Nutrition and Food Science

Volume I

Nutrition and Food Science
Volume I

Edited by **Dorothy Green**

R CALLISTO
REFERENCE

New York

Published by Callisto Reference,
106 Park Avenue, Suite 200,
New York, NY 10016, USA
www.callistoreference.com

Nutrition and Food Science: Volume I
Edited by Dorothy Green

International Standard Book Number: 978-1-63239-483-5 (Hardback)

This book contains information obtained from authentic and highly regarded sources. Reasonable efforts have been made to publish reliable data and information, but the authors, editors and publisher cannot assume responsibility for the validity of all materials or the consequences of their use.

The publisher's policy is to use permanent paper from mills that operate a sustainable forestry policy. Furthermore, the publisher ensures that the text paper and cover boards used have met acceptable environmental accreditation standards.

Printed in the United States of America.

Contents

Preface

Food is indispensable to human survival. Food is the basic necessity of every living being; be it plants, animals or humans. Food provides the energy required by humans to carry out daily chores. The right to food is a primary human right. In the early days, the primary sources of food were plants and hunting. Plants are eaten in its raw or cooked form. Today, a more rationed form of plants and animals is available in the market for consumption.

Food has a direct impact on our health. Thus, making the right choices is essential. Everything we eat adds some nutrition to our body. Our body requires a minimum amount of nutrition on everyday basis which must be fulfilled to stay healthy. In the absence of proper nutrition, the body develops certain deficiencies which can further lead to disease. In modern times, every individual is aware about the importance of nutrition. This creates a massive demand for good quality food which provides the right amount of nutrition.

The food industry has come a long way. Nowadays, detailed researches are organized to understand the nature of food, how to make it better to optimize absorption of nutrients and to discover different combinations of ingredients which can serve the diverse requirements of the human race. Along with the ingredients, a considerable attention is paid to packaging as well. There is a whole lot of incredible range of packaging available in the market, to keep the food fresh and secure.

Food and nutrition science is a wholesome science which comprises these significant yet small aspects related to food along with the major concerns. It is an interdisciplinary field which involves biology, chemical engineering and other disciplines which contribute to the better understanding of food, by delving into the details of the chemical, physical and related aspects of food compounds.

This book is a result of the consistent efforts of scientists and researchers across the globe. I would like to thank my family and publisher for their faith in me and their constant support.

Editor

The Relationship between Antioxidant and Anti-Ulcer Activities in Saudi Honey Samples Harvested from Various Regions in Different Seasons

Nabila Al-Jaber

Chemistry Department, College of Science, King Saud University, Riyadh, KSA.

ABSTRACT

The main chemical components of 13 Saudi honey samples (composed of winter and summer honeys) were identified according to anti-ulcer and antioxidant activity by using phytochemical and chromatographic analyses. Phytochemical screening of ethyl acetate and water extracts was used to detect the presence of carbohydrates, flavonoids, amino acids, and phenolic acids. High-performance liquid chromatography (HPLC) of sugar contents was used to detect the presence of galactose in all of the honey samples. The other detected sugars were sucrose, fructose, and arabinose. Fifteen amino acids were detected in all of the honey samples: Prolin is dominant in all of the honey samples. The citric acid and tartaric acids of winter and summer honey were also detected, in addition to faint traces of free oxalic acid. Both samples were tested for amino acids, phenolic compounds, and sugars. The most crucial result derived in this study is the effect of summer honey on ulcers. The anti-ulcer activities were also evaluated, showing that oral administration of the honey samples reduced the intensity of ulcer scores when compared to the control group. Similarly, there was a highly significant reduction in the values of the ulcer indices and areas in rats that received the same sample ($P < 0.05$ and $P < 0.01$) in comparison with those of the ulcer control rats. Finally, the antioxidant activity of the honey samples was evaluated, revealing a proportional relationship between the anti-oxidant and anti-ulcer activities. The results of this study could be reached that the effectiveness of honey as an anti-ulcer increase with the increasing its antioxidant activities.

Keywords: Winter and Summer Honey; Anti-Ulcer; Antioxidant; Flavonoids; Amino Acids; Phytochemical Analysis

1. Introduction

Honey, which is used in both domestic and medicinal applications, has been widely used as a sweetener since ancient times. The composition of honey varies depending on the geographical and nectar sources of a region. The quality of honey depends on its physio-chemical and sensory properties. Hence, knowledge about its constituents is essential for assessing its quality [1,2].

Certain phenolic acids and flavonoids are described in the literature as marker substances for several unifloral honeys [3]. Because the methods for extraction and determination have varied among studies, 37 phenolic acids and flavonoids are described in the literature [3,4].

Ulcerative lesions of the gastrointestinal tract are one of the major side effects associated with alcohol consumption [5]. Gastric ulcers are benign lesions occurring at a site where the mucosal epithelium is exposed to acid, alcohol, and pepsin. Several factors are implicated in the pathogenesis of gastric ulcers including stress, smoking, nutritional deficiencies and ingestion of non-steroidal anti-inflammatory drugs [6,7]. In recent years, a powerful association between gastric ulcers and infection of *Helicobacter pylori* has been adopted [8]. Gastrointestinal bleeding is the most common complication of ulcers. A sudden loss of a large amount of blood can be life-threatening, and occurs when the ulcer erodes a blood vessel. Scarring and swelling caused by ulcers may obstruct the gastric outlet.

Numerous products are used for treating gastric ulcers including H2-blockers, proton pump inhibitors that reduce acid secretion, and sucralfate, which provides mucosal protection. Although these drugs have changed ulcer therapy remarkably, their efficacy is still debatable. Reports on clinical evaluations of these drugs show that there are incidences of different adverse effects and drug interactions during ulcer therapy [9]. Thus, screening plants and other natural compounds for active drugs is still critical and might provide a useful source of new anti-ulcer products for developing pharmaceutical drugs or, alternatively, simple dietary adjuncts to existing therapies [10].

Therefore, this study assessed the efficacy of honey samples for their gastro-protective effects in rats, as well as the study of antioxidant, and conducted phytochemical and chromatographic analysis to determine their main chemical components.

2. Material and Methods

2.1. Honey Samples and Identification

Thirteen honey samples were collected from beekeepers and divided into three main groups (summer, winter, and both). None of the samples showed signs of fermentation or granulation. **Table 1** lists the 13 honey samples.

2.2. Animals

Male albino rats (220 - 230 g) and mice of both sexes (25 - 30 g) were used. The rats were maintained in cages with raised floors of wide wire mesh to prevent coprophagy, whereas the mice were housed randomly in groups in polypropylene cages. All of the animals were kept under uniform and controlled conditions of temperature and light/dark (12 h/12 h) cycles. The animals were fed a standard pellet diet and water ad libitum and were left for one week to acclimatize to the conditions of the room.

2.3. HPLC Analysis

2.3.1. For Amino Acids

1) Standards and Sample Preparation

Standards of amino acids (Sigma, St. Louis, MO) and our honey samples were prepared as described by A. Fabiani, *et al.* [11]. Identification was based on comparing the retention times of the standards amino acids with those in the honey samples and was confirmed by performing a fortification technique.

2.3.2. For Flavonoids

To identify the flavonoids, an Agilent 1100 HPLC apparatus was used. In the high-performance liquid chromatography (HPLC) analysis, a Zorbax SB-C18 column 250 mm in length, 4.6 mm i.d., and 5 μmin particle diameter was used. To detect flavonoids, a wavelength of 337 nm was set. The mobile phase was acetonitrile-water at a ratio of 48:52(v/v); the temperature was 25°C, with a flow rateof 0.3 ml/min; and the injected sample volume was 20 μl. Diluted standard solutions of rutin, quercetin, apigenin, kaempferol, acacetin, chrysin, myricetin, luteolin, tricetin, cinnamic acid, gallic acid and caffeic acid were analyzed under the same HPLC conditions. Furthermore, the detector response was calibrated, and the honey samples were dissolved in 0.5 gm/100 ml of deionized water.

2.3.3. For Carbohydrates

A hyopercarb column, 5U 100 × 4.6 mm in length was

Table 1. Types of Saudi honey.

Sample no.	Honey/location	Floral sources *species*	Family	Type of honey season
1	Farming cells/Khamis Moshiet	Mixed different flowers		Winter
2	Samra[a]/Mhail Aseer	*Acacia Etbaica*	Leguminosae	Summer
3	Sidir Aseer/Aseer Mountains	*Ziziphus Spina* Christi L	Rhammnaceae	Winter
4	Magra[a]/Rijaal Alma'a	*Hypericum Perforatum*	Gutteferae	Winter
5	Kina Kampher/Khamis Moshiet	*Eucalyptus camaldulensis*	Myrtaceae	Summer
6	Somra-Sihaiah[a]/Rijaal Alma'a	*Acacia* App.	Leguminosae	Summer
7	Sidir Tehamh/Rijaal Alma'a	*Ziziphus Spina* Christi L	Rhammnaceae	Winter
8	Mixed different mountains trees (sidir somiran, shihha, sihaiah)/Rijaal Alma'a	*Acacia* Spp., *Ziziphus* Spp., *Artemisia juncus* Spp.,	Leguminosae Rhammnaceae, Composttae Juncaceae.	Summer & Winter
9	Somiran[a]/Rijaal Alma'a	*juncus* Spp.,	Juncaceae.	Winter
10	Sihaiah/Khamis Moshiet	*Artemisia*	Composttae	Winter
11	Somiran[a]/Khamis Moshiet	*juncus* Spp.,	Juncaceae	Summer
12	Somiran/Aseer Mountains	*Ziziphus Spina* Christi L.	Rhammnaceae	Summer & Winter
13	Farming cells/Mhail Aseer	*Juniperusprocera*		Summer

The Relationship between Antioxidant and Anti-Ulcer Activities in Saudi Honey Samples
Harvested from Various Regions in Different Seasons

3

used, and the detector was RI detector, pump was GBC LC 1110 pump, software Winchrome Chromatography Ver.13, and the flow rate was 1 ml/min. The optimalmobile phase was a mixture of acetonitrile and water in the ratio of 85:15 (v/v).

Standard sugars were obtained from Sigma (St. Louis, MO). Standard solutions of sugars made up by as solving 5 g of each sugar in 1000 ML of distilled water. The honey samples were diluted in water prior to analysis (0.5 g/100 ml H$_2$O). All standards and samples were filtered and degassed by using vacuum filtration. Before injection onto the column, all syringes were fitted with syringe filters. The retention times for all standards were noted and tabulated.

2.4. Antioxidative Activity

The honey samples were analyzed for antioxidant activity by using the DPPH method. The reaction of DPPH with the honey samples was measured spectrophotometrically, at 517 nm on a Perkin Elmer, Lambda EZ Series spectrophotometer, and the data were analyzed using the corresponding UV-Vis Analyst ver. 4.67. The crude honey diluted samples, and tenth fold, were mixed with 1 mM of a DPPH ethanolic solution in 96% ethanol, and the absorbance at 517 nm was measured (the ratio of the honey sample, DPPH solution, and ethanol were 1:1:13). The antioxidative activity of the honey sample was evaluated as the relative absorbance of the sample in the presence of the DPPH solution after the constancy of the absorbance at 517 nm by using the following formula:

$$\text{Scavenging activity} = \left[\left(\text{Acontrol} - \text{Asample}\right)/\text{Acontrol}\right] \times 100$$

where A is the absorbtion at 517 nm [4].

2.5. Ulcerogenic Activity

2.5.1. Preparation of Honey Samples

A known weight of each of honey sample and the reference drug (ranitidine) was dissolved in 3% v/v Tween 80 to produce solutions of 10% concentration. The ulcerogenic agent, ethanol (Merck), was diluted to 50% (v/v, in distilled water).

2.5.2. Doses

In the absence of LD$_{50}$ values for the honey samples, an experimental dose of 100 mg·kg^{-1} was selected in this investigation to be administered orally to rats. The reference drug; ranitidine was administered orally at a certain dose (30 mg·kg^{-1}). This dose was calculated by converting the therapeutic dose used in human to rat's dose according to the findings of Paget and Barnes [11]. Ethanol (50%) was administered orally to the rats ata dose of 10 mL·kg^{-1}.

3. Phytochemical Analysis

3.1. Phytochemical Screening

The preliminary phytochemical screening of the 13 honey samples indicated the presence of phenolic compounds (flavonoids, phenolic acids, and coumarins.), amino acids, and carbohydrates. One percent of each successive ethyl acetate and water extracts were chromatographed on pre-coated silica gel plates by using the following solvent systems:

a—Chloroform—Methanol (95:5 v/v.).
b—Ethyl acetate—Methanol—Water (30:5:4 v/v/v.).
c—Benzene—Ethyl acetate (80:20 v/v/).

Solvent systems a and b achieved the most effectiveseparation, and the chromatograms from these systems were visualized by using UV lamp (short and long wave length) before and after spraying with AlCl$_3$, (a spray reagent for detecting flavonoids). The presence of amino acids was determined by spraying with ninhydrin on TLC plates.

3.2. Amino Acids

Sixteen amino acids or related compounds were identified in the honey samples by using HPLC. Proline, aspartine, threonine, serine, glutamine, glycine, alanine, valine, isoleucine, leucine, tyrocine, phenylalanine, lycine, argenine, and histidine, in addition to aspartic acid and glutamic acid, were found in all of the honey samples (**Table 2**).

3.3. Flavonoids

The flavonoids and phenolic acids of honey samples from different seasons and geographical regions were analyzed by using HPLC. More than eight flavonoids were identified in honey (**Table 3**), with an average content of 6.68 mg/100 g of honey. Lutcolin (5,7,3',4'-tetrahydroxy flavone) tricetin (5,7,3',4',5'-pentahydroxy flavone), apigenin, chrysin (5,7-dihydroxyflavone),and myricetin (3,5,7,3',4',5'-hexahydroxy flavone) were the main flavonoids identified in summer honeys, whereas quercetin (3,5,7,3',4'-pentahydroxy flavone), isorhamnetin (3,5,7,4'-tetrahydroxy flavone 3'-methyl ethyl), ruten, and kaempferol were the main flavonoids identified in winter honey. The mean content of total phenolic acids in both honey samples was arranged from 3.14 - 4.34 mg/100 g of honey, with gallic, caffic, and coumaric acids determined to be the potential phenolic acids. The content of total phenolic acids was as much as 10.08 mg/100 g of honey, with gallic acid as the main component.

Table 2. HPLC analysis of amino acids in 13 honey samples.

Honey samples amino acids	1	2	3	4	5	6	7	8	9	10	11	12	13
Proline	116.9	20.7	17.1	34.5	98.1	10.3	76.5	175.6	15.0	106.4	52.9	203.3	51.4
Threonine	14.3	1.84	6.4	-	3.2	2.1	-	-	-	-	10.8	105.5	-
Serine	15.3	6.3	5.7	3.2	5.7	7.8	4.5	53.1	34.0	3.7	21.8		14.1
Glutamine	27.9	43.8	13.1	22.2	4.4	29.0	7.9	55.1	65.4	5.1	43.7	12.4	30.8
Glycine	13.3	4.3	-	16.2	4.2	-	-	85.9	-	1.3	9.3	88.2	4.6
Alanine	26.6	15.5	-	25.7	23.3	9.6	6.8	187.6	110	6.1	41.2	216.9	25.1
Valine	17.2	16.1	7.7	3.4	14.9	18.3	9.6	47.4	65.5	0.1	25.3	58	17.8
Isoleucine	15.4	7.4	21.9	-	11.6	9.1	4.5	24.1	-	-	16.4	26.4	11.2
Leucine	28.6	10.8	15.8	4.9	28.3	15.8	7.1	72.5	46.4	-	42.7	83.4	21.3
Tyrocine	10.6	-	-	-	-	-	-	-	-	-	-	-	-
Phenylalanine	20.7	-	-	-	3.1	15.9	-	34.7	16.0	2.5	19.4	76.4	2.7
Lycine	35.6	4.7	-	6.4	10.7	6.4	-	60.7	-	0.8	13.8	79.5	4.8
Argenine	35.7	-	-	10.8	5.5	-	-	93.2	-	-	4.0	79.3	-
Histidine	2.3	-	-	-	9.7	-	-	16.1	-	0.75	7.1	22.8	-
Aspartic acid	31.1	83.7	9.0	18.4	26.7	56.33	21.4	110.9	-	36.2	52.5	105.5	38.4

Table 3. Systematic names of the standard flavonoids used in this study.

Favone

Apigenin	4',5,7-trihydroxyflavone
Luteolin	3',4',5,7-tetrahydroxyflavone
Acacetin	5,7-dihydroxy-2-(4-methoxyphenyl)chromen-4-one
Tricetin	3',4',5,5',7-pentahydroxyflavone

Flavonols

Rutin	3,3',4',5,7-pentahydroxyflavone-3-rutinoside
Quercetin	3,3',4',5,7-pentahydroxyflavone
Myricetin	3,3',4',5,5',7-hexahydroxyflavone
Kaempferol	3,4',5,7-tetrahydroxyflavone

3.4. Carbohydrates

Table 4 lists the results obtained from performing HPLC analysis to determine the carbohydrate content in the 13 honey samples. In all of the honey samples, 5 carbohydrates were evaluated and quantified: monosaccharides (fructose, glucose, arabinose, and galactose) and disaccharides (maltose). Regarding the galactose contents in all of the 13 honey samples, three of the samples (Samples 3, 5, and 8) had a higher galactose content than the others did. Significant differences were found for the car-

bohydrate content in the honey samples. The main distinguisher for these products was fructose content (found only in Sample 11, 0.06 mg/1 ml of honey) and high galactose content (more than 2.58 mg/1 ml of honey). The lowest glucose values were found in Sample 7 honey (averaging), whereas the average value for the other samples was 0.22 mg/1 ml.

4. Anti-Ulcer Activity

4.1. Determination of LD$_{50}$

LD$_{50}$ of the honey samples was estimated in albino mice (25 - 30 g) according to the method proposed by Finney [12]. In a preliminary test, animals in groups of three received one of the honey samples (*i.e.*, 100 mg·kg^{-1}, 500 mg·kg^{-1}, 1000 mg·kg^{-1}, 2000 mg·kg^{-1}, or 4000 mg·kg^{-1}) orally. The animals were observed for 24 h for signs of toxicity and death. Based on the results of the first test, doses between the minimal dose that kills all mice and the maximal dose that fails to kill any animal were administered orally to fresh groups of mice. The control animals that received the vehicle (10 mL·kg^{-1}) were kept under the same conditions without any treatment. Signs of toxicity and number of deaths per dose in 24 h were recorded, and the LD$_{50}$ of the honey samples was calculated as the geometric mean of the dose that resulted in 100% mortality and that which caused no lethality at all.

The Relationship between Antioxidant and Anti-Ulcer Activities in Saudi Honey Samples
Harvested from Various Regions in Different Seasons

5

Table 4. HPLC analysis of sugar in honey samples from Saudi.

Hony samples	Sugars	mg/Ml
1	Galactose	0.40
	Arabinose	0.44
	Glucose	0.31
	Maltose	0.27
2	Galactose	0.60
	Arabinose	0.70
	Glucose	0.33
	Maltose	0.14
3	Galactose	2.58
	Glucose	0.36
	Arabinose	1.5
	Maltose	0.80
4	Glucose	0.34
	Maltose	0.15
5	Glucose	1.02
	Galactose	1.50
6	Galactose	0.27
	Arabinose	0.48
7	Galactose	0.40
	Arabinose	0.12
	Glucose	0.22
	Maltose	0.35
8	Galactose	2.13
	Arabinose	2.19
	Glucose	0.06
9	Glucose	0.84
	Galactose	0.44
10	Galactose	0.34
	Arabinose	0.14
	Glucose	0.30
11	Fructose	0.06
	Maltose	0.16
12	Maltose	0.03
	Arabinose	0.21
13	Galactose	0.43
	Arabinose	0.21
	Glucose	0.30
	Maltose	0.16

4.2. Ethanol Induced Gastric Ulcers

The anti-ulcer activity of ranitidine and the honey samples was carried out according to the method described by Garg *et al.* (1993) [13]. Eighty-five male albino rats weighing 220 - 230 g were divided into 17 equally numbered groups. All of the animals were fasted for 48 h before use to ensure an empty stomach. During the fasting period, the rats were administered a nutritive solution of 8% sucrose in 0.2% NaCl to avoid excessive dehydration [14]. The nutritive solution was removed 1 h before the experiment. The first and second groups received the vehicle only and were kept as the normal control and ulcer control groups, respectively. The animals in the third group (reference) were administered ranitidine (30 mg·kg^{-1}). Other groups were administered honey samples in a dose of 100 mg·kg^{-1}b.wt. All medications were administered orally via a stainless steel intubation needle. Two doses were administered in the first day at 8:00 and 16:00 O' clock and a third dose was administered on the second day, 1.5 h before induction of gastric ulceration.

4.3. Induction of Gastric Ulcers

Gastric ulcers were induced in all of the rats (except for the normal control) following oral administration of 50% ethanol ata dose of 10 mL·kg^{-1} [15]. In the normal control rats, an equivolume of distilled water was dosed instead of ethanol. One hour after ethanol administration, all rats were euthanized by an overdose of ether anesthesia and the stomachs were rapidly excised. Each stomach was then opened along the greater curvature and gently rinsed under running tap water to remove gastric contents and blood clots. The stomachs were then fixed with 10% formaldehyde in saline and spread on a paraffin plate.

4.4. Assessment of Gastric Damage

Long lesions in the glandular part of the stomachs were counted and measured along their greater length by using a hand lens. Petechial lesions were counted then each five petcchial lesions were taken as 1 mm of ulcer. Ulcer scoring was performed according to Nwafor *et al.* [16].

The scores were:

0, normal tissue; 1, the presence of one ulcer and generalized erythema; 2, at least two ulcers of approximately 2 mm in length; 3, lesions that followed approximately 80% of the fold; and 4, multiple ulcers along the entire length of the gastric fold.

Theulcer index (mm) was calculated as the sum of the total length of long ulcers and petechial lesions in each rat divided by their numbers (1 - 4). The ulcer area (UA) in each stomach was assessed using a scaled glass square. The area of each cell on the glass square was 1 mm^2; thus, the number of cells was counted and the ulcer area was determined [17]. The percentage of protection (P%) availed to the animals through various treatments was calculated using the following formula:

$$P\% = \left[\left(\text{UA control} - \text{UA treatment}\right)/\text{UA control}\right] \times 100$$

4.5. Statistical Analysis

The data were analyzed by using a one-way ANOVA with SPSS 14.0 statistical software, followed by a post hoc Scheffe test. The results are expressed as the mean ± SEM. The minimal level of significance was identified at $P < 0.05$ and $P < 0.01$.

5. Results and Dissections

5.1. Chemical Constituent

In general, higher amounts of amino acids were found in the summer samples than in those of the winter samples. This may be an indication that the metabolic activities of bees were lower in winter, which also might be the result of nutritional starvation in winter. No pattern of amino acids was apparent in the samples of either the winter or summer honey, which might indicate a nutritional balance that could be used to understand these results.

In general, the honey samples from the summer season (*i.e.*, Samples 2, 6, 8, 11, and 13) showed flavonoid profiles characterized by the presence of flavones. Bycontrast, the honey samples from winter (*i.e.*, samples 1, 3, 4, 7, 9, and 10) showed only flavonoids with flavonols, whereas the mixed samples have mixed flavonoid types. These preliminary results show that flavonoid analysis in honey, could be used as an adjunct to studies on honey regarding season and geographical origin.

However, research has shown that the carbohydrates of honey are nearly identical, and feature a sugar mainly galactose in almost all samples, suggesting that the geographical environment may clearly indicate the sugar-composition of honey.

5.2. Determination of LD$_{50}$

All rats treated with different doses of the honey samples (up to 4000 mg·kg^{-1}) were alive during the 24 h of observation. The animals did not show visible signs of acute toxicity, suggesting that the oral LD$_{50}$ of each of the honey samples was higher than 4 g·kg^{-1}. The tested compounds are considered to be highly safe because substances possessing LD$_{50}$ higher than 50 mg·kg^{-1} are nontoxic [18].

5.3. Anti-Ulcerogenic Effect

Macroscopic examination of the gastric mucosa of ulcer control rats after ethanol treatment revealed edema, ulcerations, petechial lesions, and long hemorrhagic bands of different sizes. No pathological changes were observed in the normal control group, suggesting that the handling procedure did not interfere with the experimental results.

It was noticed that oral medication with the Samples 2, 6, 8, 11, and 13 (100 mg·kg^{-1}), prior to ulcer induction, significantly reduced the intensity of ulcer scores into 3.3, 3.6, 3.7, 3.5, and 3.4, respectively ($P < 0.05$ and $P < 0.01$), as compared to 4.8 in the ulcer control group (**Table 5**). Similarly, there is a high significant reduction in the values of the ulcer indices and areas in rats that received the honey samples ($P < 0.05$ and $P < 0.01$) in comparison with those in the ulcer control rats.

Conversely, Samples 3, 10, and 12 exhibited a lower anti-ulcerogenic efficacy, reducing the ulcer scores, indices, and areas ($P < 0.05$) when dosed orally to rats prior to ulcer induction. Moreover, Samples 2, 6, 8, 11, and 13 induced a protection against ulcers ranging between 27.6% and 34.2%. Toa lesser extent, oral administration of Samples 3, 10, and 12 produced protection ata ratio between 18.4% and 21.0%. The other honey samples (Samples 1, 4, 5, 7, and 9) failed to improve the gastric damage.

6. Antioxidant Activity

The results of this experiment showed that the anti-oxidative activity in all honey samples was very high it has been shown that summer samples (2, 6, 8, 11, and 13), has highly effective than the others (**Figure 1**), whereas the rest of the samples slightly lower than the summer honey samples, possibly because of the types of flavonoid compounds that have high oxygenation patterns.

7. Conclusion

The findings of this study indicate that the geographical area of harvested honey plays a role in determining the traits of honey and its activity. Previous research has

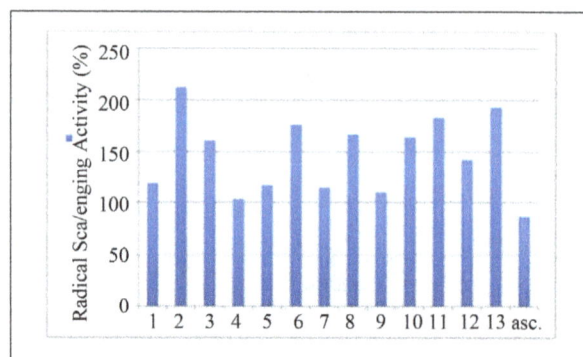

Figure 1. Antioxidant activity (%) of summer and winter honey samples.

The Relationship between Antioxidant and Anti-Ulcer Activities in Saudi Honey Samples
Harvested from Various Regions in Different Seasons

7

Table 5. Effects of honey samples on the macroscopic parameters of gastric ulcers induced by 50% ethanol in rats (n = 5).

Groups	Ulcer score (0 - 5)	Ulcer index (mm)	Ulcer area (mm^2)	Percentage of protection
Normal control	0.0 ± 0.0	0.0 ± 0.0	0.0 ± 0.0	-
Ulcer control	4.8 ± 0.37	6.1 ± 0.22	7.6 ± 0.24	-
Ranitidine	2.7 ± 0.37[b]	3.9 ± 0.18[b]	4.5 ± 0.26[b]	40.8
1	4.4 ± 0.24	5.6 ± 0.15	7.2 ± 0.55	5.3
2	3.3 ± 0.20[b]	4.3 ± 0.29[b]	5.0 ± 0.40[b]	34.2
3	3.9 ± 0.18[a]	5.2 ± 0.30[a]	6.0 ± 0.54[a]	21.0
4	4.5 ± 0.22	5.6 ± 0.21	7.3 ± 0.68	3.9
5	4.6 ± 0.40	5.8 ± 0.40	7.2 ± 0.68	5.3
6	3.6 ± 0.24[b]	4.6 ± 0.29[b]	5.4 ± 0.60[b]	28.9
7	4.6 ± 0.24	5.7 ± 0.44	7.2 ± 0.24	5.3
8	3.7 ± 0.20[b]	4.6 ± 0.29[b]	5.5 ± 0.29[b]	27.6
9	4.6 ± 0.24	5.8 ± 0.30	7.4 ± 0.56	2.6
10	3.8 ± 0.12[a]	5.1 ± 0.18[a]	6.2 ± 0.38[a]	18.4
11	3.5 ± 0.22[b]	4.5 ± 0.24[b]	5.3 ± 0.38[b]	30.3
12	3.9 ± 0.24[a]	5.1 ± 0.13[a]	6.1 ± 0.24[a]	19.7
13	3.4 ± 0.40[b]	4.5 ± 0.14[b]	5.1 ± 0.24[b]	32.9

Data are represented as mean ± SEM. [a] and [b] denote significant difference vs. control groups at $P < 0.05$ and $P < 0.01$, respectively.

shown that honey harvested in the summer has characteristics that differ from honey harvested in winter, possibly because of bees being idle in winter. Furthermore, honey that demonstrates high anti-oxidant activity and contains certain types of flavonoids can be used to treat stomach ulcers effectively.

8. Acknowledgements

This research project was supported by a grant from the "Research Center of the Center for Female Scientific and Medical Colleges", Deanship of Scientific Research, King Saudi University.

REFERENCES

[1] D. Anupama, K. K. Bhat and V. K. Sapna, "Sensory and Physico-Chemical Properties of Commercial Samples of Honey," *Food Research International*, Vol. 36, No. 2, 2003, pp. 183-119.

[2] J. Lachman, A. Hejtmánková, J. Sýkora, J. Karban, M. Orsák and B. Rygerová, "Contents of Major Phenolic and Flavonoid Antioxidants in Selected Czech Honey," *Czech Journal of Food Science*, Vol. 28, No. 6, 2010, pp. 412-426.

[3] P. Krystyna and M. Biesaga, "Analysis of Phenolic Acids and Flavonoids in Honey," *TrAC Trends in Analytical Chemistry*, Vol. 28, No. 7, 2009, pp. 893-902.

[4] S. Z. Makawi, E. A. Gadkariem and S. M. H. Ayoub, "Determination of Antioxidant Flavonoids in Sudanese Honey Samples by Solid Phase Extraction and High Performance Liquid Chromatography," *E-Journal of Chemistry*, Vol. 6, No. 1, 2009, pp. 429-437.

[5] T. Mizui, H. Sato, F. Hirose and M. Doteeuchi, "Effect of Anti-Peroxidative Drugs on Gastric Damage Induced by Ethanol in Rat," *Life Science*, Vol. 41, No. 10-11, 1987, pp. 755-763.

[6] J. Nash, L. Lynn and M. Deakin, "Histamine H2-Receptor Antagonist in Peptic Ulcer Disease. Evidence for a Prophylactic Use," *Drugs*, Vol. 6, No. 47, 1994, pp. 862-871.

[7] M. D. Basil and M. S. Howard, "Clinical Gastroenterology," 4th Edition, McGraw-Hill, New York, 1995.

[8] K. R. McQuaid, "Alimentary System," In: L. M. Tierney, S. J. McPhee and M. A. Papadakis, Eds., *Current Medical Diagnosis and Treatment*, 41st Edition, Lange Medical Books, McGraw Hill Company, New York, 2002, pp. 616-621.

[9] R. K. Goel and K. Sairam, "Anti-Ulcer Drugs from Indigenous Sources with Emphasis on Musa Sapientum, Tamrabhasma," Asparagus Racemosus and Zingiberofficinale. *Indian Journal of Pharmacology*, Vol. 34, No. 2, 2002, pp. 100-110.

[10] F. Borrelli and A. A. Izzo, "The Plant Kingdom as a Source of Anti-Ulcer Remedies," *Phytotherapy Research*, Vol. 14, No. 8, 2000, pp. 581-591.

[11] A. Fabiani, A. Versari, G. P. Parpinello, M. Castellari and S. Galassi, "High-Performance Liquid Chromatographic Analysis of Free Amino Acids in Fruit Juices Using Derivatization with 9-Fluorenylmethyl-Chloroformate," *Journal of Chromatographic Science*, Vol. 40, No. 1, 2002, pp. 14-18.

[12] D. J. Finney, "Statistical Methods in Biological Assay," Charles Griffin and Company Ltd., London, 1964, p. 597.

[13] G. P. Garg, S. S. Nigma and C. W. Ogle, "The Gastric Antiulcer Effects of the Leaves of the Neem Tree," *Planta Medica*, Vol. 59, No. 5, 1993, pp. 215-217.

[14] G. B. Glavin and A. A. Mikhail, "Stress and Ulcer Etiology in the Rat," *Physiological Behavior*, Vol. 16, No. 2, 1976, pp. 135-139.

[15] A. Alkofahi and A. AH, "Pharmacological Screening of Anti-Ulcerogenic Effects of Some Jordanian Medicinal Plants in Rats," *Journal of Ethnopharmacology*, Vol. 67, No. 3, 1999, pp. 341-345.

[16] P. A. Nwafor, F. K. Okwuasaba and L. G. Binda, "Antidiarrhoeal and Antiulcerogenic Effect of Methanolic Extracts of Asparagus Pubescensroot in Rats," *Journal of Ethnopharmacology*, Vol. 72, No. 3, 2000, pp. 421-427.

[17] S. Szabo, "Mechanisms of Mucosal Injury in the Stomach and Duodenum: Time-Sequence Analysis of Morphologic, Functional, Biochemical and Histochemical Studies," *Scandinavian Journal of Gastroenterology*, Vol. 22, No. S127, 1987, pp. 21-28.

[18] W. B. Buck, G. D. Osweiter and A. Van Gelder, "Clinical and Diagnostic Veterinary Toxicology," 2nd Edition, Kendall/Hunt Publishing Co., Iowa, 1976, p. 5211.

2

Specialty Lipids in Health and Disease

Fayez Hamam

Department of Pharmacology and Toxicology, College of Pharmacy, Taif University, Taif, Saudi Arabia.

ABSTRACT

Lipids possess a wide range of biological activities in plants, animals and humans. They also serve as important components of our daily diet and provide both energy and essential fatty acids; they also act as carriers of fat-soluble vitamins and help in their absorption. Lipids are crucial as a heating medium for food processing and affect the texture, mouth feel and flavour of foods. Structured lipids (SL) are triacylglycerols (TAG) modified to alter the fatty acid composition and/or their location in the glycerol backbone via chemical or enzymatic means. SL may offer the most efficient means of delivering target fatty acids for nutritive or therapeutic purposes as well as to alleviate specific disease and metabolic conditions. This document discusses chemistry, composition, classification, function, occurrence in food and biological activities of lipids. It also sheds light on different aspects of structured lipids, including SL applications, synthesis (chemical vs. enzymatic), SL and aquaculture and future considerations for SL.

Keywords: Structured Lipids; n-3 Fatty Acids; n-6 Fatty Acids

1. Introduction

Lipids are a chemically heterogeneous group of compounds that are insoluble or sparingly soluble in water, but soluble in non-polar solvents. They serve several important biological functions including: 1) acting as structural components of all membranes; 2) serving as storage form and transport medium of metabolic fuel; 3) serving as a protective cover on the surface of several organisms; and 4) being involved as cell-surface components concerned with cell recognition, species specificity and tissue immunity. Lipids also constitute a major component of the daily diet, and provide both energy and essential fatty acids. Lipids also act as carriers of fat-soluble vitamins A, D, E and K and help in their absorption. Finally, lipids act as a heating medium for food processing and affect the texture, mouth feel and flavour of foods [1].

2. Chemistry and Composition of Lipids

2.1. Fatty Acids

Fatty acids form the basic chemical structure of fats. The triacylglycerols (TAG) constitute 80% - 95% of lipids. One molecule of TAG consists of three fatty acids and one glycerol molecule. The physical properties of TAG differ, depending on the source they are derived from.

Those derived from animal fats are solid at room temperature (lard, butter, etc.); while those obtained from plant and marine oils are liquid at room temperature (cod liver oil, olive oil, etc.). Fatty acids fall into two main categories: saturated and unsaturated; the latter being further subdivided into monounsaturated (MUFA) and polyunsaturated (PUFA). PUFA is divided into two main classes, depending on the location of the first double bond from the methyl end group of the fatty acid; these are n-3 (also omega-3), n-6 (also omega-6). While saturated and monounsaturated fatty acids are also made in the human body, polyunsaturated fatty acids (PUFA) cannot be produced in the human body and must be obtained from dietary sources. Therefore PUFA are considered essential fatty acids (EFA).

2.2. Saturated Fatty Acids

Saturated fatty acids contain only single carbon-carbon bonds in the aliphatic chain and hydrogen atoms occupy all other available bonds. The most abundant saturated fatty acids in animal and plant tissues are usually straight chain compounds with 10, 12, 14, 16 and 18 carbon atoms. In general, saturated fats are solid at room temperature. They are found mainly in margarine, shortening, coconut and palm oils as well as foods of animal origin.

For a series of saturated fatty acids the melting point

increases as the length of the chain increases. Typically, adding double bonds to a saturated fatty acid will lower its melting point.

2.2.1. Short-Chain Fatty Acids (SCFA)

Short-chain fatty acids (SCFA) range from C2:0 to C4:0 and include acetic (C2:0), propionic (C3:0) and butyric acids (C4:0). They are the end products of carbohydrate fermentation in the human gastrointestinal tract [2]. SCFA are quickly absorbed in the stomach because of their higher solubility in water, smaller molecular size, and shorter chain length [3] and provide fewer calories than medium-chain fatty acids (MCFA) or long-chain fatty acids (LCFA) (acetic acid, 3.5 kCal; propionic acid, 5.0 kCal; butyric acid, 6.0 kCal).

2.2.2. Medium-Chain Fatty Acids (MCFA)

Medium-chain fatty acids (MCFA) comprise 6 - 12 carbon atoms that result from hydrolysis of tropical plant oils such as those of coconut and palm kernel [4, 5]. Pure medium-chain triacylglycerols (MCT) have a caloric value of 8.3 kCal /g and do not supply essential fatty acids [6,7]. MCFA are more hydrophilic than their long-chain fatty acid (LCFA) counterparts. MCFA have many distinctive features such as high oxidative stability, low viscosity and low melting point [8].

MCT exhibit unique structural and physiological characteristics; they are different from other fats and oils because they can be absorbed via the portal system without hydrolysis and reesterification because they are relatively soluble in water. MCT do not require chylomicron formation to transfer from blood stream to the cells and have a more rapid β-oxidation to form acetyl CoA end products which are further oxidized to yield CO_2 in the Kreb's cycle [9]. Absorption and metabolism of MCT is as quick as glucose and have approximately twice the caloric concentration than proteins and carbohydrates. They have little affinity to accumulate as body fat. Medium-chain triacylglycerols are not dependent on carnitine (an enzyme necessary for transport of fatty acids across the inner mitochondria membrane) to enter mitochondria. The higher solubility and smaller molecular size of MCFA make their absorption, transport and metabolism much easier than long-chain fatty acids. MCT are hydrolyzed more quickly and completely by pancreatic lipase than long-chain triacylglycerols (LCT). They may be directly absorbed by the intestinal mucosa with minimum pancreatic or biliary function [10].

Oils from tropical plant, such as those from coconut and palm kernel, contain very high amounts (approximately 50%) of lauric acid (C12:0). They also contain considerable amounts of caprylic (C8:0), capric (C10:0) and myristic (C14:0) acids.

In many medical foods, a mixture of MCT and LCT is used to provide both rapidly metabolized and slowly metabolized fuel as well as essential fatty acids. Any abnormality in the many enzymes or processes involved in the digestion of LCT can cause symptoms of fat malabsorption. Thus, patients with certain diseases (Crohn' disease, cystic fibrosis, colitis and enteritis, etc.) have shown improvement when MCT are incorporated into their diet [11]. MCT have clinical applications in the treatment of lipid malabsorption, maldigestion, obesity and deficiency of the carnitine system.

Some reports have proposed that MCT may decrease both serum and tissue cholesterol in animals and humans, even more than traditional polyunsaturated oils [12]. However, Cater et al. [13] have shown that MCT increase plasma cholesterol and TAG levels in mildly hypercholesterolemic men fed MCT, palm oil, or high oleic acid sunflower oil diets.

2.2.3. Long-Chain Fatty Acids

Most lipids consist of long-chain fatty acids (>C12) and are referred to as long-chain triacylglycerols (LCT). Palmitic acid (16:0) is a widely occurring saturated fatty acid and is found in almost all vegetable oils, as well as in fish oils and body fat of land animals. Palmitic acid is found abundantly in palm oil, cottonseed oil, lard and tallow, among others. Stearic acid (C18:0) is another important saturated fatty acids and is also a main component of cocoa butter. Triacylglycerols containing high amounts of long-chain saturated fatty acids, especially stearic acid (C18:0), are poorly absorbed in the human body partly because they have a higher melting point than the body temperature and they also display poor emulsion properties [14]. The poor absorption of long-chain saturated fatty acids makes them good candidates for the synthesis of low-calorie structured lipids (SL). SL are fats or oils modified to change the fatty acid composition and/or their location in the glycerol backbone. For example, Nabisco Food Group used this feature of C18:0 to produce a group of low-calorie SL known as Salatrium, which consist of SCFA and long-chain saturated fatty acids, mainly stearic acid [15].

2.3. Unsaturated Fatty Acids

Unsaturated fatty acids contain carbon-carbon double bonds in their aliphatic chain. In general, these fats are soft at room temperature. Monounsaturated fatty acids contain one carbon-carbon double bond. On the other hand, polyunsaturated fatty acids (PUFA) contain two or more carbon-carbon double bonds. The PUFA are liquid at room temperature due to the fact that the double bonds are rigid, thus preventing the fatty acids from packing close together. In general, they have low melting points and are susceptible to oxidation. Because most PUFA are liquid at room temperature, they are generally referred to as oils. The common sources of PUFA include grains, nuts, vegetables and seafood.

2.3.1. The n-9 Fatty Acids

The n-9 fatty acids, or monounsaturated fatty acids, contain one double bond that is located between the ninth and tenth carbon atoms from the methyl end group. They are found in vegetable oils such as olive, almond, hazelnut, canola, peanut and high-oleic sunflower as oleic acid (18:1n-9). Oleic acid is the most widely distributed and the most extensively produced of all fatty acids. Olive oil (60% - 80%), hazelnut oil (60% - 70%) and almond oil (60% - 70%) are rich sources of this fatty acid [16]. The human body can synthesize oleic acid, therefore it is not considered as an essential fatty acid. It plays a moderate role in lowering plasma cholesterol in the body [17], but increasing the uptake of oleic acid in young healthy humans is known to increase plasma high density lipoprotein (HDL) and decrease TAG [18].

2.3.2. Essential Fatty Acids (EFA)

As stated earlier PUFA with two or more double bonds in their backbone structures cannot be made in the body and hence considered EFA. There are two groups of EFA, the n-3 and the n-6 fatty acids. They are defined by the location of double bond in the molecule nearest to the methyl end of the chain. In the n-3 group of fatty acids, the first double bond occurs between the third and fourth carbon atoms and in the n-6 group of fatty acids it is situated between the sixth and seventh carbon atoms. The parent compounds of the n-6 and n-3 groups of fatty acids are linoleic acid (LA, 18:2 n-6) and α-linolenic acid (ALA, 18:3 n-3), respectively. These parent compounds are metabolized in the body via a series of alternating desaturation (in which an extra double bond is inserted by removing two hydrogen atoms) and elongation (in which two carbon atoms are added) steps.

2.3.3. The n-3 Fatty Acids

The n-3 fatty acids, such as α-linolenic acid (ALA), eicosapentaenoic acid (EPA; 20:5n-3) and docosahexaenoic acid (DHA; 22:6n-3) have many health benefits related to cardiovascular disease, inflammation, allergies, cancer, immune response, diabetes, hypertension and renal disorders [19]. Epidemiological studies have linked the low incidence of coronary heart disease in Greenland Eskimos with their high dietary intake of n-3 PUFA [20, 21]. Research studies have shown that DHA is essential for appropriate function of central nervous system and visual acuity of infants [19]. The n-3 fatty acids are essential for normal growth and development throughout the life cycle of humans and therefore should be included in the diet. The n-3 fatty acids have been extensively studied for their influence on cardiovascular disease (CVD). However, the exact mechanism by which these effects are rendered remains unknown, but research results have shown that these FA in marine oils may prevent CVD by decreasing serum TAG and acting as anti-therogenetic and antithrombotic agents [1].

Marine oils are rich sources of n-3 fatty acids, especially EPA and DHA. Cod liver, menhaden and sardine oils contain approximately 30% EPA and DHA [19]. Alpha-linolenic acid (ALA; 18:3n-3), the parent of n-3 fatty acids, can be metabolically converted to DHA via desaturation and elongation reactions. However, the efficiency of conversion of ALA to DHA in human adults is very restricted (approximately 4%) and even more restricted in infants (<1%) [22]. In certain disease conditions, the rate of conversion of ALA to DHA and/or EPA is much lower, therefore long-chain polyunsaturated fatty acids (LC PUFA) such as DHA and EPA are considered conditionally essentials and must be obtained from dietary sources [22]. ALA is a main constituent of flaxseed oil (50% - 60%). When ALA is absorbed into the animal body through the diet, it forms long-chain PUFA with an n-3 terminal structure. EPA and DHA are also important n-3 fatty acids. DHA is a major constituent of the gray matter of the brain and the retina of the eye. Human milk also contains a considerable level of DHA, therefore infants fed on mother's milk show a higher IQ and intelligence level than infants fed on formula's that lack any DHA [23]. In addition, EPA is a precursor of a series of eicosanoids and is important in protecting against heart attacks primarily due to its antithrombotic effect [24]. EPA was also shown to raise bleeding time and to decrease serum cholesterol levels [24].

In conclusion, LC PUFA exhibit multifunctional role in promotion of health and prevention of disease in the human body. However, they are highly susceptible to oxidation when stored and are known, upon consumption, to increase the body's load on natural antioxidants such as α-tocopherol. Therefore, it is very important to stabilize oils rich in LC PUFA during storage by incorporation of appropriate antioxidants and adequate packaging technologies.

2.3.4. The n-6 Fatty Acids

The n-6 fatty acids display a variety of physiological functions in the human body. The main functions of these fatty acids are related to their roles in the membrane structure and in the biosynthesis of short-lived derivatives (eicosanoids) which regulate several aspects of cellular activity. The n-6 fatty acids are responsible for maintaining the integrity of the water impermeability barrier of the skin. They are also involved in the regulation of cholesterol transport in the body.

Linoleic acid (LA; 18:2n-6) is the most common fatty

acid of this type. LA is found in all vegetable oils and is essential for normal growth, reproduction and health. LA serves as a precursor of n-6 family of fatty acids that are formed by desaturation and chain elongation, in which the terminal (n-6) structure is retained. Of these, arachidonic acid (AA; 20:4n-6) is principally important as a fundamental constituent of the membrane phospholipids and as a precursor of eicosanoids. On the other hand, γ-linolenic acid (GLA; 18:3n-6), an important intermediate in the biosynthesis of AA from LA, is a component of certain seed oils, such as borage and evening primrose, and has been a subject of intensive studies [25,26].

It is proposed that the uptake of 1% - 2% LA in the diet is adequate to protect against chemical and clinical disorders in infants. The absence of LA in the diet is associated with manifestation of several disorders, including impaired growth and reproduction, excessive water loss via the skin, scaly dermatitis and poor wound healing [27].

3. Biological Effect of Dietary Lipids

The influence of dietary lipids on the nature and constituents of adipose tissue is well recognized [28]. This lends support to the saying that "we are what we eat" for many different species tested. In a series of studies on seals and fish, Iverson and her colleagues [29] demonstrated that dietary lipids could be easily detected in their circulatory lipids and adipose tissues.

The dietary fat composition selectively affects fatty acid and TAG deposition in the adipose tissue. In turn, the composition of the fat in the adipose tissue influences lipid mobilization and release of fatty acids into the circulatory system [28]. Lipid mobilization from adipose tissue is not a random event, but instead is affected by variables, such as chain length, degree of unsaturation, and positional isomerization of fatty acids. The most readily mobilized fatty acids are those with 16 - 20 carbon atoms and four or five double bonds, while other very long unsaturated and monounsaturated fatty acids are less easily mobilized [28]. Furthermore, the reduced fat deposition in animals fed trans fatty acids may be associated with direct influence of trans isomers on fat cell metabolism [30]. Many studies provide evidence that variations in the level and type of fat incorporated in the diet can change adipose cell size (hypertrophy) and/or number (hyperplasia) [28]. It is generally accepted that a high amount of fat in the diet may induce hypertrophy and/or hyperplasia. Launay et al. [31] suggested that the multiplication rate of adipose tissue might be increased when the degree of unsaturation of the dietary lipids is increased. On the other hand, Shillabeer and Lau [32] reported a greater degree of fat cell hyperplasia with saturated rather than unsaturated dietary fat. In contrast to the studies reported above, more consistent influences on adipose cellularity were noticed with dietary n-3 PUFA that selectively restrict fat cell size and/or number in a depot-dependent manner [33].

4. Lipid Classes

Lipids are classified based on their physical characteristics at room temperature (oils are liquid and fats are solid), their polarity (polar and neutral lipids), their essentiality for humans (essential and non-essential FA), and their structure (simple, compound and fat-derived). Simple fats are made up of a glycerol, and one (monoacylglycerol), two (diacylglycerol) or three (triacylglycerol) fatty acids. The second category (compound) is the combination of simple fats with other moieties; phospholipids are one example of compound lipids. Fat-derived compounds combine simple and not contain a fatty acid, they are considered "lipid" because they are water insoluble; sterols provide a good example for this category.

Acylglycerols

The TAG consists of a glycerol backbone esterified to three fatty acids. Partial acylglycerols, such as mono- and diacylglycerols, may also be found as minor constituents in edible oils. These compounds are synthesized by enzyme systems in nature. Some 80% - 95% of lipids are generally composed of TAG. The TAG is presented in many different forms, according to the type and location of the three fatty acid components involved. Those with a single type of fatty acid in all three positions are called simple TAG and are named after their fatty acid component. However, in some cases the trivial names are more commonly used. An example of this is trioleylglycerol, which is usually referred to as triolein. The TAG with two or more different fatty acids is named by a more complex system [16].

Partial acylglycerols, such as diacylglycerols (DAG) and monoacylglycerols (MAG) are significant intermediates in the biosynthesis and catabolism of TAG and other classes of lipids. For example, 1,2-DAG is important intermediates in the biosynthesis of TAG and other lipids. On the other hand, 2-MAG is formed as intermediates or end products of the enzymatic hydrolysis of TAG.

5. Structured Lipids

5.1. Structured Lipids Applications

Structured lipids (SL) are TAG modified to change the fatty acid composition and /or their location in the glycerol backbone via chemical or enzymatic means [34]. Recently, structured lipids have attracted much attention due to their potential biological function and nutritional perspectives, including reduction in serum TAG, low-density lipoprotein (LDL) cholesterol and total choles- terol [35], improvement of immune function, protection against thrombosis [11], reduction of protein breakdown [36,37], improvement of absorption of other fats [38]), reduction of calories, preservation of reticuloendothelial system function [39], as well as improvement of nitrogen balance [4], and reduction of risk of cancer [40].

Strategies for lipid modification include genetic engineering of oilseed crops, production of oils containing high levels of polyunsaturated fatty acids, and lipase- or chemically-assisted interesterification reactions. Depending on the type of substrate available, chemical or enzymatic reactions can be used for the synthesis of SL, including direct esterification (reaction of fatty acids and glycerol), acidolysis (transfer of acyl group between an acid and ester), and alcoholysis (exchange of alkoxy group between an alcohol and an ester) [9]. However, the ordinary methods cited in the literature for production of SL are based on reactions between two triacylglycerol molecules (interesterification) or between a triacylglycerol and an acid (acidolysis) (**Figure 1**).

5.2. Synthesis of Structured Lipids

5.2.1. Chemically-Catalyzed Interesterification
Chemically-catalyzed interesterification, using alkali such as sodium methoxide, is cheap and easy to scale up. However, such reactions lack specificity and offer little or no control over the positional distribution of fatty ac-

ids in the final product [9]. In addition, the reactions carried out under harsh conditions such as high temperatures (80°C - 90°C) and produce side products which are difficult to eliminate.

5.2.2. Enzymatically-Catalyzed Interesterification
An alternative to the chemical synthesis of SL is enzymatic process using a variety of lipases. Lipase-assisted interesterification offers many advantages over chemical one. It produces fats or oils with a defined structure because it incorporates a specific fatty acid at a specific position of the glycerol moiety. It requires mild experimental conditions without potential for side reactions, reduction of energy consumption, reduced heat damage to reactants, and easy purification of products [5,41]. However, bioconversion of lipids with lipase is more expensive than chemical methods. Therefore, immobilization of lipids on suitable supports is desirable as it allows reuse of the enzymes. Screening of new lipases from organisms or production of a thermostable or sn-2 specific lipase that is rare in nature through bioengineering are desirable for industrial application.

Another approach is to produce structured lipids through bioengineering. Calgenes's Inc. of Davis (California) succeeded in production of high-laurate canola oil containing 40% lauric acid (C12:0). It is now available and marketed under the name Laurical and is used in confectionary coatings, coffee whiteners, whipped top-

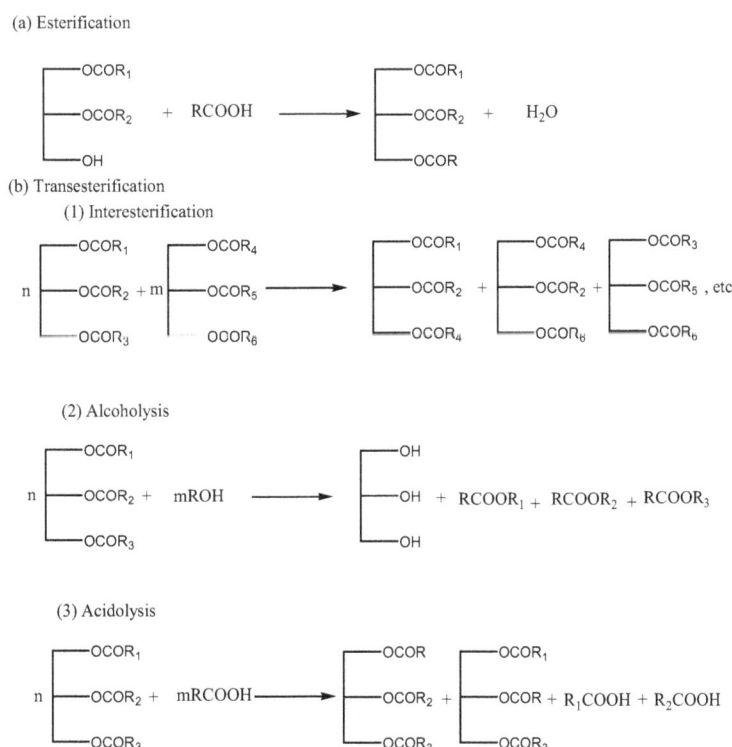

Figure 1. Schematic diagram of lipase-assisted lipid modification strategies for the synthesis of structured lipids.

pings, and filling fats. However, this genetically modified oil is deficient with essentials fatty acids. Recently Hamam and Shahidi [42-45] succeeded in enriching different kinds of high-laurate canola oil with three main kinds of n-3 fatty acids (EPA, DPA, DHA).

5.3. Structured Lipids and Aquaculture

In order to maintain an average 13 kg/person annual consumption of fish, aquaculture must continue to grow at 10 % per year. By 2010, scientists expect aquaculture to be consuming about 75% of world fish oil. The main reason for using fish oil in aqua feeds because it is a good source of n-3 fatty acids (EPA and DHA). The demand for high quality fish oil is increasing, causing the high price to remain and making vegetable oils more competitive in this market. This will result in the development of modified oils containing appropriate quality as well as providing the required amounts of n-3 fatty acids [46].

5.4. Future Considerations

Over the past two decades several research groups have successfully incorporated MCFA (caprylic or capric acids) into fish and marine oils containing PUFA [26,47-52] and into borage oil rich in γ-linolenic acid [25,26,53]. Despite their health benefits, SL containing PUFA are susceptible to rapid oxidative deterioration and thus experience stability problem. Therefore, further research is needed to stabilize these modified oils during storage by incorporation of appropriate antioxidants and adequate packaging technologies. Incorporation of SL containing n-3 PUFA into foods needs to be justified using evidence collected from animal studies and clinical trials. Further research should focus on the metabolism and medicinal importance and economic feasibility of large-scale production of SL containing a mixture of n-3 fatty acids.

Designing SL with specific fatty acids at specific locations of the TAG for use in medicine needs more studies. For example, it may be desirable to develop a SL for patients with cystic fibrosis that contains PUFA (e.g., EPA or DHA) at the sn-2 position, and MCFA at the sn-1, 3 positions.

In conclusions, fats or oils have been recognized for their nutritional, functional and sensory properties. They provide a more concentrated source of energy than do carbohydrates and proteins. There is an increasing concern about the link between a high uptake of certain types of fatty acids or an appropriate balance of the different fatty acids in the diet and certain disease conditions, such as cardiovascular disease, obesity and cancer. Thus, it is clear that there exists a need for specialty lipids that retain the physical, functional and sensory features of traditional lipids and provide specific health benefits.

REFERENCES

[1]   I. S. Newton, "Long-Chain Fatty Acids in Health and Nutrition," In: F. Shahidi and J. W. Finley, Eds, *Omega-3 Fatty Acids*: *Chemistry, Nutrition, and Health Effects*, American Chemical Society, Washington DC, 2001, pp. 14-27.

[2]   M. J. Wolin, "Fermentation in the Rumen and Large Intestin," *Science*, Vol. 213, No. 4515, 1981, pp. 463-1467.

[3]   J. Bezard and M. Bugaut, "Absorption of Glycerides Containing Short, Medium and Long Chain Fatty Acids," In: A. Kuksis, Ed., *Fat Absorption*, CRC Press, Boca Raton, 1986, pp. 119-158.

[4]   C. C. Akoh, "Structured Lipids-Enzymatic Approach," *Inform*, Vol. 6, 1995, pp. 1055-1061

[5]   C. C. Akoh, "Making New Structured Fats by Chemical Reaction and Enzymatic Modification," *Lipid Technology*, Vol. 9, 1997, pp. 61-66.

[6]   W. C. Heird, S. M. Grundy and V. S. Hubbard, "Structured Lipids and Their Use in Clinical Nutrition," *The American Journal of Clinical Nutrition*, Vol. 43, No. 2, 1986, pp. 320-324.

[7]   T. W. Lee and C. I. Hastilow, "Quantitative Determination of Triacylglycerol Profile of Structured Lipid by Capillary Supercritical Fluid Chromatography and High-Temperature Gas Chromatogaphy," *Journal of the American Oil Chemists' Society*, Vol. 76, 1999, pp. 1405-1413.

[8]   I.-H. Kim, H. Kim, K.-T. Lee, S.-H. Chung and S.-N. Ko, "Lipase-Catalyzed Acidolysis of Perilla Oil with Caprylic Acid to Produce Structured Lipids," *Journal of the American Oil Chemists' Society*, Vol. 79, No. 4, 2002, pp. 363-367.

[9]   K.-T. Lee and C. C. Akoh, "Structured Lipids: Synthesis and Application," *Food Research International*, Vol. 14, No. 1, 1998, pp. 17-34.

[10] S. J. Bell, E. A. Mascioli, B. R. Bistrian, V. K. Babayan and G. L. Blackburn, "Alternative Lipid Sources for Enteral and Parenteral Nutrition: Long- and Medium-Chain Triglycerides, Structured Triglycerides, and Fish Oils," *Journal of the American Dietetic Association*, Vol. 91, No. 1, 1991, pp. 74-78.

[11] J. P. Kennedy, "Structured Lipids: Fats of the Future," *Food Technology*, Vol. 38, 1991, pp. 76-83.

[12] J. W. Stewart, K. D. Wigges, N. C. Jacobson and P. J. Berger, "Effect of Various Triglycerides on Blood and Cholesterol of Calves," *Journal of Nutrition*, Vol. 108, 1978, pp. 561-565.

[13] N. B. Cater, J. H. Howard and M. A. Donke, "Comarison of the Effects of Medium-Chain Triacylglycerols, Palm Oil, and High Oleic Acid Sunflower Oil on Plasma Triacylglycerol Fatty Acids and Lipid and Lipoprotein Concentrations in Humans," *The American Journal of Clinical Nutrition*, Vol. 65, No. 1, 1979, pp. 41-46.

[14] A. Hashim and V. K. Babayan, "Studies in Man of

Partially Absorbed Dietary Fats," *The American Journal of Clinical Nutrition*, Vol. 31, 1987, p. S273.

[15] J. W. Finley, L. P. Klemann, G. A. Levelle, M. S. Otterburn and C. G. Walchak, "Caloric Availability of Salatrium in Rats and Humans," *Journal of Agricultural and Food Chemistry*, Vol. 42, No. 2, 1994, pp. 495-499.

[16] F. D. Gunstone, "Major Sources of Lipids," In: F. D. Gunstone and F. B. Padley, Eds., *Lipid Technologies and Application*, Marcel Dekker, Inc., New York, 1997, pp. 19-50.

[17] J. J. Gottenbos, "Nutritional Evaluation of n-3 and n-6 Polyunsaturated Fatty Acids," In J. Beare-Rogers, Ed., *Dietary Fat Requirements in Health and Development*, American Oil Chemists' Society, Chapaign, 1988, pp. 107-119.

[18] R. P. Mensink and M. B. Katan, "Effect of Mono-saturated Fatty Acids versus Complex Carbohydrates on High Density Lipoproteins in Health Men and Woman," *Lancet*, Vol. 327, No. 8525, 1987, pp. 122-127.

[19] D. J. Kyle, "The Large-Scale Production and Use of a Single-Cell Oil Highly Enriched Docosahexaenoic Acid," In: F. Shahidi and J. W. Finley, Eds., *Omega-3 Fatty Acids: Chemistry, Nutrition, and Health Effects*, American Chemical Society, Washington DC, 2001, pp. 92-105.

[20] H. O. Bang and J. Dyerberg, "Plasma Lipids and Lipoproteins in Greenlandic West-Coast Eskimos," *Acta-Medica Scandinavia*, Vol. 192, No. 1-6, 1972, pp. 85-94.

[21] H. O. Bang and J. Dyerberg, "Lipid Metabolism and Ischemic Heart Disease in Greenland Eskimos," *Advances in Nutritional Research*, Vol. 3, 1986, pp. 1-21.

[22] B. J. Holub, "Docosahexaenoic Acid in Human Health, Omega-3 Fatty Acids: Chemistry, Nutrition, and Health Effects," In: F. Shahidi and J. W. Finley, Eds., *ACS Symposium Series* 788, American Chemical Society, Washington DC, 2001, pp. 54-65.

[23] F. Shahidi and J. W. Finley, "Omega-3 Fatty Acids: Chemistry, Nutrition, and Health Effects, In: F. Shahidi and J. W. Finley, Eds., *ACS Symposim Series* 788, American Clinical Society, Washington DC, 2001.

[24] H. O. Bang, J. Dyerberg and E. Stofferson, "Eicosap-entaenoic Acid and Prevention of Thrombosis and Athero-sclerosis," *Lancet*, Vol. 312, No. 8081, 1978, pp. 117-122.

[25] S. P. Senanayake and F. Shahidi, "Enzyme-Assisted Acidolysis of Borage (*Borago officinalis* L.) and Evening Primrose (*Oenothera biennis* L.) Oils: Incorporation of Omega-3 Polyunsaturated Fatty Acids," *Journal of Agri-cultural and Food Chemistry*, Vol. 47, No. 8, 1999, pp. 3105-3112.

[26] S. P. Senanayake and F. Shahidi, "Structured Lipids via Lipase-Catalyzed Incorporation of Eicosapentaenoic Acid into Borage (*Borago officinalis* L.) and Evening Primrose (*Oenothera biennis* L.) Oils," *Journal of Agricultural and Food Chemistry*, Vol. 50, No. 3, 2002, pp. 477-483.

[27] C. C. Akoh, "Structured Lipids," In: C. C. Akoh and D. B. Min, Eds., *Food Lipids*, Marcel Dekker, Inc., New York, 2002, pp. 877-908.

[28] D. B. Hausman, D. R. Higbee and B. M. Grossman, "Dietary Fats and Obesity," C. C. Akoh and D. B. Min, Eds., *Food Lipids*, Marcel Dekker, Inc., New York, 2002, pp. 663-694.

[29] S. M. Budge, S. J. Iverson, W. D. Bowen and R. G. Ackman, "Among- and Within-Species Variability in Fatty Acid Signatures of Marine Fish and Invertebrates on the Scotian Shelf, George Bank, and Southern Gulf of St. Lawrence," *Canadian Journal of Fisheries and Aquatic Sciences*, Vol. 59, No. 5, 2002, pp. 886-898.

[30] K. D. Cromer, T. C. Jenkins and E. J. Thies, "Replacing Cis Ocatdecenoic Acid with Trans Isomers in Media Containing Rat Adipocytes Stimulates Lipolysis and Inhibits Glucose Utilization," *Journal of Nutrition*, Vol. 125, 1995, pp. 2394-2399.

[31] M. Launay, N. Vodovar and J. Raulin, "Development Du Tissue Adipeux: Nombre et Taille Des Cellules En Function De la Valent Energetique et de l'insaturation Des Lipids du Regime," *Bulletin de la Société Chimique Biologies*, Vol. 50, 1986, pp. 439-445.

[32] G. Shillabeer and D. C. W. Lau, "Regulation of New Formation in Rats: The Role of Dietary Fats," *Journal of Lipids Reseach*, Vol. 35, 1994, pp. 592-596.

[33] T. Raelot, R. Groscolas, D. Langin and P. Ferre, "Site-Specific Regulation of Gene Expression by n-3 Poly-unsaturated Fatty Acids in Rat White Adipose Tissue," *Journal of Lipids Reseach*, Vol. 38, 1997, pp. 963-1967.

[34] K.-T. Lee and C. C. Akoh, "Structured Lipids: Synthesis and Application," *Food Reseach International*, Vol. 14, No. 1, 1998, pp. 17-34.

[35] I. Ikeda, Y. Tomari, M. M. Sugano, S. Watanabe and J. Nagata, "Lymphatic Absorption of Structured Glycerides Containing Medium-Chain Fatty Acids and Linoleic Acid, and Their Effect on Cholestrol Absorption in Rats," Lipids in Health and Disease, Vol. 26, 1991, pp. 369-373.

[36] V. K. Babayan, "Medium Chain Triglycerides and Structured Lipids," *Lipids*, Vol. 22, No. 6, 1987, pp. 417-420.

[37] S. J. DeMichele, M. D. Karlstad, V. K. Babayan, N. Istfan, G. L. Blackburn and B. R. Bristrain, "Enhanced Skeletal Muscle and Liver Protein Synthesis with Structured Lipid in Enterally Fed Burned Rats," *Metabolism*, Vol. 37, No. 8, 1988, pp. 788-795.

[38] I. Ikeda, *et al.*, "Lymphatic Absorption of Structured Glycerolipids Containing Medium-Chain Fatty Acids and Linoleic Acid, and Their Effect on Cholesterol Ab-sorption in Rats," *Lipids*, Vol. 26, No. 5, 1991, pp. 369-373.

[39] R. Sandstrom, *et al.*, "Structured Triglycerides to Posto-perative Patients: A Safety and Tolerance Study," *Journal of Parenteral and Enteral Nutrition*, Vol. 17, No. 2, 1993, pp. 153-157.

[40] L. E. Crosby, E. S. Swenson, V. K. Babayan, N. Istfan, G. L. Blackburn and B. R. Bistrain, "Effect of Structured

Lipid-Enriched Total Parental Nutrition in Rats Bearing Yoshida Sarcoma," *Journal of Nutrition*, Vol. 1, No. 1, 1990, pp. 41-47.

[41] C. C. Akoh, "Lipid-Based Fat Substitutes," *Critical Reviews in Food Science and Nutrition*, Vol. 35, No. 5, 1995, pp. 405-430.

[42] F. Hamam and F. Shahidi, "Synthesis of Structured Lipids Containing Medium-Chain and Omega-3 Fatty Acids," *Journal of Agricultural and Food Chemistry*, Vol. 54, No. 12, 2006, pp. 4390-4396.

[43] F. Shahidi and F. Hamam, "Structured Lipids Containing Medium-Chain and Omega-3 Fatty Acids," *Inform*, Vol. 17, 2006, pp. 178-181.

[44] F. Hamam, J. Daun and F. Shahidi, "Lipase-Catalyzed Acidolysis of High-Laurate Canola Oil with Eicosapentaenoic Acid," *Journal of the American Oil Chemists' Society*, Vol. 82, 2005, pp. 875-879.

[45] F. Hamam and F. Shahidi, "Structured Lipids from High-Laurate Canola Oil and Long-Chain Omega-3 Fatty Acids," *Journal of the American Oil Chemists' Society*, Vol. 82, No. 10, 2005, pp. 731-736.

[46] A. P. Bimbo, "Fishmeal and Oil: Update Turmoil and Transition," In: F. Shahidi, Ed., *Seafood in Health and Nutrition*, Sci. Tech Publishing Company, St. John's, 2000, pp. 45-67.

[47] C. C. Akoh and C. O. Moussata, "Characterization and Oxidative Stability of Enzymatically Produced Fish and Canola Oil-Based Structured Lipids," *Journal of the American Oil Chemists' Society*, Vol. 78, No. 1, 2001, pp. 25-30.

[48] B. H. Jennings and C. C. Akoh, "Enzymatic Modification of Triacylglycerols of High Eicosapentaenoic and Docosahexaenoic Acids Content to Produce Structured Lipids," *Journal of the American Oil Chemists' Society*, Vol. 76, No. 10, 1999, pp. 1133-1137.

[49] A. Kawashima, Y. Shimada, M. Yamamoto, A. Sugihara, T. Nagao, S. Komemushi and Y. Tominaga, "Enzymatic Synthesis of High-Purity Structured Lipids with Caprylic Acid at 1,3-Postions and Polyunsaturated Fatty Acid at 2-Position," *Journal of the American Oil Chemists' Society*, Vol. 78, No. 6, 2001, pp. 611-616.

[50] Y. Shimada, A. Sugihara, K. Maruyama, T. Nagao, S. Nakayama, H. Nakano and Y. Tominaga, "Production of Structured Lipids Containing Docosahexaenoic and Caprylic Acids Using Immobilized *Rhizopus delemar*," *Journal of Fermentation and Bioengineering*, Vol. 81, No. 4, 1996, pp. 299-303.

[51] F. Hamam and F. Shahidi, "Synthesis of Structured Lipids via Acidolysis of Docosahexaenoic Acid Single Cell Oil (DHASCO) with Capric Acid," *Journal of Agricultural and Food Chemistry*, Vol. 52, No. 10, 2004, pp. 2900-2906.

[52] F. Hamam and F. Shahidi, "Enzymatic Incorporation of Capric Acid into a Single Cell Oil Rich in Docosahexaenoic Acid (DHA) and Docosapentaenoic Acid (DPA)," *Food Chemistry*, Vol. 91, No. 4, 2005, pp. 583-591.

[53] C. C. Akoh and C. O. Moussata, "Lipase-Catalyzed Modification of Borage Oil: Incorporation of Capric and Eicosapentaenoic Acids to Form Structured Lipids," *Journal of the American Oil Chemists' Society*, Vol. 75, No. 6, 1998, pp. 697-701.

List of Abbreviations

AA	Arachidonic acid		EFA	Essential fatty acids
ALA	α-linolenic acid		LA	Linoleic acid
DAG	Diacylglycerol		LCFA	Long-chain fatty acids
DHA	Docosahexaenoic acid		LCT	Long-chain triacylglycerols
DPA	Docosapentaenoic acid		MAG	Monoacylglycerols
EPA	Eicosapentaenoic acid		MCFA	Medium-chain fatty acids
FA	Fatty acid		SL	Structured lipids
FFA	Free fatty acids		TAG	Triacylglycerols

3

Fatty Acids and Associated Cardiovascular Risk

Caroline Le Goff[1], Jean-François Kaux[2], Ludovic Leroy[1], Joël Pincemail[3], Jean-Paul Chapelle[1], Etienne Cavalier[1]

[1]Department of Clinical Chemistry, University and University Hospital of Liege, Liege, Belgium; [2]Department of Clinical Sciences, University and University Hospital of Liege, Liege, Belgium; [3]CREDEC, Department of Cardiovascular Surgery, University Hospital of Liege, Liege, Belgium.

ABSTRACT

Introduction: A fatty acid (FA) is a carboxylic acid with a long aliphatic chain, which is either saturated or unsaturated. Recently, the role of FA and particularly omega-3 and -6 has emerged as cardiovascular risk factor in the literature. The aim of our study was to establish reference values for these FA and to compare them with data obtained in a population of acute myocardial infarction (AMI) patients. **Materials and methods:** Hundred thirty five healthy subjects (59.38 ± 27.12 yo, 75 men) were selected as reference population. We also evaluated FA in thirty three patients (55 ± 9 yo, 23 men) admitted in the Emergency Department of our Institution for AMI. The fasting whole blood was drawn in vacutainer containing EDTA. Before analysis, samples were washed and transmethylated. We performed the quantification of different FA by gas chromatography associated with flame ionization detector (GCFID). **Results:** We obtained results in control healthy patients to be used as reference values. In the AMI group, levels of omega-6 were significantly higher ($p < 0.05$) for C18:2n6 and C18:3n6 than the reference population and omega-3 values were significantly lower ($p < 0.01$) compared to reference value for C22:6n3. The omega-3 index was lower and the ratio omega-6/omega-3 was higher in AMI group compared to reference values. **Conclusions:** We have established reference value for FA and have compared these values with the results obtained in AMI population. FA determination is a new tool we are able to use and to process in our laboratory which can help the clinician to screen patients with the highest cardiovascular risks because of the implication of FA in the etiopathogeny of atherosclerosis.

Keywords: Fatty Acids; Acute Myocardial Infarction; Gas Chromatography

1. Introduction

A fatty acid (FA) is a carboxylic acid with a long aliphatic chain. This chain can either be saturated or unsaturated. Most naturally occurring FA have a chain of 4 to 28 carbon atoms. FA are produced by hydrolysis of the ester linkages in a fat or biological oil (both of which are triglycerides), with the removal of glycerol. There are many families of FA: "essential FA" with omega-3, omega-6, omega-7, omega-9, "saturated" FA (SFA) and "unsaturated" (cis and trans) FA. **Table 1** shows the fatty acids with their common name, the International Union of Pure and Applied Chemists (IUPAC) name and the formula [1]. For many years, research has focused on the involvement of fatty acids and their repercussions on the cardiovascular system.

The implication of FA, particularly, omega-3 and -6 in different diseases has been widely published in literature [2-9]. SFA, abundantly found in our diet, participate in the development of certain metabolic abnormalities such as insulin resistance or atherosclerosis [10,11]. However, all SFA have different metabolic effects. For example, palmitic acid (which consists of 50% palm oil versus a maximum of 17% for other oils such as peanut oil), unlike other SFA such as stearic acid, is a potent inhibitor of the intracellular pathway for insulin in some tissues such as skeletal muscle [10,12]. SFA are also considered as pro-inflammatory agents because they stimulate the inflammatory response in different cell types, such as adipocytes, particularly during the postprandial phase [12, 13]. Finally, this class of fatty acids probably contributes to the development of atherosclerotic plaques because it causes increased plasma concentrations of LDL-cholesterol [12]. All of this knowledge means that the current intake of SFA in western food, and in particular, palmitic acid, is a key player in the expansion of certain metabolic diseases such as type 2 diabetes, cardiovascular disease

Table 1. Fatty acids with their common name, the International Union of Pure and Applied Chemists (IUPAC) name and the formula.

Symbol	Chemical structure	Acid's common name	IUPAC name
	Saturated fatty acid		
C14:0	$CH_3(CH_2)_{12}COOH$	Myristic acid	tétradécanoïque
C15:0	$CH_3(CH2)_{13}COOH$	Pentadecanoic acid	Pentadecanoic
C16:0	$CH_3(CH2)_{14}COOH$	Palmitic acid	hexadecanoic
C17:0	$CH_3(CH2)_{15}COOH$	Margaric acid	heptadecanoic
C18:0	$CH_3(CH2)_{16}COOH$	Stearic acid	octadecanoic
C20:0	$CH_3(CH2)_{18}COOH$	Arachidic acid	eicosanoic
	Unsatured fatty acid ω-3		
C18: 3 ω-3	$CH_3(CH_2CH=CH)_3(CH_2)_7COOH$	α-linolenic acid	cis,cis,cis-9,12,15-octadécatrienoic
C20: 5 ω-3	$CH_3CH_2(CH=CHCH_2)_4CH=CH(CH_2)_3COOH$	Eicosapentaenoic acid	cis,cis,cis,cis,cis-5,8,11,14, 17-eicosapentaenoic
C22: 5 ω-3	$CH_3CH_2CH=CH(CH_2)_2CH=CHCH_2(CH=CH(CH_2)_2)_3COOH$	Docosapentaenoic acid	cis,cis,cis,cis,cis-4,8,12,15, 19-docosapentaenoic
C22: 6 ω-3	$CH_3CH_2(CH=CHCH_2)_5CH=CH(CH_2)_2COOH$	Docosahexaenoic acid	cis,cis,cis,cis,cis,cis-4,7,10,13,16, 19-docosahexaenoic
	Unsatured fatty acid ω-6		
C18: 2 ω-6	$CH_3(CH_2)_4CH=CHCH_2CH=CH(CH_2)_7COOH$	Linoleic acid	cis,cis-9,12-octadecadienoic
C18: 3 ω-6	$CH_3(CH_2)_3(CH_2CH=CH)_3(CH_2)_4COOH$	γ-linolenic acid	cis,cis,cis-6,9,12-octadecatrienoic
C20: 3 ω-6	$CH_3(CH_2)_4(CH=CHCH_2)_3(CH_2)_5COOH$	di-homo-γ-linolenic acid	cis,cis,cis-8,11,14-eïcosatrienoic
C20: 4 ω-6	$CH_3(CH_2)_4(CH=CHCH_2)_3CH=CH(CH_2)_3COOH$	Arachidonic acid	cis,cis,cis,cis-5,8,11, 14-eicosatetraenoic
	Others unsatured fatty acid		
C16: 1 ω-7	$CH_3(CH_2)_5CH=CH(CH_2)_7COOH$	Palmitoleic acid	cis 9 hexadécénoïc
C18: 1 ω-9	$CH_3(CH_2)_7CH=CH(CH_2)_7COOH$	Oléic acid	cis-9-octadécénoic
C20: 1 ω-9	$CH_3(CH_2)_7CH=CH(CH_2)_9COOH$	Gondoic acid	cis-9-eicosénoic
	Trans unsatured fatty acid		
C18: 1 ω-9	$CH_3(CH_2)_7CH=CH(CH_2)_7COOH$	Elaidic acid	trans-9-octadecenoic
C18: 2 ω-6	$CH_3(CH_2)_4CH=CHCH=CH(CH_2)_8COOH$	Linolelaidic acid	trans,cis-10,12-octadecadienoic

or metabolic syndrome [8,14]. Dietary long-chain, poly-unsaturated fatty acid (PUFA) also benefit the blood vessel wall by enhancing cell membrane fluidity and regulating membrane receptors [15].

Hundreds of epidemiologic studies, either of mechanisms of action, or experimental in animals have shown that dietary intake of omega-3 FA presented anti atherosclerotic potential [16-24]. The role of FA in the atherosclerossis process was already discussed by Sinclain in 1956 in the Lancet [25].

Some authors have demonstrated that the protective effect of omega-3 PUFAs in sudden cardiac death which was most often due to ventricular arrhythmia [5,26]. However, in patients with implantable cardioverter defibrillators and cardiac failure, no benefit of supplementation with omega-3 could be established [27]. The erythrocyte fatty acid profile is modified; the omega-3 PUFA concentrations are elevated and predict the risk of ventricular arrhythmia. These results highlight an altered cardiac metabolism in patients with heart failure and may have clinical implications for early identification of subjects at high risk [28].

In general, prospective epidemiological studies have shown a relationship between intake of trans fatty acids and increased morbidity and mortality from cardiovascular disease in Europe and North America [29,30]. The incomes of trans fatty acids compared to the intake of cis fatty acids, increase levels of LDL-cholesterol, triglycerides, lipoprotein Lp (a) and decrease levels of HDL-cholesterol and the actual particle size of LDL-cholesterol. Each point may increase the risk of coronary heart disease [31].

An important concept in the field of fatty acids is the balance omega-6/omega-3. An ideal relationship for dietary intake would be 1:1 but it is rather around the ratio of 15:1, linked to deficiency of omega-3 PUFA intakes. The increase in the proportion of omega-6 PUFA is considered as a risk factor for many diseases such as heart disease, inflammatory, autoimmune and various cancers [2]. Freije compared the fatty acid composition of erythrocyte membranes in healthy patients with those of patients with coronary heart disease in Bahrain [2]. The results mainly showed a low level of docosahexaenoic acid which increases the risk of coronary heart disease. Moreover, there was also an increased ratio of omega-6/omega-3. An omega-3 index decrease (which is the sum of the percentages of docosahexaenoic acid and eicosapentaenoic acid), was also observed in these patients [3].

The aim of our study was to establish reference values for these FA and to compare the results obtained with data observed in patients with acute myocardial infarction (AMI) seen the protector role of PUFA particularly, omega-3, and the implication of SFA in the development of atherosclerosis [32].

2. Materials and Methods

Hundred thirty five healthy subjects (59.38 ± 27.12 yo, 75 men) were selected as reference population (patient without cardiac antecedent or diabetes). We also evaluated FA in thirty three patients (55 ± 9 yo, 23 men) admitted in the Emergency Department of our Institution for AMI. All patients gave their informed consent.

Fasting whole blood samples were drawn in vacutainer tubes containing EDTA. One millilitre of whole blood was centrifuged at 3500 rpm for 5 min at 4°C. Plasma was carefully separated and the red blood cells (RBC) pellet was washed three times with equal volumes of standard saline. All the washings were carried out at 4°C. The white viscous material after each washing was carefully removed to minimize contamination by non-erythrocytic cells. The samples were finally stored at −20°C until analyzed (personal data).

Before analysis, the samples were washed and an extraction was done.

Briefly, 0.3 ml packed RBCs were placed in a 5 ml clean glass vial with 100 μl of C28 (Internal Standard). RBCs were then treated with 2 ml of methanolic-HCl.95%/5%. The vials were sealed and incubated at 80°C for 2 h. After incubation the mixture was treated with 5 ml of hexane and centrifuged at 3500 rpm for 10 min at room temperature using a swinging rotor. The top aqueous layer comprised mainly of methylated fatty acids. Hexane mixture was then carefully collected in a separate vial. This step was repeated three times and all extractions were pooled. The methylated-fatty acids hexane mixture was then completely dried at 55°C under nitrogen. Methylated fatty acids were dissolved in 100 μl hexane and 1 μl was identified and quantified by gas chromatography with flame ionization detector (GC/FID) performed on a Shimadzu GC2010, using a capillary column SP-2380 of 30 mm × 0.25 mm × 0.20 μm dimensions (Supelco). A FID was used with an oven temperature set at 250°C. The temperature of the injector was 250°C whereas the detector was set at 260°C.

The quantification of FA is realized according to the calibration curve that we have established with standards for all peaks that we must identify.

First, we expressed the results in mg/L and after we chose to express results in percent (%) from the total of results. The sum of the FA found in a patient equals 100% and with a simple calculation we expressed them in % in relation to the sum of FA.

For the statistical analysis, we used an independent sample T-test with the statistical software Medcalc 9.1 (Mariakerk, Belgium).

3. Results

Figure 1 shows the chromatographic profile of the different FA analyzed. The **Table 2** presents the comparison of the results observed in AMI and in healthy subjects (reference values).

We obtained differences between AMI group and our healthy population.

In AMI, the levels of omega-6 were significantly higher ($p < 0.05$) for C18:3n6. The level of omega-3 was lower in comparison with reference values but not significantly. We also calculated the omega-3 index and the ratio omega-6/omega-3. The omega-3 index was significantly lower ($p < 0.0001$) in AMI compared to the reference value and the omega-6/omega-3 ratio was significantly higher ($p < 0.005$) in AMI than in reference patients. The sum of SFA was also significantly higher in AMI group ($p < 0.05$). We also noted a significantly difference ($p < 0.005$) between both types of patients for C18:1n9.

4. Discussion

The incidence of cardiovascular disease is different between the countries and increases from Southern to the Northern countries (Europe) [33,34].

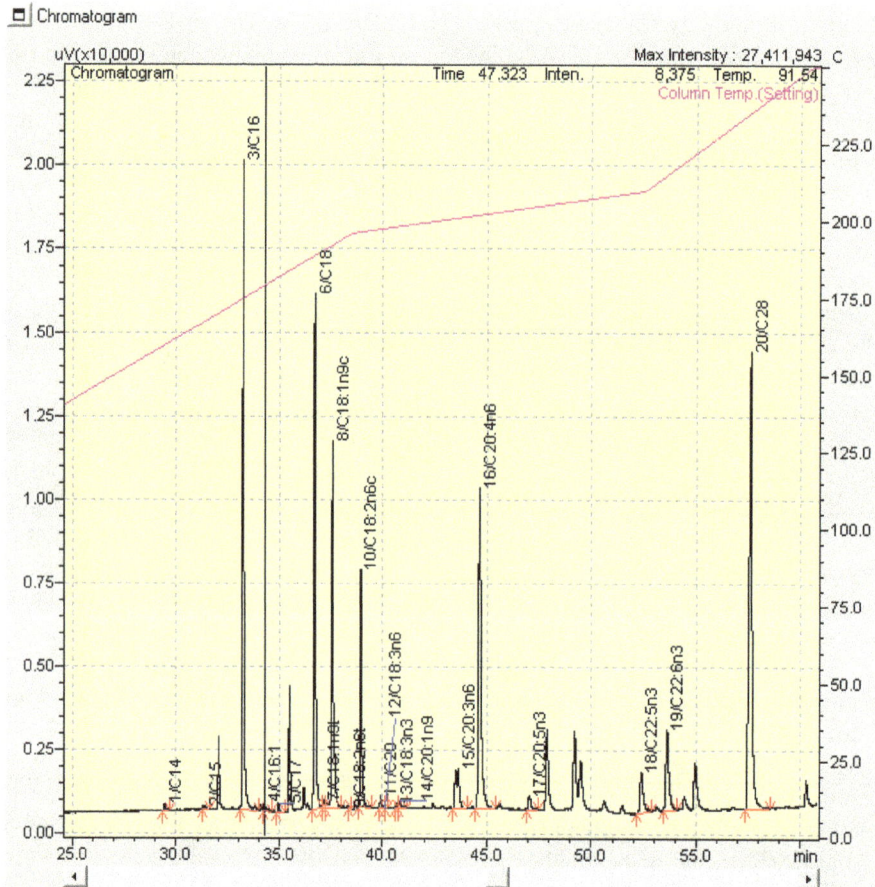

Figure 1. Example of a chromatographic profile of the different FA analyzed by gas chromatography in a healthy subject.

Cardiovascular diseases are the first cause of mortality and morbidity in the developped countries. They are the world's largest killers, claiming 17.1 million lives a year [35]. This incidence has increased with time. One explanation is probably found in changes of life style: aging, tobacco use, unhealthy diet, stress, physical inactivity and harmful use of alcohol increase the risk of heart attacks and strokes [35]. Atherosclerosis initiation and early progression is almost invariability a result of abnormal interactions between circulating oxidized lipids and blood vessel walls. Endothelial dysfunction has been linked to diverse vascular abnormalities like atherosclerosis. The first detectable stage in atherosclerosis is a slightly raised, fatty "streak" in the arterial wall. Among Western populations, these can be found in most individuals beginning as early as the teenage years [15]. In Northern of Europe, it is well known that people do not follow the Mediterranean diet, which has long been celebrated as the gold standard of healthy diet for its favourable impact on the prevention of chronic diseases, promotion of greater longevity and quality of life [36].

We have studied AMI patients because they are at risk of a second myocardial infarction due to their cardiovascular risk phenotype. Moreover, the implication of SFA

and PUFA in the etiopathogeny of atherosclerosis is well known. We found thus very interesting to be able to make the quantification of these different FA to achieve on the dietary prevention (supplementation in omega-3, decrease of intake of SFA…).

For the SFA, we observed a statistical increase between AMI and control subjects. These results are in accordance with Walrand et al. study [12], which explained that SFA was a key player in the expansion of cardiovascular disease. This class of fatty acids is most likely involved in the development of atherosclerosis probably due to the induced increasing of plasma concentrations of LDL-cholesterol. However, one of the limitations of our study is without any doubt the age difference between both patient groups as well as daily hygiene which we were able to define with our control group but not with the AMI group.

Our study also showed a statistically significant difference between the percentages of oleic acid (C18: 1n 9c) in the 2 groups. This increase in oleic acid has also been observed in the study of Harris et al. [4] who studied the lipid profile and erythrocytic plasma after myocardial infarction. It must be said that this fatty acid is often considered beneficial as it benefits from the posi-

Table 2. Comparison between AMI and reference group for FA (FA in %). p-values in red are statistically significative (p < 0.05).

	FA AMI % (n = 33)	FA reference % (n = 135)	p-value
C14	1.00 ± 0.21	0.9 ± 0.19	0.024
C15	0.34 ± 0.07	0.37 ± 0.15	<0.001
C16	25.45±1.51	31.05 ± 7.7	<0.001
C16: 1	0.56 ± 0.21	0.52 ± 0.23	0.329
C17	0.42 ± 0.09	0.41 ± 0.07	0.085
C18	20.53 ± 1.35	20.16 ± 1.56	0.342
C18: 1ω9t	0.23 ± 0.13	0.19 ± 0.09	0.003
C18: 1ω9c	15.43 ± 1.38	14.68 ± 1.03	0.005
C18: 2ω6t	0.07 ± 0.02	0.07 ± 0.04	0.513
C18: 2ω6c	9.15 ± 1.34	9.64 ± 0.89	0.043
C20	0.54 ± 0.09	0.57 ± 0.08	0.063
C18: 3ω6	0.11 ± 0.04	0.08 ± 0.02	0.034
C18: 3ω3	0.22 ± 0.04	0.24 ± 0.06	0.212
C20: 1ω9	0.27 ± 0.08	0.26 ± 0.05	0.553
C20: 3ω6	2.14 ± 0.50	2.26 ± 0.51	0.301
C20: 4ω6	15.45 ± 1.54	14.99 ± 1.15	0.112
C20: 5ω3	0.78 ± 0.26	0.72 ± 0.34	0.079
C22: 5ω3	1.94 ± 0.34	2.00 ± 1.08	0.749
C22: 6ω3	5.37 ± 1.36	6.28 ± 3.04	0.094
Sum of SFA	48.27 ± 1.38	47.57 ± 1.35	0.022
ω6/ω3	3.39 ± 0.86	2.86 ± 0.58	0.001
ω3 index	6.15 ± 1.51	7.60 ± 1.62	<0.001

tive image of olive oil, containing 70% oleic acid, and no less favourable than the Mediterranean diet. However, a portion of beneficial effects in olive oil on endothelial functions, inflammation, platelet aggregation, and perhaps even on HDL cholesterol is the lipid fraction of the olive oil [37].

In regards to trans fatty acids some statistically significant difference (p < 0.005) was observed between the 2 groups although many articles [31,38] show the harmful effects related to the consumption of trans fatty acids on the development of cardiovascular risk factors. Indeed, they are proven to increase levels of LDL-cholesterol, triglycerides, lipoprotein Lp (a), and lower levels of HDL-cholesterol as well as particle size of LDL-cholesterol. It is for all these reasons that the political authorities try to regulate the possible content of trans fat in food.

In this study, we observed differences between the two studied groups in terms of docosahexaenoic acid (C22: 6 n3), omega-6/omega-3 ratio and the omega-index, but also a significant difference for linoleic acid (C18: 2 n6c). These differences are consistent with both studies [3,39], which highlighted the same differences. However in the Freije study, the percentage of C22: 6 n3 decreased from 2.81% to 0.41% and the ω3 index from 3.14% to 0.64% in control subjects and AMI patients [2]. These results were obtained in the kingdom of Bahrain and the authors believe that these low values of C22: 6 n3 could be due to very low omega-3 fatty acids in fish inhabiting nearby waters [2]. The attempts to corroborate the higher results were obtained.

5. Conclusion

We have established reference values for FA implicated in different pathologies like cardiovascular diseases. This is a new tool we are able to use in our laboratory which can help clinicians to highlight and target patients with a higher cardiovascular risks. The clinicians could then supplement the patients with, for example, omega-3 to reduce the risk of developing cardiac diseases because of the implication of FA in the etiopathogeny of atherosclerosis.

REFERENCES

[1] A. D. McNaught and A. Wilkinson, "IUPAC Compendium of Chemical Terminology," 2nd Edition, Blackwell Science, Oxford, 1997.

[2] A. Freije, "Fatty Acid Profile of the Erythrocyte Membranes of Healthy Bahraini Citizens in Comparison with Coronary Heart Disease Patients," *Journal of Oleo Science*, Vol. 58, No. 7, 2009, pp. 379-388.

[3] G. Durand and J. L. Beaudeux, "Biochimie Médicale: Marqueurs Actuels et Perspectives," Lavoisier, Paris, 2008.

[4] W. S. Harris and C. Von Schacky, "The Omega-3 Index: A New Risk Factor for Death from Coronary Heart Disease?" *Preventive Medicine*, Vol. 39, No. 1, 2004, pp. 212-220.

[5] C. M. Albert, H. Campos, M. J. Stampfer, P. M. Ridker, J. E. Manson, W. C. Willett, *et al.*, "Blood Levels of Long-Chain n-3 Fatty Acids and the Risk of Sudden Death," *The New England Journal of Medicine*, Vol. 346, No. 15, 2002, pp. 1113-1138.

[6] F. Paganelli, J. M. Maixent, M. J. Duran, R. Parhizgar, G. Pieroni and S. Sennoune, "Altered Erythrocyte n-3 Fatty Acids in Mediterranean Patients with Coronary Artery Disease," *International Journal of Cardiology*, Vol. 78, No. 1, 2001, pp. 27-32.

[7] H. Rupp, D. Wagner, T. Rupp, L. M. Schulte and B.

Maisch, "Risk Stratification by the 'EPA + DHA Level' and the 'EPA/AA Ratio' Focus on Anti-Inflammatory and Antiarrhythmogenic Effects of Long-Chain Omega-3 Fatty Acids," *Herz*, Vol. 29, No. 7, 2004, pp. 673-685.

[8]   A. M. Hodge, D. R. English, K. O'Dea, A. J. Sinclair, M. Makrides, R. A. Gibson, *et al.*, "Plasma Phospholipid and Dietary Fatty Acids as Predictors of Type 2 Diabetes: Interpreting the Role of Linoleic Acid," *The American Journal of Clinical Nutrition*, Vol. 86, No. 1, 2007, pp. 189-197.

[9]   G. Mamalakis, M. Kiriakakis, G. Tsibinos, E. Jansen, H. Cremers, C. Strien, *et al.*, "Lack of an Association of Depression with n-3 Polyunsaturated Fatty Acids in Adipose Tissue and Serum Phospholipids in Healthy Adults," *Pharmacology Biochemistry and Behavior*, Vol. 89, No. 1, 2008, pp. 6-10.

[10]   S. M. Hirabara, R. Curi and P. Maechler, "Saturated Fatty Acid-Induced Insulin Resistance Is Associated with Mitochondrial Dysfunction in Skeletal Muscle Cells," *Journal of Cellular Physiology*, Vol. 222, No. 1, 2010, pp. 187-194.

[11]   J. Lovergrove, L. Brady, S. Lesauvage, A. M. Minihane and C. Williams, "Platelet Membrane Phospholipid Fatty Acids Are Weakly Associated with Markers of Insulin Resistance and Some CVD Risk Factors," ISSFAL2008.

[12]   S. Walrand, F. Fisch and J. M. Bourre, "Do Saturated Fatty Acids Have the Same Metabolic Effect?" *Nutrition Clinique et Metabolisme*, Vol. 24, No. 2, 2010, pp. 63-75.

[13]   S. Sierra, F. Lara-Villoslada, M. Comalada, M. Olivares and J. Xaus, "Dietary Eicosapentaenoic Acid and Docosahexaenoic Acid Equally Incorporate as Decosahexaenoic Acid but Differ in Inflammatory Effects," *Nutrition*, Vol. 24, No. 3, 2008, pp. 245-254.

[14]   R. B. Moore and S. H. Appel, "Methylation of Erythrocyte Membrane Phospholipids in Patients with Myotonic and Duchenne Muscular Dystrophy," *Experimental Neurology*, Vol. 70, No. 2, 1980, pp. 380-391.

[15]   P. R. Kidd, "Cell Membranes, Endothelia, and Atherosclerosis—The Importance of Dietary Fatty Acid Balance," *Alternative Medicine Review*, Vol. 1, No. 3, 1996, pp. 148-165.

[16]   C. Von Schacky, "Prophylaxis of Atherosclerosis with Marine Omega-3 Fatty Acids. A Comprehensive Strategy," *Annals of Internal Medicine*, Vol. 107, No. 6, 1987, pp. 890-899.

[17]   T. A. Dolecek, "Epidemiological Evidence of Relationships between Dietary Polyunsaturated Fatty Acids and Mortality in the Multiple Risk Factor Intervention Trial," *Proceedings of the Society for Experimental Biology and Medicine*, Vol. 200, No. 2, 1992, pp. 177-182.

[18]   C. M. Albert, C. H. Hennekens, C. J. O'Donnell, U. A. Ajani, V. J. Carey, W. C. Willett, *et al.*, "Fish Consumption and Risk of Sudden Cardiac Death," *JAMA*, Vol. 279, No. 1, 1998, pp. 23-28.

[19]   D. S. Siscovick, T. E. Raghunathan, I. King, S. Weinmann, K. G. Wicklund, J. Albright, *et al.*, "Dietary Intake and Cell Membrane Levels of Long-Chain n-3 Polyunsaturated Fatty Acids and the Risk of Primary Cardiac Arrest," *JAMA*, Vol. 274, No. 17, 1995, pp. 1363-1367.

[20]   M. L. Daviglus, J. Stamler, A. J. Orencia, A. R. Dyer, K. Liu, P. Greenland, *et al.*, "Fish Consumption and the 30-Year Risk of Fatal Myocardial Infarction," *The New England Journal of Medicine*, Vol. 336, No. 15, 1997, pp. 1046-1053.

[21]   J. Dyerberg and H. O. Bang, "Haemostatic Function and Platelet Polyunsaturated Fatty Acids in Eskimos," *Lancet*, Vol. 2, No. 8140, 1979, pp. 433-435.

[22]   W. E. Kaminski, E. Jendraschak, R. Kiefl and C. von Schacky, "Dietary Omega-3 Fatty Acids Lower Levels of Platelet-Derived Growth Factor mRNA in Human Mononuclear Cells," *Blood*, Vol. 81, No. 7, 1993, pp. 1871-1879.

[23]   S. Endres and C. von Schacky, "n-3 Polyunsaturated Fatty Acids and Human Cytokine Synthesis," *Current Opinion in Lipidology*, Vol. 7, No. 1, 1996, pp. 48-52.

[24]   C. Von Schacky, "Cardiovascular Effects of n-3 Fatty Acids," In: J. C. Frölich and C. Von Schacky, Ed., *Fish, Fish Oil, and Human Health*, Zuckschwerdt Verlag, Munich, 1992, pp. 167-178.

[25]   H. M. Sinclair, "Deficiency of Essential Fatty Acids and Atherosclerosis, Etcetera," *Lancet*, Vol. 270, No. 6919, 1956, pp. 381-383.

[26]   M. Wilhelm, R. Tobias, F. Asskali, R. Kraehner, S. Kuly, L. Klinghammer, *et al.*, "Red Blood Cell Omega-3 Fatty Acids and the Risk of Ventricular Arrhythmias in Patients with Heart Failure," *American Heart Journal*, Vol. 155, No. 6, 2008, pp. 971-977.

[27]   P. Saravanan, N. C. Davidson, E. B. Schmidt and P. C. Calder, "Cardiovascular Effects of Marine Omega-3 Fatty Acids," *Lancet*, Vol. 376, No. 9740, 2010, pp. 540-50.

[28]   F. B. Hu, M. J. Stampfer, J. E. Manson, E. Rimm, G. A. Colditz, B. A. Rosner, *et al.*, "Dietary Fat Intake and the Risk of Coronary Heart Disease in Women," *The New England Journal of Medicine*, Vol. 337, No. 21, 1997, pp. 1491-1499.

[29]   W. C. Willett, M. J. Stampfer, J. E. Manson, G. A. Colditz, F. E. Speizer, B. A. Rosner, *et al.*, "Intake of Trans Fatty Acids and Risk of Coronary Heart Disease among Women," *Lancet*, Vol. 341, No. 8845, 1993, pp. 581-585.

[30]   D. Mozaffarian, "Trans Fatty Acids—Effects on Systemic Inflammation and Endothelial Function," *Atherosclerosis Supplements*, Vol. 7, No. 2, 2006, pp. 29-32.

[31]   A. P. Simopoulos and L. G. Cleland, "Omega-6/Omega-3 Essential Fatty Acid Ratio: The Scientific Evidence," Karger AG, Basel, 2003.

[32]   Q. Sun, J. Ma, H. Campos, S. E. Hankinson and F. B. Hu,

"Comparison between Plasma and Erythrocyte Fatty Acid Content as Biomarkers of Fatty Acid Intake in US Women," *The American Journal of Clinical Nutrition*, Vol. 86, No. 1, 2007, pp. 74-81.

[33] W. C. Smith and H. Tunstall-Pedoe, "European Regional Variation in Cardiovascular Mortality," *British Medical Bulletin*, Vol. 40, No. 4, 1984, pp. 374-379.

[34] J. Muller-Nordhorn, S. Binting, S. Roll and S. N. Willich, "An Update on Regional Variation in Cardiovascular Mortality within Europe," *European Heart Journal*, Vol. 29, No. 10, 2008, pp. 1316-1326.

[35] World Health Organization, "Prevention of Cardiovascular Disease : Guidelines for Assessment and Management of Total Cardiovascular Risk," World Health Organization, Geneva, 2007.

[36] J. B. Brill, "The Mediterranean Diet and Your Health," *American Journal of Lifestyle Medicine*, Vol. 3, No. 1,

2009, pp. 44-56.

[37] D. R. Morgan, L. J. Dixon, C. G. Hanratty, N. El-Sherbeeny, P. B. Hamilton, L. T. McGrath, *et al.*, "Effects of Dietary Omega-3 Fatty Acid Supplementation on Endothelium-Dependent Vasodilation in Patients with Chronic Heart Failure," *American Journal of Cardiology*, Vol. 97, No. 4, 2006, pp. 547-551.

[38] Y. Park, J. Lim, Y. Kwon and J. Lee, "Correlation of Erythrocyte Fatty Acid Composition and Dietary Intakes with Markers of Atherosclerosis in Patients with Myocardial Infarction," *Nutrition Research*, Vol. 29, No. 6, 2009, pp. 391-396.

[39] D. Mozaffarian, T. Pischon, S. E. Hankinson, N. Rifai, K. Joshipura, W. C. Willett, *et al.*, "Trans-Fatty Acid Intake and Systemic Inflammation among Women," *Circulation*, Vol. 109, No. 7, 2004, p. 195.

4

Acidity/Rancidity Levels, Chemical Studies, Bacterial Count/Flora of Fermented and Unfermented Silver Catfish (*Chrysichthys nigrodigitatus*)

O. A. Oyelese, O. M. Sao, M. A. Adeuya, J. O. Oyedokun

Department of Aquaculture and Fisheries Management, Faculty of Agriculture and Forestry, University of Ibadan, Ibadan, Nigeria.

ABSTRACT

The keeping quality and shelf life of fermented and unfermented *Chrysichthys nigrodigitatus* were monitored in this study. Four kilograms of fresh *Chrysichthys nigrodigitatus* was minced into fine particles (with an initial pH of 7.2 before distribution into 8 samples). Samples 1-4 are unfermented cooked while Samples 5-8 were fermented, not cooked. All the 8 prepared samples barely lasted for two weeks, while samples 1, 3 and 7 lasted for six weeks. Total Volatile Base (TVB) ranged higher (24.12 - 29.43) mg/100gm in Samples 1-4 than (14.23 - 18.09) mg/100gm recorded in Samples 5-8. In Samples 1-4, FFA values were not significantly ($P > 0.05$) different; also followed a narrow range of (6.14 - 6.45)% while higher range of (6.42 - 12.27)% recorded in samples (5-8). Peroxide values (PV) increased in all the 8 samples in the second, fourth and sixth week, however higher values were recorded in Samples 5-8. Acidity generally increased with length (weeks) of fermentation with a gradual drop in pH from 7.2 (in the fresh fish) to pH 4.5 (sample 7), the worst sample at six weeks. Sample 4 with bacteria load of 5.05×10^5 at second week and sample 7 (8.2×10^5) at sixth week became unfit for consumption having exceeded the 5.0×10^5 ICMSF standard for safe fish product. Five bacteria species (*Lactobacillus sp*, *Proteus spp*, *Staphylococcus aureus*, *Staphylococcus epidermis*, *Bacillus sp*) with the exception of *Proteus sp* were not represented in sample 1 (due to salt content). Strong positive correlation ($r = 0.97$, $P < 0.01$) exists between PV and FFA. Acidity of the fermented products increased over the weeks with strong negative correlation ($r = -0.121$, $P < 0.01$) exists between pH and FFA. Acidity (*l.e* drop in PH) with increasing rancidity since ($r = -0.313$, $P < 0.05$) exists between PV and pH.

Keywords: Acidity; Bacteria; Fermented; Unfermented; Rancidity

1. Introduction

Fermentation is one method of fish curing in which the development of a distinctive flavor in the final product is the principal objective. Therefore, this product is mainly used as a condiment in the preparation of traditional sources. Fermentation alone as a curing process does not preserve fish because it results in the breakdown of fish muscle. For this reason, fermentation is often combined with salting and/or drying in order to reduce water activity and retard or eliminate the growth of proteolytic and putrefying bacteria.

Fermentation method is an ancient method of processing food products dating back thousands of years. It is a process by which beneficial bacteria (like *Lactobacillus*) are encouraged to grow and increase the acidity of food, prevent spoilage and food poisoning, bacteria growth, hence preserving it. Fermented fish can be described as any fishery product that has undergone degradation changes through enzymatic or microbiology activities either in the presence or absence of salt [1]. Fermented fish products have for many years been considered a South East Asian product [2].

Fermented fish have for many years been considered as a value added product. However, since the raw fish are sometimes from poor quality or underutilized species which are very cheap, product price are affordable to many low income consumers. Some examples of fermented fish products include shushi of Japan, Patis of Phillipines, nuocmam of Vietnam and Cambodia, nampla of Thailand and Budu of Malaysia [3].

Acidity/Rancidity Levels, Chemical Studies, Bacterial Count/Flora of Fermented and Unfermented
Silver Catfish (Chrysichthys nigrodigitatus)

25

In Africa, the popularity of fermented fish products has been influenced by the fish consumption pattern which has been reported to be relatively higher in the coastal countries due to proximity to fish source than in the hinterland which is far from source of fish, hence less fish supply [4]. Many Africans have a strong preference for fresh fish when it is available. However in the absence of fresh fish, cured fish products such as smoked, salted, sundried and fermented fish are predominant and popular.

Also in Africa, fish preservation is accompanied by partial fermentation within a few days during which flavor can be developed in the fish. [5] noted that in Africa, fermented fish products are used as condiments especially in the rural areas. In some Africa countries such as Ghana, Gambia, Uganda, Sierra Leone, Chad, Cote d' Ivoure, Mali and Sudan, there has been relative popularity of fermented fish products [1,6-9].

In Nigeria, little has been documented on fermented fish products. However, Azeez, N. I. [10] reported that fermented fish had a ready market in the Lake Chad region of Northern Nigeria. Though fish fermentation is not common and really appreciated in Nigeria, the benefit inherent in the consumption of this product will be enormous, more especially in salvaging the fish farmer and marketers from post-harvest losses.

Fish fermentation in the Southeast Asian sub-region normally lasts for several months (three to nine months) and the fish flesh may liquefy or turn into a paste [2]. No African fermented fishery products are mentioned in the FAO Fisheries Report No. 100 on fermented fish; however, Fessiokh from the Sudan is mentioned as a Mediterranean product. Fermentation is characterized by a strong odour and for this reason, various authors have described the product as "stink" fish. In Africa, fermentation is usually accomplished with salting and drying. Fish in its natural environment has its own microflora in the slime on its body, in its gut and in its gills. These micro-organisms, as well as the enzymes in the tissues of the fish, bring about putrefactive changes in fish when it dies. Micro-organisms require water for growth and metabolism, while growth may be inhibited at a water activity (Aw) below 0.60.

Several studies have been carried out to study the biochemical pathways followed during the degradation process of fish fermentation. [11] reported that the strong odour in spoilt fish may be a reaction between TMAD and lactic acid. Tomiyasu, Y. et al. [12] also incriminated organic acids in deteriorated fish. Pearson, D. [13] identified the following five chemical changes in deteriorating fish:

1) Enzymic degradation of nucleotides and nucleosides in the flesh leading to the formation of inosine, hy-poxanthine, ribose etc.

2) Bacterial reduction of trimethylamine oxide (TMAO), a non-volatile and non-odoriferous compound, to volatile trimethylamine (TMA) which has ammonical smell.

3) Formation of dimethylamine (DMA).

4) Breakdown of protein with subsequent formationof ammonia (NH_3), indole, hydrogen sulphide.

5) Oxidative rancidity of the fat.

Therefore, the choice of Chrysichthys nigrodigitatus for fermentation in this study is based on the known fact that fatty fishes produce a more acceptable texture and flavor on fermentation. Hence the keeping quality and shelf life of fermented and unfermented Chrysichthys nigrodigitatus is monitored in this study through Acidity (pH levels) and Rancidity (PV) peroxide value levels along with other chemical parameters (TVB, FFA) and bacteria load of sample preparations for this fermentation study.

2. Materials and Methods

2.1. Collection of Samples

Twenty kilograms of live Chrysichthys nigrodigitatus of average weight 500 gm per fish was procured from Asejire Dam stored in plastic coolers containing ice cubes and transported to the Department of Aquaculture and Fisheries Management.

2.2. Preparation of Samples

The fresh fish were gutted and washed thoroughly, 4kg of fresh Chrysichthys nigrodigitatus was minced into fine particles (with an initial pH 7.2 recorded at 0 week before distribution into 8 different samples).

The procedures for fermented and unfermented fish paste preparation used in this study were according to the method of Oyelese and Odubayo [14]. A total of eight samples (four unfermented and four fermented samples) were prepared for fish paste experiment as follows:

1) Unfermented Chrysichthys nigrodigitatus: two kilograms fresh Chrysichthys nigrodigitatus was gutted and washed thoroughly in clean water. The fish was ground into fine paste (with a little quantity of water added) using an electric grinder. The fish was reweighed after grinding, cooked for 30 min, then weighed again after cooking; was shared into 4 equal parts of 0.45 kg each and were treated as follows:

Sample 1: This sample was salted with 45 g (i.e 10% salting) and frozen at −25°C (freezer temperature) and then covered.

Sample 2: This sample was salted with 45 g salt (i.e 10% salting) but left at room (ambient temperature of 26°C) on a shelf opened.

Sample 3: This sample was not salted but frozen at −25°C (freezer temperature) and then covered.

Sample 4: This sample was not salted, but left at room (ambient temperature of 26°C) on a shelf opened.

2) Fermented *Chrysichthys nigrodigitatus*: another 2kg of fresh *Chrysichthys nigrodigitatus* fish was gutted and washed thoroughly in clean water. The fish was ground into fine paste (with a little quantity of water added) using anelectric grinder. The fish was reweighed after grinding but this product was not cooked. It was then divided into four equal parts of 0.45 kg each, the uncooked fish paste was treated as follows:

Sample 5: This sample was salted with 45 g salt (*i.e* 10% salting) covered and left at room (ambient temperature of 26°C).

Sample 6: This sample was salted with 45 g salt (*i.e* 10% salting), opened and left at room (ambient temperature of 26°C).

Sample 7: This sample was not salted, covered and left at room (ambient temperature of 26°C).

Sample 8: This sample was salted, opened and left at room (ambient temperature of 26°C).

The proximate analysis which are the crude protein, moisture content, fat, crude fibre, ash of the initial fresh fish will be carried out and also the bacteria count and an identification will be carried out every two weeks and also the chemical analysis will be done bi-weekly for a six week period in other to determine d shelf life of the 8 sample preparation..

2.3. Chemical Analysis

Free Fatty Acid content (FFA), Peroxide Value (PV) (Rancidity) and Acidity (pH level) of the unfermented and fermented fish were determined using A.O.A.C. [15] method while Total Volatile Base content was determined by volumetric method for the determination of volatile bases in fish as described by Pearson, D [13].

2.3.1. Determination of Free Fatty Acid—Pearsons Method

Mix 25 ml diethyl either with 25 ml alcohol neutralizes with 0.1m alkali. Dissolve 1 - 10 g of the oil or melted fat in the mixture neutral solvent and titrate with aqueous 0.1 m alkali. Dissolve 1 - 10 m NaOH shaking constantly until pink colour is obtained which persists for 15 seconds. The titration should not exceed about 10 ml or there is a danger of 2 phases separating. This can be avoid by using hot neutral alcohol as solvent or alcoholic for titration.

$$\text{Acid value} = \frac{\text{vol. of 0.1 N alkali 5.61}}{\text{Samples wt}}$$

The FFA Figure is usually calculated as oleic acid.

1.00 ml 0.1 N alkali = 0.0282 g Oleic acid *i.e* AV = 2 × FFA.

2.3.2. Determination of TVB

100 g of flest of fresh fish sample would be weighed and blended with 300 ml of 5% tricholoroacetic acid. The blend will then be centrifuged at 3000 × g for 1 h to obtain clear extract 5 ml of the extract was pipette into the Markhan apparatus and 5 ml of 2 M NaOH added. This would be steam distilled into 15 ml of standard 0.01 M HCl containing 0.1 ml rosolic indicator. After distillation, the excess acid was then titrated in the receiving flask using standard 0.01 M NaOH to a pale pink end point. A procedural blank would be done using 5 ml trichloroacetic acid with no sample and titrated as before. The concentration of TVB (in mg/100g sample) would be computed as follows:

$$\text{TVB (mg/100g sample)} = \frac{(M)(VB-VS)(14)(300+W)}{5}$$

where VB = ml NaOH used for blank titration, W = water content of sample in g/100g, M = molarity of NaOH standard solution and VS + ml NaOH used for sample titration. The water content (W) of the sample was obtained by drying an initial weight of fish sample at 77°C in an oven to constant weight. This temperature is used to dehydrate the material completely and to limit the vaporization of volatile materials.

2.3.3. pH Determination

PH was measured with a standard pH meter (Hanna Instruments, USA) by dipping the pH probe into each of the sample preparations to measure the pH reading directly every week.

2.3.4. Rancidity (Peroxide Value)—PV Determination

The fish sample was weighed into a clear dry boiling tube and 1.0 g of powdered potassium iodide and 20 ml of solvent mixture (2 volume glacial acetic acid plus 1 volume of chloroform) was added. The tube was placed in boiling water so that the sample boils within 30 seconds and allowed to boil vigorously for not more than 30 seconds. Pour the contents quickly into a flask containing 20 ml of potassium iodide solution (5%), wash out the tube twice with 25 ml of water and titrate with 0.002 M sodium thiosulphate solution using starch as an indicator (1%). A blank should be performed at the same time.

$$\text{P.V (meq/100g)} = \frac{\text{Titre value} \times \text{Normality of acid used} \times 100}{\text{Weight of sample used 1}}$$

(Peroxide value is in milliequivalent of peroxide per

Acidity/Rancidity Levels, Chemical Studies, Bacterial Count/Flora of Fermented and Unfermented
Silver Catfish (Chrysichthys nigrodigitatus)

27

100 g of sample.)
(A.O.A.C., 1990 Edition 11 and 12, Washington DC.)

2.4. Proximate Composition

Proximate analysis was done on the processed samples of *Chrysichthys nigrodigitatus* were analyzed chemically according to the official methods of analysis described by the Association of official Analytical Chemist (A.O.A.C) [15] on a dry matter basis to determine a general proximate analysis initially before processing the fish for cold smoking during the experiment. The crude protein content, crude fibre content, moisture content, ash content, fat content and nitrogen free extract are the parameters determined after which the sample was also analyzed at the first day of smoking and subsequently done bi-weekly.

2.4.1. Protein Content

The protein content of the samples was determined. The procedure involves digesting the material with conc. H_2SO_4 to dehydrate and char the sample (carbonisation) and H_2O_2 to complete sample decomposition by providing a reducing environment, which helps in converting the nitrogen to ammonium salts. On treatment with a dispersing agent (polyvinyl alcohol (PVA)), the ammonium salt decomposes to liberate ammonia, which in the presence of Nessler's reagent gives an orange color which is read at 460 nm according to the specification of the Hach procedure manual.

0.25 g sample was weighed into Hach digestion flask and 4ml of conc. Sulphuric acid was added. The sample was transferred to the fume hood and heated for 5 minutes at 440°C. To the charred sample was added 16 ml of H_2O_2 to clear off the brown fumes and make the digest colorless. The flask was taken off the heater, allowed to cool and the contents made up to 100 ml mark with demonized water and mixed. To 1 ml of the digest, was added 3 drops of mineral stabilizer and 3 drops of polyvinyl alcohol dispersing agent. It was mixed, made up to 25 ml and 1 ml of Nessler's reagent was added. The color was read within 5 minutes at 460 nm on the Hach spectrophotometer against deionized water blank. The absorbance gives mg/l apparent Nitrogen. The true Kjeldahl nitrogen is calculated as follows:

$$\% \text{ N} = \frac{0.005625 \times A}{B \times C}$$

where A = mg/l (reading displayed), B = ml or g sample digested and C = ml digest analyzed.

2.4.2. Crude Fat, Automated Method (Soxtec System HT2)

PROCEDURE: Grind and dry the sample properly. Load each thimble with about 2 - 3 g of the sample and plug with cotton wool. Dry the thimbles. Insert the thimbles into the Soxtec. HT. Dry and weigh the extraction cups (with boiling chips). Add 25 - 50 ml of the solvent into each cup. Insert the cup into the Soxtec HT. Extract for 15 mins in boiling position and for 30 - 45 mins in "Rinsing" position. Evaporate the solvent. Release the cups and dry at 100°C for 30 mins. Cool the cups in a desiccator and weigh.

Weight of the cup with the extracted oil = W_3.
Weight of the empty cup = W_2
Weight of sample = W_1

$$\% \text{ fat/oil} = \frac{(W_3 - W_2)}{W_1} \times 100.$$

2.4.3. Ash Content Determination

The sample was weighed into a porcelain crucible. This was transferred into the muffle furnace set at 550°C and left for 4hours. About this time, it had turned white Ash. The crucible and its content were cooled to about 100°C in air, at room temperature in desiccator and weighed.

The percentage Ash content was calculated from the formula below:

$$\text{Percentage Ash content } (\%)$$
$$= \frac{\text{Weight of Ash} \times 100}{\text{Original sample weight 1}}$$

2.4.4. Moisture Content and Dry Matter

The sample was weighed into a previously weighed crucible W_0. The crucible plus sample taken was then transferred into the oven set at 100°C to dry to a constant weight for 24 hours. At the end of 24 hours, the crucible plus sample was removed from the oven and transferred to the desiccator, cooled for 10 minutes and then weighed.

$$\text{Percentage Dry matter } (\%) = \frac{W_2 - W_0 \times 100}{W_1 - W_0 \, 1}$$

where W_0 is weight of empty crucible, W_1 is weight of crucible plus sample and W_2 is weight of crucible plus oven dried sample.

$$\text{Percentage moisture content } (\%) = \frac{W_1 - W_2 \times 100}{W_1 - W_0 \, 1}$$

Or % moisture = 100% Dry matter

2.4.5. Crude Fibre Determination (H₂SO₄-Method)

PRINCIPLE: The method involves the digestion of the food material in boiled dilute acid to hydrolysed the carbohydrate and protein. This is followed by digestion in dilute alkali to effect saponification of the fat in the food material.

APPARATUS: 600 ml long beaker, Buchner funnel, muffle furnace, crude fiber refluxing apparatus, filter paper, Porcelain crucible.

REAGENTS: Weigh 2 g sample into a 600 ml long beaker, Add 200 ml hot 1.25% H_2SO_4, Place beakers on digestion apparatus with preheated plates. Boil and reflux for 30 minutes, Filter through Whitman GF/A paper by gravity or with the aid of vacuum/air pressure pump. Rinse the beakers with distilled water. Wash the residue on the paper with distilled water until the filtrate is neutral. Transfer the residue from the paper back to the beaker with the aid of hot 1.25% NaOH to 200 ml. Return the beakers to the digestion apparatus, boil and reflux for 30 minutes. Repeat steps 4 and 5. Transfer paper with residue into a crucible. Dry samples at 100°C overnight. Cool in a desiccator's and weigh. (Weight A). Put samples in furnance at 600°C for 6 hours. Cool in a desiccator's and reweigh (Weight B).

The loss in weight during incineration represents the weight of Crude fibre in the sample.

$$\% \text{ Crude fibre} = \frac{(\text{Weigh A}) - (\text{Weight B})}{\text{Sample weight}} \times 100$$

2.4.6. Nitrogen Free Extract (N.F.E) Determination

The nitrogen free extract (N.F.E) calculation was made after the completion of analysis of the Crude Protein, Crude Fibre, Moisture content, Ash content and Ether extract by adding the percentage values in dry basis of these analysis contents and subtracting them from 100% gives N.F.E.

$$\begin{aligned} \text{N.F.E} = {} & 100 - \big(\%\text{Crude protein} \big) + \big(\%\text{Crude fibre} \big) \\ & + \big(\%\text{Moisture content} \big) + \big(\%\text{Ash content} \big) \\ & + \big(\%\text{Ether extract} \big) \end{aligned}$$

2.5. Microbiological Analysis

1) The standard plate count (SPC) was determined by pour-plating appropriate dilution on stardard count agar (Biokar, France), the plates were incubated at 30°C for 48 hours.

2) Enterobacteriaceae counts were determined by plating dilutions from 10^{-1} to 10^{-6} on Violet Red Bile Glucose Agar. The plates were incubated at 37°C, for 24 hours.

3) Salmonella were determined on 25 g of the sample added to 125 ml of sterile buttered peptone water and incubated at 35°C for 18 hours for enrichment in buffered peptone water overnight, tubes of Rappaport-Vassiliadis broth (Oxoid 669) and Selenite-cystein broth (Merck, Germany) were inoculated with 1 ml from culture in buffered peptone water overnight and incubated at 37°C for 24 hours. The positive tubes of broth media were

streaked on Hektoen agar (merck, Germany) and salmonella shigella agar medium, plates were incubated at 37°C for 24 hours. Noncoloured colonies with and without a dark centre were purified and streaked on trypticase soya agar (Biokar, France) slants and stored at 4°C.

2.6. Statistical Analysis

Analysis of Variance (ANOVA) in completely randomized design was performed on the data obtained using SPSS (2006). Significant means were compared at 5% probability level using Duncan's New Multiple Range Test (DMRT) as provided in the same SPSS (2006).

3. Results

Table 1 shows the proximate composition of the fresh *Chrysichthys nigrodigitatus* used in this study for the fermented and unfermented samples. It shows high Crude protein (55.64%), Crude Fat (5.89%), Crude Fiber (2.40%), Ash (18.75%) and Moisture Content (16.46%).

Table 2 shows the pH of the freshly minced *Chrysichthys nigrodigitatus* before the above eight sample preparations to be pH 7.2. There was a significant drop in the pH of the fermented fish products ranging from 4.8 - 5.8 (Samples 5-8), while pH of Samples 1-4 ranged from 6.0 - 6.8 at the end of 2 weeks.

Also TVB values were higher ranging from (24.12 - 29.43) mg/100gm fish in samples (1-4) unfermented (as shown in **Figure 1-4**), compared to (14.23 - 18.09) mg/100gm fish in the fermented product (Samples 5 - 8) also higher rancidity seen as high Peroxide Value (PV) range of (13.29 - 34.28) meq/kg were recorded in the fermented product of Samples 5-8 (as shown in **Figures 5-8**) compared to a range of (14.23 - 17.12) meq/kg in the unfermented (Samples 1-4) at the end of two weeks.

The free fatty acids values were lower and followed a narrow range of (6.14 - 6.45)% in the unfermented (Samples 1-4) compared to fermented (Samples 5-8) with a higher value and higher range of (6.42 - 12.27)% at the end of two weeks.

Table 1. Proximate composition of fresh *Chrysichthys nigrodigitatus*.

Parameters	%Level
%Ash	18.75
%Moisture	16.46
%Crude Fiber	2.40
%Crude Fat	5.89
%Crude Protein	55.64
NFE	0.86

Acidity/Rancidity Levels, Chemical Studies, Bacterial Count/Flora of Fermented and Unfermented
Silver Catfish (Chrysichthys nigrodigitatus)

29

Table 2. Chemical analysis at week two for the eight sample preparations (unfermented (1-4) and fermented (5-8)) of *Chrysichthys nigrodigitatus*.

Samples	TVB (mg/100gm)	%FFA	PV (meq/kg)	PH	PH of fresh *C. nigrodigitatus* before sample preparation
1	24.12	6.34	16.15	6.8	7.2
2	25.89	6.21	14.23	6.0	7.2
3	28.03	6.14	15.34	6.8	
4	29.43	6.45	17.12	6.5	
5	14.23	6.42	13.29	5.8	
6	15.78	9.03	30.45	5.2	
7	17.34	12.27	32.93	5.0	
8	18.09	10.17	34.28	4.8	

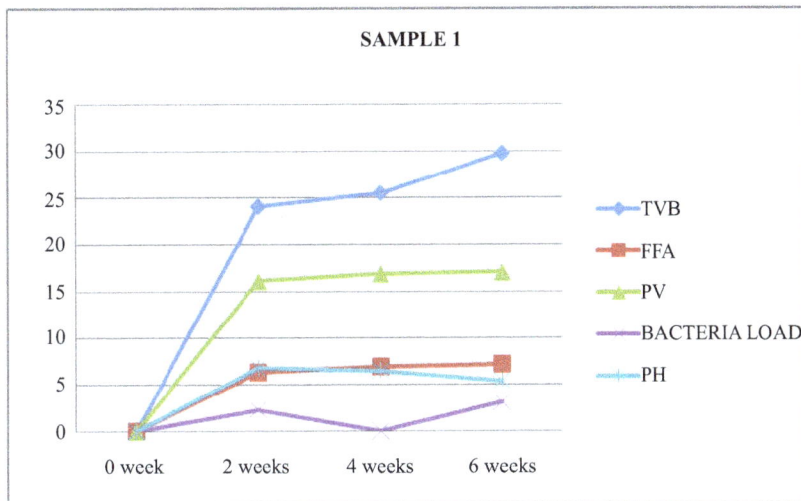

Figure 1. Graphically representation showing bacteria load, ph, pv, ffa and tvb of unfermented *Chrysichthys nigrodigitatus* for Sample 1.

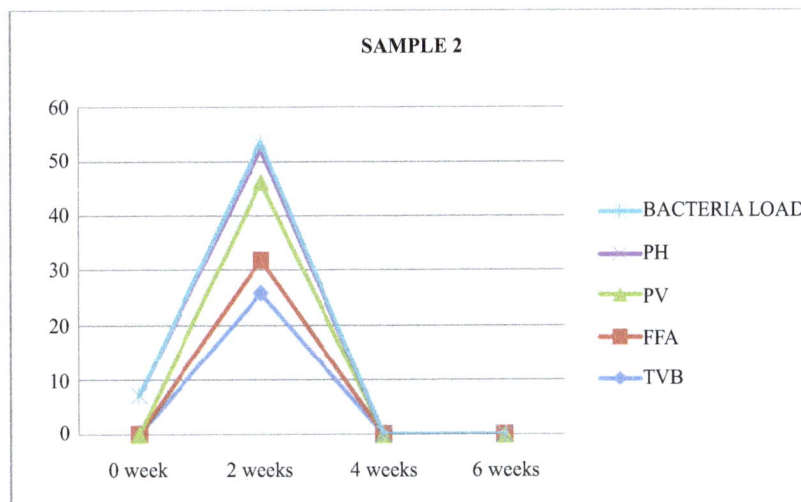

Figure 2. Graphically representation showing bacteria load, ph, pv, ffa and tvb of unfermented *Chrysichthys nigrodigitatus* for Sample 2.

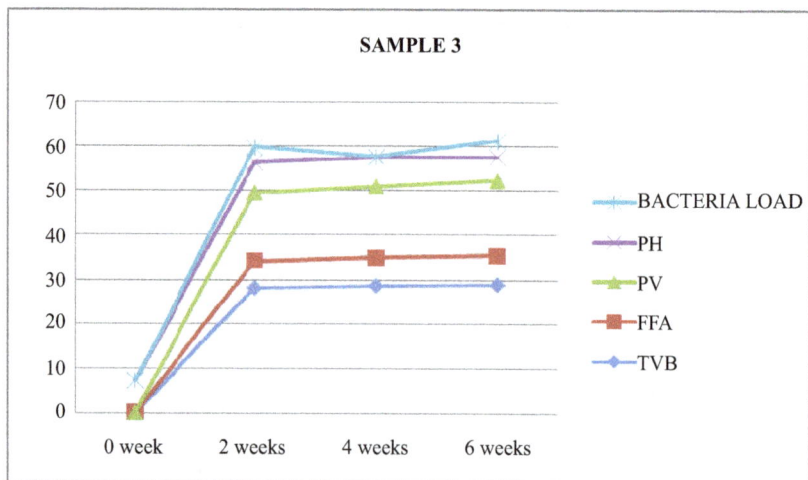

Figure 3. Graphically representation showing bacteria load, ph, pv, ffa and tvb of unfermented *Chrysichthys nigrodigitatus* for Sample 3.

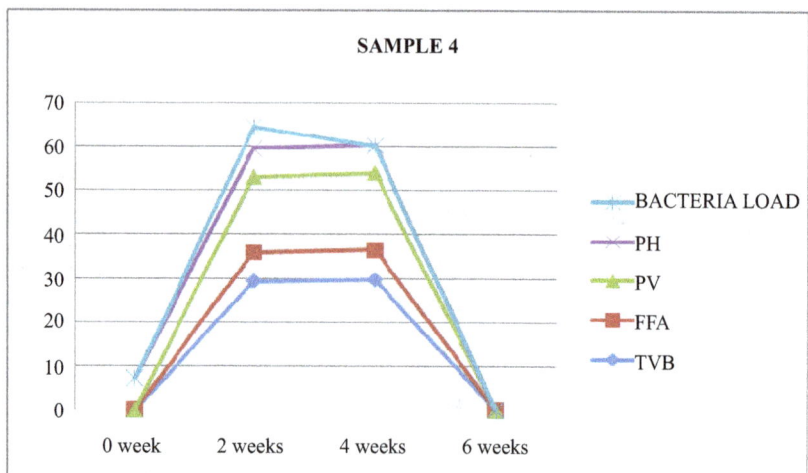

Figure 4. Graphically representation showing bacteria load, ph, pv, ffa and tvb of unfermented *Chrysichthys nigrodigitatus* for Sample 4.

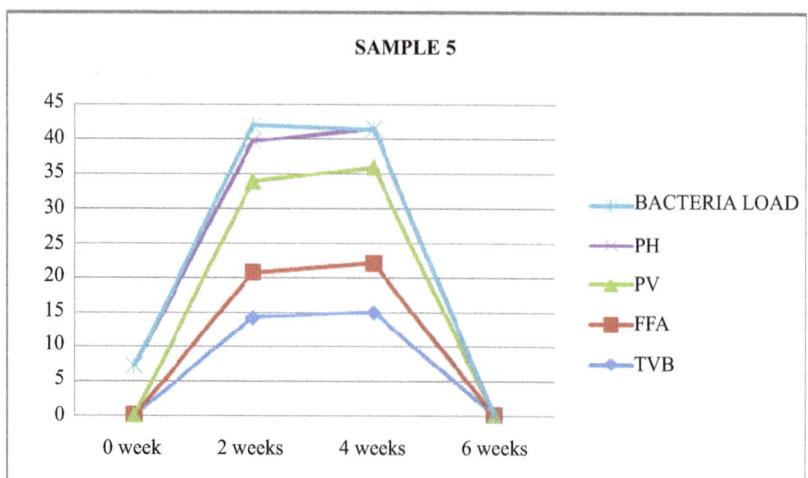

Figure 5. Graphically representation showing bacteria load, ph, pv, ffa and tvb of fermented *Chrysichthys nigrodigitatus* for Sample 5.

Acidity/Rancidity Levels, Chemical Studies, Bacterial Count/Flora of Fermented and Unfermented
Silver Catfish (Chrysichthys nigrodigitatus)

31

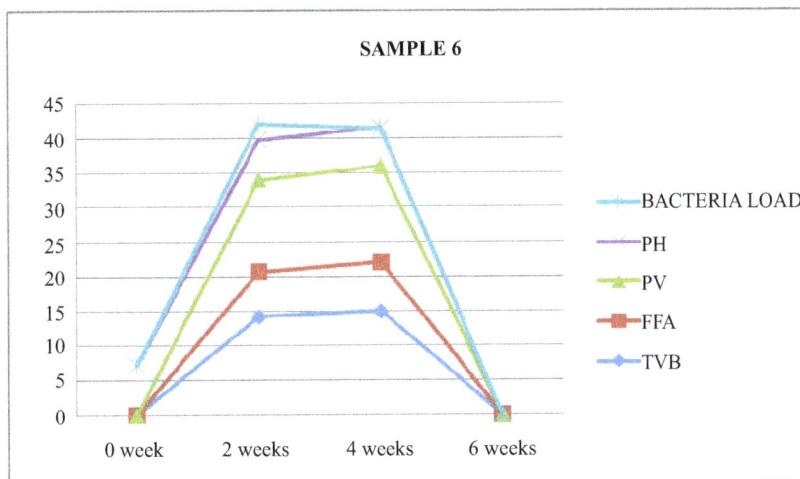

Figure 6. Graphically representation showing bacteria load, ph, pv, ffa and tvb of fermented *Chrysichthys nigrodigitatus* for Sample 6.

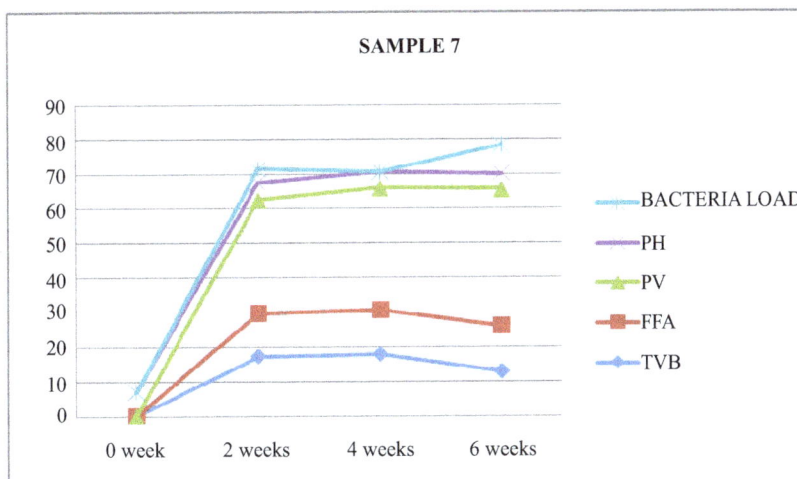

Figure 7. Graphically representation showing bacteria load, ph, pv, ffa and tvb of fermented *Chrysichthys nigrodigitatus* for Sample 7.

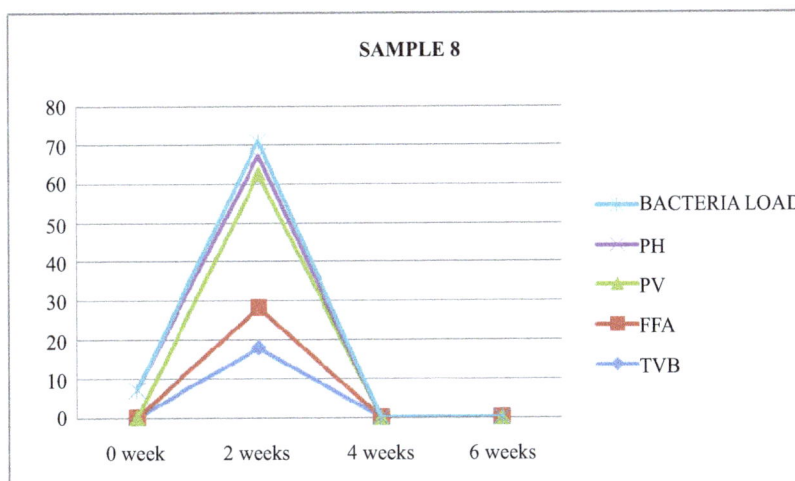

Figure 8. Graphically representation showing bacteria load, ph, pv, ffa and tvb of fermented *Chrysichthys nigrodigitatus* for Sample 8.

It should be noted that all the eight samples were left for two weeks, although Samples 2 and 8 were spoilt at the end of one week with maggots, offensive odours and colour changes.

Table 3 shows that Sample 2 (unfermented) and Samples 8 (fermented) were terminated at the end of two weeks as a result of spoilage. As shown in Figures 2 and 8.

Sample 2 is unfermented, salted, cooked, opened and maintained under ambient temperature (26°C), its physical appearance changed to brownish in colour with some maggot growing in it and had a very pungent odour and swelling was noticeable after a week and was terminated at the end of two weeks.

Sample 8 is fermented, not salted, not cooked, opened and left at ambient temperature (26°C), was totally spoilt and maggot were seen on the product, which has turned nearly to liquid. It was evident that this product was contaminated with flies and was terminated at the end of two weeks.

As shown at the end of four weeks FFA (free fatty acids) increased with length of storage of the products this ranged from (6.45 - 12.73) meq/kg in (Samples 1-7). This also increased with PV, ranging from (15.89 - 35.40) meq/kg and TVB (14.96 - 29.80) mg/100 gm. However with the increase in all the measured chemical parameters a corresponding drop in pH from (6.7 - 4.6) was recorded which implies acidity and rancidity increases with length of storage/spoilage; especially with PV ranges of (15.89 - 35.40) meq/kg compared to a corresponding drop in pH from 6.7 - 4.6.

Table 4 shows that only 3 out of the 8 sample preparations lasted the six weeks experimental period. These are unfermented (Samples 1 and 3) and only one fermented (Sample 7). As shown in Figures 1, 3 and 7.

Sample 1 is unfermented, cooked, salted and maintained at freezer temperature of −25°C, had no physical change in appearance, still retained its colour and appearance, no pungent odour and was terminated at the end of six weeks (no maggot found).

Sample 3 is unfermented, cooked, covered, not salted and maintained at freezer temperature of −25°C, had no physical change in appearance, no strong odour and still maintained its sea weedy odour, with no colour change, was terminated at the end of six weeks (no maggot found).

Sample 7 is fermented, not cooked, not salted, covered and left under room (ambient) temperature had more moisture content, no maggot was found. Had a very strong offensive odour black and whitish colourmoulds were visible through the transparent nylon: it was terminated after six weeks. Sample 7 is the most acidic (out of the 8 sample preparations) with a pH of 4.5 and it also had the highest rancidity of 35.66 meq/kg Peroxide Value, and it is also the only fermented fish product that lasted six weeks without maggot seen. While higher TVB values (29.84 and 28.79) lower FFA (7.18 and 6.73)%, lower PV (17.09 and 16.85) and higher pH (5.4 and 5.2) were recorded in samples 1 and 3 compared to sample 7.

Table 5, shows that all the 8 sample preparations (fermented or unfermented) are best consumed within two weeks of preparations; with the exception of sample 4

Table 3. Chemical analysis at week four for the six sample preparations remaining (unfermented (1, 3, 4) and fermented (5, 6, 7)) of *Chrysichthys nigrodigitatus*.

Samples	TVB (mg/100 gm)	%FFA	PV (meq/kg)	PH	PH of fresh *C. nigrodigitatus* before sample preparation
1	25.56	6.95	16.89	6.5	7.2
3	28.56	6.45	15.89	6.7	"
4	29.80	6.77	17.44	6.3	"
5	14.96	7.12	13.87	5.6	"
6	16.38	9.28	31.60	5.0	"
7	17.84	12.73	35.40	4.6	"

Table 4. Chemical analysis at week six for the three sample preparations remaining (unfermented (1 & 3) and fermented (7)) of *Chrysichthys nigrodigitatus*.

Samples	TVB (mg/100gm)	%FFA	PV (meq/kg)	PH	PH of fresh *C. nigrodigitatus* before sample preparation
1	29.84	7.18	17.09	5.4	7.2
3	28.79	6.73	16.85	5.2	"
7	12.97	12.97	35.66	4.5	"

Acidity/Rancidity Levels, Chemical Studies, Bacterial Count/Flora of Fermented and Unfermented
Silver Catfish (Chrysichthys nigrodigitatus)

33

Table 5. Initial bacteria load at 2nd week for the eight sample preparations, unfermented (1-4) and fermented (5-8) of *Chrysichthys nigrodigitatus*.

Samples	Bacteria load (Log10 cfug)
1	2.30×10^5
2	1.4×10^5
3	3.5×10^5
4	5.05×10^5 not fit for consumption exceeded the ICMSF standard value of 5.0×10^5 cfug recommended for safe fish products consumption [16]
5	2.35×10^5
6	1.10×10^5
7	3.90×10^5
8	3.80×10^5

(unfermented) which is not fit consumption because of the heavy bacteria load of 5.05×10^5 which exceeded the recommended safe value of 5.05×10^5 for fish product comsumption [16].

Sample 4, is unfermented, cooked, not salted, opened and maintained under ambient (room) temperature, had green mould seen through transparent nylon, state with strong offensive odour emanating. It was terminated after four weeks of preparation.

Table 6 shows that the cooked unfermented *Chrysichthys nigrodigitatus* minced sample at six weeks, (samples 1 and 3) had the lowest bacteria count of 3.1×10^5 cfug and 3.7×10^5 cfug respectively below the ICMSF standard of 5.0×10^5. Hence may be fit consumption as fish paste, sauces etc. however to avoid food poisoning they are better consumed before two weeks.

Sample 7 which is not cooked, not salted fermented covered at room (ambient) temperature of 26°C is not fit for consumption at all because of heavy bacteria load of 8.2×10^5 which far exceeded the recommended ICMSF standard of 5.0×10^5 [16]. Although Sample 7 lasted for six weeks, without maggot, with high moisture content, it had a very strong offensive odour, black with whitish mould visible through the transparent nylon.

Table 7 shows that only Proteus sp (−) was not present in samples implicated in Samples 1, 3 and 7 at the end of six weeks. Very few bacteria sp were isolated in this study possibly because of the increasing acidity with length of storage and fermentation. For example in Sample 7, pH (acidity) dropped from 7.2 (in fresh minced *Chrysichthys nigrodigitatus* to 4.5 at the end of six weeks. Also increasing rancidity (Sample 7) with highest PV (35.66) meq/kg coupled with high acidity of pH (4.5) is likely to be responsible for the low number of bacteria (5 bacteria sp) isolated.

Table 8 shows that pH is (negatively) inversely re-

lated to TVB (r = −0.32), FFA (r = −0.12), PV (r = −0.31) and Bacteria Load (r = −0.43). as shown in **Table 7** while very few bacteria species (5 bacteria sp) were represented due to the high rancidity (e.g. as in Sample 7 with highest PV (35.66 meq/kg) and low pH (high acidity) of 4.5 had the highest bacteria load of 8.2×10^5 (**Table 6**).

This high value is possibly due to the incidence of Acidiophillies bacteria; which multiplied rapidly (with increasing length of fermentation) above the ICMSF sage limit of 5.0×10^5. Peroxide Value is strongly positively correlated with FFA (r = 0.97), TVB (r = 0.45) and bacteria load (r = 0.73). Total Volatile Base (TVB) is also positively correlated with FFA (r = 0.51), PV (r = 0.45) and bacteria laod (r = 0.85). Free Fatty Acid also posi-

Table 6. Initial bacteria load at 6th week for the eight samples 1, 3 (unfermented) and 7 (fermented) preparations of *Chrysichthys nigrodigitatus*.

Samples	Bacteria load (Log10 cfug)
1	3.1×10^5
3	3.7
7	8.2×10^5 Not fit for consumption exceeded the ICMSF standard value of 5.0×10^5 cfug recommended for safe fish products consumption [16].

Table 7. Spatial occurrence of bacteria isolates in samples 1, 3 and 7 at the end of week six of *Chrysichthys nigrodigitatus*.

Isolated organisms	Samples		
	1	2	3
Lactobacillus sp	+	+	+
Proteus spp	-	+	+
Staphylococcus aureus	+	+	+
Staphylococcus epidermis	+	+	+
Bacillus spp	+	+	+

Table 8. Correlation (r) table between tvb, ffa, pv, ph and bacterial load.

	TVB	FFA	PV	PH	Bacteria load
TVB	1	0.51	0.45	−0.32	0.85
FFA	0.51	1	0.97	−0.12	0.38
PV	0.45	0.97	1	−0.31	0.73
PH	−0.32	−0.12	0.31	1	−0.43
Bacteria load	0.85	0.38	0.73	−0.43	1

*Correlation is significant at the 0.05 level; **Correlation is significant at the 0.05.

tively correlated with TVB (r = 0.51), PV (r = 0.45) and bacteria load (r = 0.85).

4. Discussion

The result of proximate composition (**Table 1**) shows the silver catfish (*Chrysichthys nigrodigitatus*) is very high in crude protein (55.64%) and fat (5.89%), while %Ash (18.75%), %Moisture (16.46%), %crude fiber (2.40%) and NFE (0.86%) were also recorded.

The quality characteristics of the 8 samples used for the fermentation processes are as discussed below at the point of termination of the fermented sample products.

Sample 1: is unfermented, cooked, salted, covered and maintained at freezer temperature of −25°C, had no physical change in appearance, still retained its colour and appearance and was terminated after six weeks.

Sample 2: is unfermented, salted, cooked, opened and maintained under ambient temperature (26°C). Its physical appearance changed to brownish in colour with some maggot growing in it and had a very strong odour and swelling was noticeable after a week; it was terminated at the end of two weeks.

Sample 3: is unfermented, cooked, covered not salted and maintained at freezer temperature (−25°C) had no physical change in appearance and still maintained its sea weedy odour. It was terminated after six weeks.

Sample 4: is unfermented, cooked, not salted, opened, maintained under ambient temperature (26°C), had green mould seen through transparent nylon, stale with strong offensive odour emanating. It was terminated after four weeks.

It should be noted that Samples 1-4 are unfermented cooked, while Samples 1 and 2 are salted, Samples 3 and 4 are not salted, Samples 1 and 3 are covered and frozen at −25°C. Samples 2 and 4 are maintained at ambient temperature.

Sample 5: is fermented, not cooked, salted, covered, maintained under room temperature (26°C) (ambient) turned brownish in colour, slimy, swollen with maggot growing on it, was terminated after four weeks.

Sample 6: is fermented, not cooked, salted, opened, maintained at ambient (room) temperature (26°C), had no maggot, no swelling, but sickly sweet odour was present and was terminated after four weeks

Sample 7: is fermented, not cooked, not salted, covered and left under ambient (room) temperature (26°C), had more moisture content, no maggot was found. It had a very strong offensive odour, black and whitish colour-moulds were visible through the transparent nylon. It was terminated after six weeks.

Sample 8: is fermented not cooked, not salted, opened and left at room temperature (26°C), was totally spoilt and maggot were seen moving on the products which has

turned nearly to liquid. It was evident that this product was contaminated by houseflies, and was terminated at the end of two weeks.

Samples 5-8 were fermented, not cooked; 5 and 6 were salted, Samples 7 and 8 not salted, Samples 5 and 7 covered, while 6 and 8 were opened. All fermented samples (5-8) were left open, not frozen.

The samples under cold storage (Samples 1 and 3) maintained at freezer temperature −25°C, cooked, salted (Sample 1) and not salted (Sample 3) did not change in physical appearance (both samples lasted six weeks) and still retained its sea weedy odour, remains the best. A peculiar characteristic of fermentation is a strong, sometimes offensive odour.

Since all the 8 prepared samples barely lasted for two weeks without visible signs of spoilage at (least minimal) fermented products are best consumed within one or two weeks. This observation is in line with Watanabe, K. [9] findings that the characteristics smell of fermented fish is the result of enzymatic and microbial activity in the fish muscle. Fermented fish is, therefore, any fishery product which has undergone degradative changes through enzymatic or microbiological activity either in the presence or absence of salt [1].

Total Volatile Base (TVB) were much higher in Samples 1-4 (unfermented) ranging from (24.12 - 29.43) mg/100gm fish, than in Samples 5-8 (fermented) with a range of (14.23 - 18.09) mg/100gm fish. This observation is in line with Hiltz [17] report that the volatile bases particularly TMA, DMA and NH_3 are associated with changes in the organoleptic and textural quality of fish.

In samples (1-4) unfermented, cooked FFA values were not significantly (P < 0.05) difference from each other and also followed a narrow range of (6.14 - 6.45)%. Comparatively much higher value and higher range (6.42 - 12.27)% were recorded in samples (5-8) fermented and uncooked. The increasing FFA levels observed in the fermenting 8 samples is as a result of oxidation of poly unsaturated fatty acids (PUFA) contained in the fish tissue to products such as peroxides, aldehydes and ketones and free fatty acids [18].

Also, in line with Horner, W.F.A [18] observations, PV values increased in all the 8 samples in the second, fourth and sixth week (for the 8, 6 and 3 samples) respectively, however higher values 13.29 - 34.28 (second week), 13.87 - 35.40 (fourth week) and 16.85 - 35.66 (sixth week) were recorded in the fermented (Samples 5-8) uncooked products as shown in **Tables 2-4**.

Acidity generally increased with length (weeks) of fermentation with a gradual drop in pH from 7.2 in the fresh *C. nigrodigitatus* to pH 4.5 (Sample 8) at six weeks. This clearly shows that as rancidity increases in the fermented/uncooked and unfermented/cooked samples acid-

Acidity/Rancidity Levels, Chemical Studies, Bacterial Count/Flora of Fermented and Unfermented
Silver Catfish (Chrysichthys nigrodigitatus)

35

ity increases seen as a drop in pH values generally, ranging from pH 7.2 (in the fresh state) and pH 4.5 (in sample 8) the worst sample.

Table 5 shows that at the second week, Sample 4 became unfit for consumption with a bacteria load of 5.05×10^5 and also Sample 7 (**Table 6**) became unfit for consumption at the end of the sixth week with a bacteria load of 8.2×10^5. This is because these values exceeded the 5.0 recommended standards for safe, edible fish products [16].

Table 7 shows the spatial distribution of bacteria load a total of five bacteria sp (*Lactobacillus sp, Proteus spp, Staphylococcus aureus, Staphylococcus epidermic, Bacillus sp*) were present in all the 3 samples (1, 3, 7) at the end of six weeks. This fewer species represented in probably due to the increasing acidity of the fermented products. However, *Proteus sp* is absent (not acid loving) in sample 1.

As shown in **Table 8** the strong positive correlation (r = 0.97, P < 0.01) exist between PV and FFA shows that the rate of oxidative rancidity is very high in fermented fish products. Also the strong negative correlation (r = −0.121, P < 0.01) between pH and FFA shows that acidity (seen as drop in pH) of the fermented products increased with increasing oxidative rancidity of the fermented product. This is also confirmed with increasing acidity (drop in pH) (r = −0.313, P < 0.05) between PV and pH.

5. Conclusions

Fermented products are best consumed with in 1 or 2 weeks, since all the 8 samples prepared lasted for barely 2 weeks (in order to prevent food poisoning) before maggot infestation, turning greenish, blackish or brownish before losing their sea weedy odour.

However, samples 4 unfit at two weeks and sample 7 unfit at six weeks because of heavy bacteria load. Acidity (seen as drop in pH) increases with rancidity in fermented products over the weeks. Not all samples lasted six weeks, with the exception of samples 1, 3 and 7.

REFERENCES

[1] N. Zakhia and J. L. Cuq, "Apercu sur la Qualite du Tilapia Secheetcommercialise au Mali," *Proceedings of the FAO Expert Consultation on Fish Technology in Africa*, Accra, 22-25 October 1991.

[2] H. H. Huss and G. Valdimarson, "Micro-Biology of Salted Fish," *FAO Food Tech News*, Vol. 10, No. 1, 1990, p. 190.

[3] B. Klinic, "Fish Sauce Technology," *EUJ Fish Aquat. Sci*, Vol. 20, No. 1-2, 2003, pp. 263-273.

[4] FAO, "Fermented Fish Products," Prepared by I. M. Mackie, R. Hardy and G. Hobbs, *FAO Fish Rep.*, Vol. 100, 1971, p. 54.

[5] K. M. Essuman, "Fermented Fish in Africa. A Study on Processing," Marketing and Consumption FAO Fisheries Technical Paper No 329, FAO, Rome, 1992, 80 p.

[6] K. K. Eyeson, "Composition of Foods Commonly Used in Ghana," Food Research Institute (CSIR), Accra, 1975.

[7] A. A. Eyo, "Fish Processing Technology in the Tropic National Institute for Freshwater Fisheries Research (NIFFR)," New Bussa, 2001, pp. 10-170.

[8] J. C. D. Watts, "Some Observations on the Preservation of Fish in Sierra Leone," Bulletin de MPANTXXVII Sci. of No. 1, 1965.

[9] K. Watanabe, "Fish Handling and Processing in Tropical Africa," *Proceedings of the FAO Expert Consultation on Fish Technology in Africa*, Casablanca, 7-11 June 1982, FAO Fish. Rep, Vol. 268, 1982, pp. 1-5.

[10] N. I. Azeez, "The Problem of Choice and Safer Methods of Reducing Post Harvest Losses in Lake Chad Processed Fish," *Proceedings of the FAO Expert Consultation on Fish Technology in Africa*, Lusaka, 21-25 January 1985, FAO Fish, Rep., Vol. 329, 1986, pp. 340-348.

[11] G. M. Reay and J. M. Shewan, "The Spoilage of Fish and Its Preservation by Chilling," *Advances in Food Research*, Vol. 2, 1949, pp. 343-389.

[12] Y. Tomiyasu and W. B. Zenitani, "Spoilage of Fish and Its Preservation by Chemical Agents," *Advances in Food Research*, Vol. 7, 1957, pp. 41-82.

[13] D. Pearson, "Chemical Analysis of Foods," 7th Edition, Churchill Livingstone, Edinburg, 1962.

[14] Oyelese and Odubayo, "Shelf Life of Fish Meal Paste and Cake of *Tilapia niloticus* and Supplementation of Convertional Fishmeal with Tilapia Fishmeal in the Diet of *Clariasgariepinus* Fingerlings," *Journal of Food Processing and Preservation*, Vol. 34, Suppl. 1, 2010, pp. 149-163.

[15] A.O.A.C., "Official Methods of Analysis of AOAC International," 17th Edition, 1st Revision, Association of Analytical Communities, Gaithersburg, 2002.

[16] ICMSF, "International Commission on Microbiological Specifications for Foods. Micro-Organisms in Foods 7. Microbiological Testing in Food Safety Management," Kluwer Academic Plenum Publishers, New York, 2002, p. 199.

[17] D. F. Hiltz, "Deteriorative Changes during Frozen Storage in Fillets and Minced Flesh of Silver Lake (*Mertuucciusbilineans*) Processed from Round (Whole) Fish Held in Ice and Refrigerator Sea Water," *Journal of the Fisheries Research Board of Canada*, Vol. 33, No. 11, 1976, pp. 2560-2567.

[18] W. F. A Horner, "Preservation of Fish by Curing (Drying, Salting and Smoking)," In: G. M Hall, Ed., *Fish Processing Technology*, 2nd Edition, Chapman and Hall, New York, 1997, pp. 21-39.

5

Nutritional Status of Patients with Cutaneous Leishmaniasis from a Tropical Area of Bolivia, and Implications for Zinc Bioavailability

Claudia E. Lazarte[1,2*], Claudia Alegre[3], Ernesto Rojas[4], Yvonne Granfeldt[1]

[1]Department of Food Technology, Engineering and Nutrition, Lund University, Lund, Sweden; [2]Food and Natural Products Center, San Simon University, Cochabamba, Bolivia; [3]Department of Nutrition, San Simon University, Cochabamba, Bolivia; [4]Biomedical Research Institute, San Simon University, Cochabamba, Bolivia.

ABSTRACT

Macro and micronutrient deficiencies are a significant problem among people in rural areas in developing countries. Deficiencies may lead to an impaired immune system making the organism vulnerable to infections and diseases. In this paper, the dietary intake, anthropometric measurements, zinc and copper levels in serum, of patients with cutaneous leishmaniasis (CL) are compared with a group of healthy controls and reference values. Results showed no significant differences in most nutrient intake or anthropometrics between patients and controls. However, serum zinc level of patients (80 μg/dl) was significantly lower ($P < 0.001$) than in controls (85 μg/dl), likely explained by the presence of leishmaniasis. The median serum zinc level in both groups was below the reference values, even though their median zinc intake met the zinc recommendations from WHO. Consequently, besides the presence of leishmaniasis, serum zinc levels may be compromised by inhibitory components in their diet, such as phytates, as it is shown by the molar ratio phytate:zinc (Phy:Zn) that was between 11 and 19, while 15 is the level said to compromise zinc status. There were significant ($P < 0.05$) negative correlations between serum zinc and Phy:Zn, for patients ($r = -0.413$) and controls ($r = -0.410$). In conclusion this study shows that patients with CL in Chapare, Bolivia had low serum zinc levels due to the leishmaniasis per se and the decreased zinc bioavailability in their diet. CL infection was not in direct association with the nutritional status indicated by the anthropometric or dietary assessments. However, dietary intake showed 5 essential nutrients below the nutrient recommendation in both groups.

Keywords: Nutritional Status; Leishmaniasis; Dietary Intake; Anthropometrics; Zinc Bioavailability; Phytates

1. Introduction

In tropical areas, especially in developing countries, nutritional deficiencies play an important role in the chronicity of some parasitic infections. Malnutrition is the most common cause of immunodeficiency worldwide; thus, epidemiological and clinical evidence suggests that nutritional deficiencies lead to an increased risk for development of infectious diseases due to impaired immunological responses [1].

Leishmaniasis is an infectious disease, endemic in 21 countries in America, and 39 million people in America are at risk for acquiring the disease [2]; it is a protozoan parasitic disease transmitted to humans by several phlebotomine sandflies. Depending on the strain of the parasite *Leishmania*, and the immunological status of the host, the infection may vary and develop from a skin ulcer (CL) to mucosal leishmaniasis (ML) leading to deformations, or to visceral leishmaniasis (VL) a severe form of systemic infection with hepatosplenomegaly [3]. Data about the prevalence of leishmaniasis in Chapare, a tropical area of Bolivia are scarce; most cases (85%) of CL are caused by the strain *Leishmania (Viannia) braziliensis* [4,5].

During the leishmaniasis infection, the impaired immune system of the host may cause an uncontrolled parasite replication that delays the healing of CL leading to a diffuse CL, ML or VL [6]. Even though it has been shown that leishmaniasis may occur in individuals in endemic areas independently of the nutritional status [7], studies with children have shown a link between poor

---

Nutritional Status of Patients with Cutaneous Leishmaniasis from a Tropical Area of Bolivia, and Implications for Zinc Bioavailability

37

nutritional status, growth retardation and iron deficiency with CL [8,9]. Furthermore, malnourished children are at a greater risk for developing severe VL than well-nourished children [10]. Malnutrition and micronutrient deficiencies are likely to interfere with several important functions of the immune system resulting in an impaired capability to overcome the leishmaniasis infection; nutritional status of the host is a key factor for the outcome of infection [10-12].

Nutritional status is evaluated primarily by dietary assessment methods that are widely used in both developed and developing countries for measuring the risks of nutrient deficiencies and excesses and evaluating the effects of nutrition interventions. Inadequate levels of nutrients originate either from primary deficiencies due to low levels of the nutrients in the diet or secondary deficiencies due to other factors like drugs, disease states or dietary components that inhibit the absorption of nutrients [13]. Secondly, anthropometric methods involving measurements of the body to provide an indirect evaluation of body composition are applied for the assessment of nutritional risks [13,14]. Furthermore, biochemical measurements of nutrients in biological fluids or tissues are used to detect subclinical deficiency states [13].

There is a lack of data on the nutritional status of rural populations in Bolivia, and no available data about nutritional status of patients with leishmaniasis. Recently we have studied the dietary patterns among a healthy population in Chapare; the results showed that their food consumption is mainly based on starchy tubers, cereals, and legumes providing 72E% from carbohydrates, with small portions of meat and eggs for protein 13E%, and with oil or tallow as a source of fat 15E% [15]. This type of diet has been associated with micronutrient deficits, notably iron, zinc, and calcium [16,17], due to the small amount of animal-source foods and the presence of mineral inhibitors like phytates, which is a strong chelator of minerals, reducing their bioavailability [18,19]. The inhibitory effect of phytates in mineral absorption appears to follow a dose dependent response, and the molar ratios Phy:Zn, Phy:Fe (phyate:iron) and Phy:Ca (phytate:calcium) in the diet have been used to predict the proportion of absorbable minerals [16,20,21].

Zinc and copper are essential trace elements of great importance for many enzymes and biological processes and their deficits or excesses may lead to different health problems [22]. Zinc deficiency in particular has a great impact on the defense mechanisms of the body and the immune response to infections. These have been well documented [23,24], but there is limited information about zinc and copper status and CL. Although a few studies have reported alterations in the status of these minerals during leishmaniasis [25,26], there has been no reported information about their relation to the dietary components.

This paper presents the dietary and anthropometric assessments of a group of patients with CL from the tropical area Chapare in Bolivia compared with healthy subjects. Additionally, in order to provide an estimate of the relative bioavailability of zinc, iron and calcium in the diet of the studied population, the content of phytate and minerals in their diet, were calculated and the molar ratios phytate: mineral are presented. Furthermore, biochemical indicators of zinc and copper were studied and correlated with the anthropometric and dietary features in order to contribute to the knowledge concerning the zinc and copper status in adult patients and gain some insight into the effect that phytates from the diet and the presence of the leishmaniasis infection may have on the absorption and metabolism of these minerals. The results of this nutritional evaluation are aimed to be used as a baseline in a further intervention study of zinc supplementation during leishmaniasis treatment.

2. Subjects and Methods

2.1. Study Participants and Design

The study was carried out in a tropical region located approximately 160 km east of Cochabamba, Bolivia, including the rural villages named Villa Tunari, Eterazama, San Gabriel, Aroma, Chimore, Shinaota and Ivirgarzama. Patients were recruited by contact with the local health centers; CL diagnosis was confirmed by microscopic examination of lesion smears and by isolation of parasites by culture according to the procedure previously described [27]; 34 patients were enrolled but complete data were collected for 32 patients.

The exclusion criteria for patients considered: patients with skin ulcers by another etiology (negative CL diagnosis), patients with previous leishmaniasis episodes, patients currently receiving leishmaniasis treatments or other drugs, patients with additional ML, patients with multiple CL lesions, patients taking mineral-vitamin supplements and pregnant or lactating women. In the aforementioned area 32 healthy control participants, of the same sex and approximately the same age (±5 years) as the patients, were enrolled. The exclusion criteria for healthy controls were: subjects presenting any disease at the moment, subjects with previous leishmaniasis episodes, subjects taking drugs or mineral-vitamin supplements and pregnant or lactating women.

All patients and controls signed a letter of consent prior to their participation. The study follows a case-control design, approved by the Ethics Committee of the Faculty of Medicine at Lund University and Faculty of Medicine at San Simón University.

2.2. Anthropometric Measurements

The anthropometric indicators, body mass index (BMI), mid-upper-arm muscle area (AMA), and mid-upper-arm fat area (AFA), were evaluated in patients and controls lightly dressed and without shoes. Measurements of weight were done with a digital electronic scale (Omron HBF400), 150 kg ± 0.1 kg, height with a portable staidometer ±1 mm, mid-upper-arm circumference in the left arm with a flexible non-stretch tape ±1 mm, and triceps-skin-fold in the left arm with a caliper ±0.2 mm (Harpender Skinfold Caliper, Baty International, United Kingdom). The indicators BMI, AMA and AFA were calculated with equations from the WHO committee [14], and evaluated according to the WHO and Frisancho classification [14,28].

2.3. Assessment of Dietary Intake

The dietary intake of the patients and controls was assessed during three consecutive days by Food Photography 24-hours Recall Method (FP24h-R) previously evaluated and described in detail [15]. Briefly explained the method is a 24-h recall supported by a photographic food record; subjects take digital photographs of all their meals and beverages consumed over a period of time, then nutritionists visit the subjects after each 24-h period to fill in a 24-h recall questionnaire with the detailed information of all the consumed foods. The portion sizes are estimated using the digital photographs taken by the subjects compared with standard food portions depicted in a photo atlas. Food consumption data were extracted from the questionnaires and the nutrient calculation was done in an excel file with a food data base from National Nutrient Data Base for standard reference [29] and a few items from the Bolivian Food Composition Table [30].

The calculations were performed for the intake of energy, protein, fat, carbohydrates, fiber, calcium, iron, phosphorus, zinc, copper, thiamin, riboflavin, niacin, pantothenic acid, folate, β-carotenoids, and vitamins A, C, E, B6, and B12. These nutrients were selected to elucidate differences in the nutrient intake between patients and controls and to shed light on possible deficiencies presented in this rural population; thus the subjects' median daily dietary intake results were compared with the recommended nutrient intake (RNI) from World Health Organization (WHO) [31], according to sex and age of each patient and control. Additionally, data of the phytates content were included in the database according to a literature review [32-35]. The intake of phytates was calculated and the molar ratios Phy:Zn, Phy:Fe and Phy:Ca are presented to give some insight into the relative bioavailability of iron, zinc and calcium in the diet of patients and controls.

2.4. Trace Elements Indicators

After the diagnosis of CL was confirmed, blood samples (5 ml) were drawn from fasting patients and controls from the antecubital vein, into free trace element tubes; the samples were immediately centrifuged (5000 g at 4°C for 10 min) in order to separate the serum, which was divided into aliquots and stored at −20°C until zinc and copper analyses.

Serum zinc was quantified by flame atomic absorption spectrometry (Model 2280, Perkin Elmer Corporation, Norwalk, CT, USA), and serum copper by a graphite furnace atomic absorption spectrometry (Model SIMAA 6100, Perkin Elmer Corporation, Norwalk, CT, USA). Before analysis the samples were diluted 10 times with deionized water [36], and a calibration curve for each mineral was prepared from certified Atomic Absorption Standard solutions (Perkin Elmer Corp.). The reference material Seronorm[TM] trace elements serum L-1-2 (SERO AS, Norway) was used to validate the mineral analysis.

2.5. Statistical Analysis

The normal distribution of the data was evaluated for all the parameters by Shapiro-Wilk test, and by measurements of skewness and kurtosis. Most of the parameters did not have normal distribution and thus all the results are presented as medians and percentiles 25th and 75th. First, the data were evaluated for continuous variables in the whole group of patients (n = 32) and controls (n = 32) using the statistic tests for matched data Wilcoxon rank test, and Chi-Square analysis was used to test the group differences in categorical variables, such as BMI, AMA and AFA classification. Later on, the groups were divided in female patients (n = 12) and controls (n = 12), male patients (n = 20) and controls (n = 20) and compared with Wilcoxon rank test to evaluate differences between same gender patients and controls.

Spearman's correlations were computed to evaluate the association between the anthropometric indices (BMI, AMA, AFA) with energy and macronutrient intake, and between the biochemical measurements of zinc and copper with the corresponding intakes and with the anthropometric variables. Correlations between serum zinc with phytates and Phy:Zn were also calculated to elucidate the effect of phytates on the serum zinc. Statistical analysis was performed using the Statistical Package for Social Sciences (SPSS) version 18.0 (SPSS Inc., IBM corporation 2010, www.spss.com). The significance level was set up at P values < 0.05.

3. Results

3.1. Anthropometric Measurements

Thirty-two patients and 32 controls participated in the

Nutritional Status of Patients with Cutaneous Leishmaniasis from a Tropical Area of Bolivia, and Implications for Zinc Bioavailability

39

study; the age range was between 14 and 50, and each group consisted of 12 females and 20 males. There were no significant differences in anthropometric results of BMI, AMA and AFA between patients and controls (**Table 1**). According to the WHO classification [14], most of the patients (50.0%) and controls (59.5%) were in the normal weight classification and there were no subjects in the underweight classification in any of the groups; hence 50.0% of the patients and 40.5% of the controls were overweight or obese and differences were not statistically different ($P = 0.340$). The AMA and AFA indicated that most of the patients and controls were in the average muscle and fat status according to the Frisancho classification [28] and not significantly different ($P = 0.485$ and 0.192 respectively). Wilcoxon rank test to compare groups of female patients (n = 12) with female controls (n = 12), and male patients (n = 20) with male controls (n = 20) showed no significant differences (results not shown).

3.2. Dietary Intake

Results of the dietary intake of patients and controls, were calculated as the nutrient density (nutrient/MJ) and

Table 1. Anthropometric characteristics of patients with cutaneous leishmaniasis and healthy controls.

	Patients (n = 32)				Controls (n = 32)				P value[c]
	Median	Percentiles		% (n)	Median	Percentiles		% (n)	
		25th	75th			25th	75th		
Age (y)	23.5	18.3	31.8		24.5	19.3	30.0		0.340
Weight (kg)	64.1	54.4	69.5		61.2	58.5	68.8		0.779
Height (m)	1.62	1.51	1.65		1.60	1.51	1.68		0.550
BMI[a] (kg/m²)	25.0	22.0	27.8	100 (32)	24.1	23.0	27.2	100 (32)	0.701
Frequencies BMI									0.340
Normal weight				50.0 (16)				59.5 (19)	
Pre-obese				38.0 (12)				28.0 (9)	
Obese class I				6.0 (2)				12.5 (4)	
Obese class III				6.0 (2)					
AMA[b] (cm²)	49.5	39.9	58.2	100 (32)	51.02	42.0	56.9	100 (32)	0.822
Frequencies AMA									0.485
Low, muscle wasted				3.0 (1)				6.0 (2)	
Below average				12.5 (4)				12.5 (4)	
Average				47 0 (15)				50.0 (16)	
Above average				12.5 (4)				22 (7)	
High				25.0 (8)				9.5 (3)	
AFA[b] (cm²)	15.8	13.4	23.2	100 (32)	14.95	10.3	23.2	100 (32)	0.432
Frequencies AFA									0.192
Lean				6.0 (2)				-	
Below average				12.5 (4)				25.0 (8)	
Average				63.0 (20)				69.0 (22)	
Above average				12.5 (4)				6.0 (2)	
Excess fat				6.0 (2)				-	

[a]BMI classification according WHO [14]; underweight (<18.5), normal weight (18.50 - 24.99), pre-obese (25.00 - 29.99), obese class I (30.00 - 34.99), obese class II (35.00 - 39.99) and obese class III (>40); [b]AMA-Muscle status and AFA-Fat status, classification according Frisancho [28]; [c]P-value, Wilcoxon rank test for continuous variables, and Chi Square for categorical variables (BMI, AMA and AFA frequencies).

compared by Wilcoxon rank test, which showed significant differences in fat and carbohydrates intake, but energy and most of the micronutrients were not statistically different ($P > 0.05$) between the two groups, excepting for that of Vitamin C (**Table 2**).

The median energy intake was generally low in both groups, consequently, the results were compared with their corresponding energy expenditure, calculated by equations from FAO/WHO [31]. The energy intake of patients met 78% and of controls 76% of their energy expenditure. The intake of calcium, folate, vitamin A and E, were low, patient's intake met between 38% and 65%

of the RNI and control's intake met between 36 and 59 of RNI; the other nutrients showed values close to the recommendations (**Table 2**).

Besides, the median value of iron intake was lower in the group of female patients (12.5 mg/d) and female controls (12.6 mg/d), meeting only the 52% of RNI. Furthermore, in order to gain knowledge about food components affecting mineral absorption, phytates intake and molar ration phytate:mineral were evaluated (**Table 2**). Phy:Zn ratios were between 11 and 19 (25th - 75th), Phy: Fe was between 7 and 9, and Phy:Ca between 0.14 and 0.35.

Table 2. Nutrient intake of patients with cutaneous leishmaniasis and healthy control group.

	Patients (n = 32)				Controls (n = 32)				P value
	Median	Percentiles		% RNI[b]	Median	Percentiles		% RNI[b]	
		25th	75th			25th	75th		
Energy (MJ/d)	7.44	6.64	8.58	78[c]	7.33	6.73	8.71	76[c]	0.736
Protein (g/d) (%E)[a]	56.8 (12.9)	50.6	69.1		60.9 (14.2)	52.5	74.1		0.501
Fat (g/d) (%E)[a]	38.9 (18.1)	29.6	44.9		48.0 (22.7)	37.3	56.2		0.018*
Carbohydrates (g/d) (%E)[a]	311 (69.3)	256	361		275 (62.8)	246	332		0.020*
Fiber (g/d)	17.4	13.5	20.3		15.6	12.9	20.4		0.347
Calcium (mg/d)	366	228	477	38	346	285	428	36	0.831
Iron (mg/d)	13.6	12.2	14.9	105	13.1	11.1	15.0	100	0.197
Magnesium (mg/d)	261	239	294	112	251	226	285	108	0.151
Phosphorus (mg/d)	899	801	1096	141	965	821	1157	139	0.81
Zinc (mg/d)	8.09	6.85	9.95	117	8.49	7.56	10.81	124	0.336
Copper (mg/d)	1.21	1.06	1.40	140	1.15	0.94	1.39	134	0.179
Vitamin C (mg/d)	64	44	104	167	43	33	53	109	0.003*
Thiamin (mg/d)	1.12	0.90	1.37	103	0.99	0.81	1.29	92	0.071
Riboflavin (mg/d)	0.99	0.75	1.14	87	0.94	0.74	1.23	84	0.963
Niacin (µg/d)	16.1	13.6	22.8	122	18.3	14.3	25.2	123	0.686
Pantothenic acid (mg/d)	4.95	4.52	5.97	108	4.87	4.08	5.55	99	0.453
Vitamin B6 (mg/d)	1.83	1.48	2.17	151	1.65	1.39	2.25	139	0.133
Folate (µg/d)	221	181	257	58	221	179	275	59	0.8
Vitamin B12 (µg/d)	1.43	1.15	2.7	86	1.92	1.53	2.85	94	0.426
Vitamin A (µgRAE/d)	324	242	464	65	242	191	326	51	0.079
β-Carotenoids (µg/d)	2356	1570	3399		1579	989	2214		0.066
Vitamin E (mg/d)	3.03	2.34	3.84	38	3.33	2.74	4.50	42	0.161
Phytates (g/d)	1.24	1.02	1.44		1.21	0.94	1.48		0.707
Molar ratio Phy:Zn	14.3	10.7	18.5		13.2	11.6	16.8		0.286
Molar ratio Phy:Fe	7.7	6.7	8.9		8.6	6.8	9.2		0.594
Molar ratio Phy:Ca	0.21	0.14	0.35		0.20	0.16	0.28		0.695

*Difference is statistically significant at 0.05 level. Wilcoxon rank test, calculated for the nutrient intake as nutrient density (nutrient units/MJ). [a]Median values of protein, fat and carbohydrates and corresponding percentage of energy intake in parenthesis (%E). [b]%RNI Percentage of recommended nutrient intake met by the diet. Calculated as: %RNI = (Estimated nutrient intake from the diet/Recommended nutrient intake)*100. [c]Percentage of energy expenditure which is covered by the energy intake, calculated as (EI/EE)*100. The energy expenditure was calculated with equations from FAO/WHO [31], and Goldberg cut-off [37] according to age and sex.

Nutritional Status of Patients with Cutaneous Leishmaniasis from a Tropical Area of Bolivia, and Implications for Zinc Bioavailability

41

3.3. Trace Elements Indicators

Serum zinc in patients (80 µg/dl) was significantly lower ($P = 0.033$) than in controls (85 µg/dl) and serum copper was not significantly different (**Table 3**). The results of serum zinc were compared with reference values from NHANES III (90 µg/dl for females and 98 µg/dl for males and lower cut-off 70 and 74 µg/dl for females and males respectively) [38]. The median values of serum zinc were below the average reference values, in 88% of the patients and 79% of the controls, furthermore, 29% of patients and 15% of controls showed zinc serum levels

below the lower cut-off, indicating that they are at risk of zinc deficiency. Values of serum copper were within the range of reference values (70 to 140 µg/dl) [39].

Correlations between anthropometric indicators BMI, AMA, AFA and intake of energy, protein, fat, and carbohydrates are presented for the groups of all patients, all controls, male patients, male controls, female patients and female controls (**Table 4**). BMI was positively correlated ($P < 0.001$) with the muscle and fat status for all groups; the correlations were stronger for females patients and controls ($P < 0.001$) than for males patients

Table 3. Zinc and copper serum levels of patients with cutaneous leishmaniasis and controls.

	Patients (n = 32)					Controls (n = 32)					P value
	Median	Percentiles		% ≤ ref[a]	% ≤ cut-off[a]	Median	Percentiles		% ≤ ref[a]	% ≤ cut-off[a]	
		25th	75th				25th	75th			
Serum Zn (µg/dl)	80	70	89	88	29	85	80	95	79	15	0.033[*]
Serum Cu (µg/dl)	104	85	119	-	6	105	96	121	-	-	0.191

[*]Difference is statistically significant at the 0.05 level; [a]Percentage of patients and controls below the values and lower cut-offs of Zn and Cu. References derived from NHANES II for zinc [22] and copper [39].

Table 4. Correlations of BMI, muscle and fat status with energy and macronutrients intake.

	AMA	AFA	Energy intake	Protein intake	Fat intake	Carbohydrates intake
BMI						
All Patients (n = 32)	0.755[*]	0.541[**]	0.290	0.250	0.247	0.273
All Controls (n = 32)	0.441[*]	0.694[**]	0.198	0.212	0.076	0.183
Male Patients (n = 20)	0.893[*]	0.432	0.451[*]	0.365	0.421	0.445
Male Controls (n = 20)	0.493[*]	0.602[**]	0.400	0.621[**]	0.465[*]	0.203
Female Patients (n = 12)	0.914[**]	0.913[**]	0.614[*]	0.213	0.239	0.577[*]
Female Controls (n = 12)	0.878[**]	0.803[**]	0.592	0.108	−0.102	0.497[*]
AMA, Muscle status						
All Patients (n = 32)		0.139	0.221	0.200	0.238	0.129
All Controls (n = 32)		0.090	0.975[**]	0.443[*]	0.271	0.316
Male Patients (n = 20)		0.308	0.159	0.186	0.293	0.186
Male Controls (n = 20)		0.520[*]	0.448	0.555[*]	0.569[**]	0.166
Female Patients (n = 12)		0.805[**]	0.472	0.189	0.316	0.374
Female Controls (n = 12)		0.706[*]	0.154	−0.091	−0.448	0.343
AFA, Fat Status						
All Patients (n = 32)			−0.046	−0.074	−0.050	0.077
All Controls (n = 32)			−0.223	−0.045	−0.161	−0.188
Male Patients (n = 20)			0.003	−0.107	−0.063	0.089
Male Controls (n = 20)			0.093	0.456[*]	0.341	−0.101
Female Patients (n = 12)			0.665[*]	0.226	0.291	0.611[*]
Female Controls (n = 12)			0.063	−0.203	−0.266	0.224

[**]Correlation is significant at the 0.001 level; [*]Correlation is significant at the 0.05 level.

and controls ($P < 0.05$). The associations between BMI and energy intake were positive for all groups, significantly so for male and female patients but not for male and females controls. A similar tendency is shown for correlation between BMI with protein and fat. AMA was positively related to protein and fat intake for male controls ($r = 0.555$ and 0.569, $P < 0.05$) but not significantly for male patients not for females.

There were no significant correlations between AFA with energy and macronutrient intake for any of the groups. Also no significant correlations with serum zinc and copper for any of the anthropometric variables for either group were found (results not shown).

Serum zinc was correlated with zinc intake at level $P < 0.05$ for the groups of all patients and female patients (**Table 5**). Further, serum zinc presented a negative significant correlation ($P < 0.05$) with copper intake for the groups of female controls ($r = -0.657$). Negative correlation with phytates intake was found for all groups; significant ($P < 0.05$) for the group of male patients ($r = -0.488$), male controls ($r = -0.460$). In addition, the molar ratio Phy:Zn for daily intake showed negative correlations ($P < 0.05$) for all groups except for male patients. Serum copper was negatively correlated with zinc intake for all the groups, significantly ($P < 0.05$) for all patients ($r = -0.361$). The correlations between serum copper and copper intake were not completely conclusive; some of them were negative and others positive.

4. Discussion

One of the most interesting findings of the present study is the significantly lower serum zinc concentrations found in patients with CL compared with the healthy controls; this might be associated with the presence of leishmaniasis. Another interesting finding was the apparently lower serum zinc status of both patients and controls in spite of a zinc intake according to the recommended values. This was most likely due to zinc absorption being impaired for phytates content in their diet.

The zinc dietary intake of patients and controls met the dietary recommendations (7.8 mg/d for females and 7.0 mg/d for males) [40], however, the zinc in serum of both groups was below the average reference value (90 - 98 μg/dl) [38]. Furthermore, the serum zinc in patients was significantly lower than in healthy controls. The results are consistent with previous studies [25,26,41] where serum zinc was significantly lower in CL patients.

Diminished serum zinc was also seen in patients with ML [41] and, to a greater extent, in patients with VL [41,42]. The decreased serum zinc levels in patients with leishmaniasis are probably due to the redistribution of zinc from plasma to the liver. Cytokines (IL-1) released during the acute-phase response of the host's immune system activates the synthesis of metallothionein in the liver and other tissues; metallothionein participates in the process of energy production and protection against reactive oxygen species that may be generated during the

Table 5. Correlations of zinc and cooper serum levels with the corresponding dietary intakes.

	Zn Intake	Cu Intake	Phytate Intake	Phy:Zn
Serum Zn				
All Patients (n = 32)	0.393*	−0.165	−0.202	−0.413*
All Controls (n = 32)	0.106	−0.105	−0.332	−0.410*
Male Patients (n = 20)	0.402	−0.215	−0.488*	−0.243
Male Controls (n = 20)	0.025	−0.098	−0.460*	−0.454*
Female Patients (n = 12)	0.652*	−0.257	−0.529	−0.624*
Female Controls (n = 12)	0.552	−0.657*	−0.568	−0.658*
Serum Cu				
All Patients (n = 32)	−0.361*	−0.195		
All Controls (n = 32)	−0.298	0.094		
Male Patients (n = 20)	−0.398	−0.065		
Male Controls (n = 20)	−0.136	−0.081		
Female Patients (n = 12)	−0.318	0.192		
Female Controls (n = 12)	−0.576	0.582*		

*Correlation is significant at the 0.05 level.

Nutritional Status of Patients with Cutaneous Leishmaniasis from a Tropical Area of Bolivia, and Implications for Zinc Bioavailability

43

infection, and it is a metal-binding protein which appears to alter the hepatic uptake of zinc [43]. A study with mice has demonstrated that, besides metallothionein, the zinc transporter Zip14 contributes towards the reduction of the zinc levels during inflammation and infection [44].

In recent years, there has been a great interest in the study of the zinc status and supplementation, since the demonstration of its critical role in reducing the risk and severity of diverse infections [21], and of CL in particular, because it is known that wound healing is impaired in zinc deficiency [1], and that oral zinc administration in the treatment of acute CL has shown a good response in the healing of the lesions caused by the infection [45].

Besides the changes of serum zinc during leishmaniasis, it has been reported that levels of serum copper were higher in patients with CL [25,26,41,46], ML and VL [41]. The increased level of serum copper could be attributed to inflammation due to the presence of the leishmaniasis parasite [25]. In the present study the levels of serum copper were not significantly different between patients and controls. Further studies are needed in order to investigate the changes of copper during the course of the infection, and the use of more sensitive indicators to detect changes in copper status [47].

Complex mechanisms might be involved in bringing about the differences in serum zinc and copper in patients with CL and control subjects, most of them produced as a consequence of the acute-phase response of the defense strategies of the host's immune system as mentioned above. Immune cells, just as any other cells, require an adequate supply of trace elements, so there is a redistribution of essential minerals like zinc and copper and an increase in the hepatic synthesis of acute-phase proteins like ceruloplasmin [48]. Another enzyme involved in the immune response is superoxide dismutase (SOD) which requires both zinc and copper for its normal activity; copper is necessary for catalysis and zinc stabilizes the enzyme. In this sense there is a competition between the two minerals to reach the enzyme, which may cause the imbalance of the minerals [49].

The low levels of serum zinc were not only found in the patients with CL but also in healthy controls with an adequate zinc intake according to the RNI from WHO/FAO [31]. The zinc intake of patients and controls met the zinc requirements in 117% and 124% respectively. However, 88% of the patients showed values below the reference serum zinc value and 29% of them presented values below the lower cut-off, indicating that they were at risk for zinc deficiency. In the healthy controls, 79% of them presented values below the reference and 15% of them were at risk for zinc deficiency. These results drew attention to the need to investigate other factors besides the leishmaniasis that could decrease the levels of serum

zinc in both patients and healthy controls.

Among the causes of the low serum zinc concentrations are; a low dietary zinc intake and a low absorption of dietary zinc as a result of other components in the food and the physicochemical interactions in the intestine; low serum zinc levels may often be due to a combination of these factors [13,22,38]. In the present study the zinc intake was not the cause of the low serum zinc, so it was most likely, due to the presence of zinc inhibitors in the diet, such as phytates, which are strong chelators of divalent minerals and are primarily to be found in cereal grain, legumes, seeds and tubers [17,22], which are the principal components of the diet in this area as we have previously shown [15].

The diet of the studied population showed Phy:Zn between 11 and 19 (25th -75th), indicating that zinc absorption may be compromised for the level of phytates content. According to the WHO committee [40] Phy:Zn higher than 15 are considered to inhibit zinc absorption and even molar ratios between 5 and 15 may have a certain negative effect on the absorption of zinc. It was reported that diets in rural areas following similar dietary patterns with high Phy:Zn impair mineral bioavailability, leading to zinc deficiencies [16,17,50-52]. Besides, the Phy:Fe (7 to 9) was much higher than 1, which is the level considered adequate for iron bioavailability [53]. Furthermore, Phy:Ca (0.14 to 0.35) was higher than the desirable value 0.17 [54], indicating that it is also likely that phytates compromised iron and calcium absorption in this diet.

The association between serum zinc and dietary zinc, phytates, and Phy:Zn (**Table 5**) showed positive correlations ($P < 0.05$) between serum zinc and zinc intake, indicating that serum zinc is a good indicator to reflect dietary zinc, as has been demonstrated in other studies [55,56]. Correlations between serum zinc and phytates intake were inverse and significant ($P < 0.05$) for male patients ($r = -0.488$) and male controls ($r = -0.460$). Stronger significant inverse correlations ($P < 0.05$) were found between serum zinc and Phy:Zn for all the groups (except for male patients), indicating that the zinc absorption was impaired by the presence of phytates in the diet. In a study of vegetarian and omnivorous diets, a similar inverse significant association was reported between serum zinc and Phy:Zn in women with low levels of serum zinc [57]; the same findings were reported in a study with women from New Zealand [58]. Associations between serum copper and the corresponding copper intake were not conclusive; most probably because this biomarker is not sensitive enough to reflect copper intake, except when a severe deficiency is present or the intake is very low [22,59].

In relation to the anthropometric characteristics, there

were no significant differences between the median values of BMI, AMA and AFA between patients and controls (**Table 1**). None of the patients or controls was underweight. However, 50.0% of patients and 40.5% of controls were overweight or obese and similar results were presented by Ferreira da Cunha *et al.* [60]. Overweight and obesity is not an indication that the patients have a better nutritional status; they may have micronutrient deficiencies and an impaired immunity associated with the consumption of imbalanced diets [61,62]. The increased prevalence of overweight is a consequence of the dietary transition reported in Latin America with a reduced consumption of fruits and vegetables and an increase in fats and sugar [63].

The analysis of the dietary intake showed that the basic diet of patients and controls present a composition of macronutrients as energy percentage, within the dietary recommendations from WHO [31]; 69.3 and 62.8 E% from carbohydrates, 12.9 and 14.2 E% from protein and 18.1 and 22.7 E% from total fat for patients and controls respectively. These results are consistent with those reported in a study of dietary intake carried out in the same area, where the contribution from carbohydrates was 72E%, proteins 13E%, and total fat 15E% [15]. The energy distribution is, in accordance with the dietary patterns of the area, based primarily on carbohydrates from cereals, tubers and legumes, protein from small portions of meat or eggs and fat from oil and tallow; the vegetables and fruits are present in small portions. The diet of the subjects varied very little from one day to another constituting a monotonous diet. The median energy intake was, in general, low: 7.44 MJ/d for patients and 7.33 MJ/d for controls. Similar dietary and energy intake patterns were found in a rural population in Argentina, where energy intake was between 6.65 to 7.77 MJ/d [64].

Furthermore, the nutrient intake of patients and controls in this studied population indicates the existence of deficits of several essential nutrients; calcium, iron (for women), folate, Vitamin A, and E, with median values below the 65% of the RNI. Micronutrient deficiencies have been seen among rural populations, especially in rural areas in developing countries with dietary patterns containing small amounts of animal foods, which may lead to a micronutrient malnutrition or "hidden hunger" [65,66].

The results in this study indicate that the infection of CL does not stand in direct association with the nutritional status shown by the anthropometric and dietary assessments. Thus, the infection may randomly affect the exposed individuals, independently of their current nutritional status. However, further studies are needed in order to determine whether the development of the disease is exacerbated by a low nutritional status. There are few studies about the association of CL and nutritional status; in children a more severe clinical manifestation of leishmaniasis was found when chronic energy-protein malnutrition was present [9], and it was also reported that in patients older than 22 years of age the risk of severe manifestations of leishmaniasis increases when the nutritional status decreases [67]. Studies of the nutritional status and the outcome of VL have shown that malnourished children presented a more aggravated state of the infection, creating a circle of malnutrition and infection, causing low growth-rate and nutritional deficiencies [7,10].

5. Conclusions

The present paper shows that the serum zinc levels of patients with CL were significantly lower than those of healthy subjects. Furthermore, it was found that even though the zinc intake of patients and controls met the dietary recommendations, the serum zinc levels were below the average reference values, indicating a low absorption of dietary zinc in both healthy subjects and patients with CL. Indeed, the results showed that zinc absorption and metabolism might be compromised by inhibitory components in the diet, such as phytates, and by the presence of the CL infection. Additionally, CL was found to not be directly associated with the nutritional status observed in the anthropometric and dietary results. However, the results of dietary patterns and nutrient intake shed some light on the existence of deficits of several essential micronutrients, which are below the recommended intake.

Studies of the nutritional status of the population at risk for acquiring leishmaniasis are of great importance for the design and implementation of new strategies, both nutritional and therapeutic. In order to prevent complications in the outcome of leishmaniasis, as well as other adverse effects of imbalanced diets and nutritional deficiencies among rural populations in developing countries, the nutritional status of the host should be appropriately considered.

6. Acknowledgements

We thank Miguel Guzmán of Biomedical Research Institute, San Simon University, Cochabamba, Bolivia, for the collaboration providing blood samples. Financial support from the Swedish International Development Agency (SIDA/SAREC) is gratefully acknowledged.

REFERENCES

[1] R. K. Chandra, "Nutrition and the Immune System: An Introduction," *The American Journal of Clinical Nutrition*, Vol. 62, No. 2, 1997, pp. 460S-463S.

[2] R. W. Ashford, P. Desjeux and P. Deraadt, "Estimation of Population at Risk of Infection and Number of Cases of Leishmaniasis," *Parasitology Today*, Vol. 8, No. 3, 1992, pp. 104-105.

[3] R. Reithinger, *et al.*, "Cutaneous Leishmaniasis," *The Lancet Infectious Diseases*, Vol. 7, No. 9, 2007, pp. 581-596.

[4] A. L. García, *et al.*, "Leishmaniases in Bolivia: Comprehensive Review and Current Status," *The American Journal of Tropical Medicine and Hygiene*, Vol. 80, No. 5, 2009, pp. 704-711.

[5] E. Rojas, *et al.*, "Leishmaniasis in Chaparé, Bolivia," *Emerging Infectious Diseases*, Vol. 15, No. 4, 2009, pp. 678-680.

[6] L. Soong, C. A. Henard and P. C. Melby, "Immunopathogenesis of Non-Healing American Cutaneous Leishmaniasis and Progressive Visceral Leishmaniasis," *Seminars in Immunopathology*, Vol. 34, No. 6, 2012, pp. 735-751.

[7] A. J. M. Caldas, *et al.*, "Leishmania (Leishmania) Chagasi Infection in Children from an Endemic Area of American Visceral Leishmaniasis on Sao Luis Island, Maranhao, Brazil," *Revista da Sociedade Brasileira de Medicina Tropical*, Vol. 34, No. 5, 2001, pp. 445-451.

[8] D. F. D. Cunha, *et al.*, "Retardo do Crescimento em Crianças com Reação Intradérmica Positiva para Leishmaniose: Resultados Preliminares," *Revista da Sociedade Brasileira de Medicina Tropical*, Vol. 34, No. 1, 2001, pp. 25-27.

[9] M. M. Weigel, *et al.*, "Nutritional Status and Cutaneous Leishmaniasis in Rural Ecuadorian Children," *Journal of Tropical Pediatrics*, Vol. 41, No. 1, 1995, pp. 22-28.

[10] B. L. L. Maciel, *et al.*, "Association of Nutritional Status with the Response to Infection with *Leishmania chagasi*," *The American Journal of Tropical Medicine and Hygiene*, Vol. 79, No. 4, 2008, pp. 591-598.

[11] H. Perez, I. Malave and B. Arredondo, "Effects of Protein-Malnutrition on the Course of Leishmania-Mexicana Infection in C57B1-6 Mice-Nutrition and Susceptibility to Leishmaniasis," *Clinical and Experimental Immunology*, Vol. 38, No. 3, 1979, pp. 453-460.

[12] G. Malafaia, "Protein-Energy Malnutrition as a Risk Factor for Visceral Leishmaniasis: A Review," *Parasite Immunology*, Vol. 31, No. 10, 2009, pp. 587-596.

[13] R. S. Gibson, "Principles of Nutritional Assessment," 2nd Edition, Oxford University Press, New York, 2005.

[14] WHO, "Physical Status: The Use and Interpretation of Anthropometry in Technical Report Series No.8541995," World Health Organization, Geneva.

[15] C. Lazarte, *et al.*, "Validation of Digital Photographs, as a Tool in 24-h Recall, for the Improvement of Dietary Assessment among Rural Populations in Developing Countries," *Nutrition Journal*, Vol. 11, No. 1, 2012, p. 61.

[16] R. S. Gibson, *et al.*, "A Review of Phytate, Iron, Zinc, and Calcium Concentrations in Plant-Based Complementary Foods Used in Low-Income Countries and Implications for Bioavailability," *Food & Nutrition Bulletin*, Vol. 31, No. S2, 2010, pp. 134-146.

[17] R. S. Gibson and C. Hotz, "Dietary Diversification/Modification Strategies to Enhance Micronutrient Content and Bioavailability of Diets in Developing Countries," *British Journal of Nutrition*, Vol. 85, 2001, pp. S159-S166.

[18] B. Lönnerdal, "Dietary Factors Influencing Zinc Absorption," *The Journal of Nutrition*, Vol. 130, No. 5, 2000, pp. 1378S-1383S.

[19] B. Sandstrom, "Bioavailability of Zinc," *European Journal of Clinical Nutrition*, Vol. 51, 1997, pp. S17-S19.

[20] E. L. Ferguson, *et al.*, "Dietary Calcium, Phytate and Zinc Intakes and the Calcium, Phytate and Zinc Molar Ratios of the Diets of a Selected Group of East-African Children," *American Journal of Clinical Nutrition*, Vol. 50, No. 6, 1989, pp. 1450-1456.

[21] C. Hotz and K. H. Brown, "International Zinc Nutrition Consultative Group (IZiNCG) Technical Document #1. Assessment of the Risk of Zinc Deficiency in Populations and Options for Its Control," *Food and Nutrition Bulletin*, Vol. 25, No. 1, 2004, pp. S94-S203.

[22] C., Hotz, *et al.*, "Assessment of the Trace Element Status of Individuals and Populations: The Example of Zinc and Copper," *The Journal of Nutrition*, Vol. 133, No. 5, 2003, pp. 1563S-1568S.

[23] A. S. Prasad, "Effects of Zinc Deficiency on Immune Functions," *Journal of Trace Elements in Experimental Medicine*, Vol. 13, No. 1, 2000, pp. 1-20.

[24] P. Shetty, "Zinc Deficiency and Infections. Nutrition, Immunity and Infection," Cabi Publishing, Wallingford, 2010, pp. 101-113.

[25] A. Kocyigit, *et al.*, "Alterations of Serum Selenium, Zinc, Copper, and Iron Concentrations and Some Related Antioxidant Enzyme Activities in Patients with Cutaneous Leishmaniasis," *Biological Trace Element Research*, Vol. 65, No. 3, 1998, pp. 271-281.

[26] M. Faryadi and M. Mohebali, "Alterations of Serum Zinc, Copper and Iron Concentrations in Patients with Acute and Chronic Cutaneous Leishmaniasis," Iranian *Journal of Public Health*, Vol. 4, 2003, p. 53.

[27] M. F. Zubieta Durán and M. C. Torrico Rojas, "Manual de Normas y Procedimientos Técnicos de Laboratorio (para Leishmaniasis 2010: Ministerio de Salud y Deportes Bolivia,", 2010.

[28] A. R. Frisancho, "Triceps Skin Fold and Upper Arm Muscle Size Norms for Assessment of Nutritional Status," *The American Journal of Clinical Nutrition*, Vol. 27, No. 10, 1974, pp. 1052-1058.

[29] USDA, "USDA National Nutrient Database for Standard Reference," 2001.
http://www.nal.usda.gov/fnic/foodcomp/search

[30] INLASA, "Tabla de Composicion de Alimentos Bolivianos," C. Edicion, Ed., Ministerio de Salud y Deportes, La Paz, 2005.

[31] WHO and FAO, "Dietary Recommendations in the Report of a Joint WHO/FAO Expert Consultation on Diet," Nutrition and the Prevention of Chronic Diseases (WHO Technical Report Series 916), 2003.

[32] N. R. Reddy, "Occurrence, Distribution, Content, and Dietary Intake of Phytate," *Food Phytates*, CRC Press, 2001.

[33] R. M. García-Estepa, E. Guerra-Hernández and B. García-Villanova, "Phytic Acid Content in Milled Cereal Products and Breads," *Food Research International*, Vol. 32, No. 3, 1999, pp. 217-221.

[34] B. Q. Phillippy, M. Lin and B. Rasco, "Analysis of Phytate in Raw and Cooked Potatoes," *Journal of Food Composition and Analysis*, Vol. 17, No. 2, 2004, pp. 217-226.

[35] J. Ruales and B. M. Nair, "Saponins, Phytic Acid, Tannins and Protease Inhibitors in Quinoa (Chenopodium Quinoa, Willd) Seeds," *Food Chemistry*, Vol. 48, No. 2, 1993, pp. 137-143.

[36] A. Taylor, "Measurement of Zinc in Clinical Samples," *Annals of Clinical Biochemistry*, Vol. 34, No. 2, 1997, pp. 142-150.

[37] G. R. Goldberg, *et al.*, "Critical-Evaluation of Energy-Intake Data Using Fundamental Principles of Energy Physiology. Derivation of Cutoff Limits to Identify Under-Recording," *European Journal of Clinical Nutrition*, Vol. 45, No. 12, 1991, pp. 569-581.

[38] C. Hotz, J. M. Peerson and K. H. Brown, "Suggested Lower Cutoffs of Serum Zinc Concentrations for Assessing Zinc Status: Reanalysis of the Second National Health and Nutrition Examination Survey Data (1976-1980)," *The American Journal of Clinical Nutrition*, Vol. 78, No. 4, 2003, pp. 756-764.

[39] M. A. Knovich, *et al.*, "The Association between Serum Copper and Anaemia in the Adult Second National Health and Nutrition Examination Survey (NHANES II) Population," *British Journal of Nutrition*, Vol. 99, No. 6, 2008, pp. 1226-1229.

[40] WHO, "Trace Elements in Human and Health Nutrition," 1996.
http://whqlibdoc.who.int/publications/1996/9241561734_eng.pdf

[41] J. Van Weyenbergh, *et al.*, "Zinc/Copper Imbalance Reflects Immune Dysfunction in Human Leishmaniasis: An *ex Vivo* and *in Vitro* Study," *BMC Infectious Diseases*, Vol. 4, No. 1, 2004, p. 50.

[42] J. Mishra, S. Carpenter and S. Singh, "Low Serum Zinc Levels in an Endemic Area of Visceral Leishmaniasis in Bihar," *Indian Journal of Medical Research*, Vol. 131, No. 6, 2010, pp. 793-798.

[43] J. J. Schroeder and R. J. Cousins, "Interleukin 6 Regulates Metallothionein Gene Expression and Zinc Metabolism in Hepatocyte Monolayer Cultures," *Proceedings of the National Academy of Sciences*, Vol. 87, No. 8, 1990, pp. 3137-3141.

[44] J. P. Liuzzi, *et al.*, "Interleukin-6 Regulates the Zinc Transporter Zip14 in Liver and Contributes to the Hypozincemia of the Acute-Phase Response," *Proceedings of the National Academy of Sciences of the United States of America*, Vol. 102, No. 19, 2005, pp. 6843-6848.

[45] K. E. Sharquie, *et al.*, "Oral Zinc Sulphate in the Treatment of Acute Cutaneous Leishmaniasis," *Clinical and Experimental Dermatology*, Vol. 26, No. 1, 2001, pp. 21-26.

[46] G. Culha, E. Yalin and K. Sanguen, "Alterations in Serum Levels of Trace Elements in Cutaneous Leishmaniasis Patients in Endemic Region of Hatay (Antioch)," *Asian Journal of Chemistry*, Vol. 20, No. 4, 2008, pp. 3104-3108.

[47] I. M. Goldstein, *et al.*, "Ceruloplasmin: A Scavenger of Superoxide Anion Radicals," *Journal of Biological Chemistry*, Vol. 254, No. 10, 1979, pp. 4040-4045.

[48] C. A. Dinarello, "Interleukin-1 and the Pathogenesis of the Acute-Phase Response," *New England Journal of Medicine*, Vol. 311, No. 22, 1984, pp. 1413-1418.

[49] M. Panemangalore and F. N. Bebe, "Effect of High Dietary Zinc on Plasma Ceruloplasmin and Erythrocyte Superoxide Dismutase Activities in Copper-Depleted and Repleted Rats," *Biological Trace Element Research*, Vol. 55, No. 1-2, 1996, pp. 111-126.

[50] Y. Abebe, *et al.*, "Phytate, Zinc, Iron and Calcium Content of Selected Raw and Prepared Foods Consumed in Rural Sidama, Southern Ethiopia, and Implications for Bioavailability," *Journal of Food Composition and Analysis*, Vol. 20, No. 3-4, 2007, pp. 161-168.

[51] E. I. Adeyeye, *et al.*, "Calcium, Zinc and Phytate Interrelationships in Some Foods of Major Consumption in Nigeria," *Food Chemistry*, Vol. 71, No. 4, 2000, pp. 435-441.

[52] K. C. Menon, *et al.*, "Concurrent Micronutrient Deficiencies Are Prevalent in Nonpregnant Rural and Tribal Women from CENTRAL INDIA," *Nutrition*, Vol. 27, No. 4, 2011, pp. 496-502.

[53] R. F. Hurrell, "Phytic Acid Degradation as a Means of Improving Iron Absorption," *International Journal for Vitamin and Nutrition Research*, Vol. 74, No. 6, 2004, pp. 445-452.

[54] M. Umeta, C. E. West and H. Fufa, "Content of Zinc, Iron, Calcium and Their Absorption Inhibitors in Foods

Nutritional Status of Patients with Cutaneous Leishmaniasis from a Tropical Area of Bolivia,
and Implications for Zinc Bioavailability

47

Commonly Consumed in Ethiopia," *Journal of Food Composition and Analysis*, Vol. 18, No. 8, 2005, pp. 803-817.

[55] K. H. Brown, "Effect of Infections on Plasma Zinc Concentration and Implications for Zinc Status Assessment in Low-Income Countries," *American Journal of Clinical Nutrition*, Vol. 68, No. 2, 1998, pp. 425S-429S.

[56] S. Y. Hess, *et al.*, "Use of Serum Zinc Concentration as an Indicator of Population Zinc Status," *Food & Nutrition Bulletin*, Vol. 28, No. 3, 2007, pp. 403S-429S.

[57] U. M. Donovan and R. S. Gibson, "Iron and Zinc Status of Young-Women Aged 14 to 19 Years Consuming Vegetarian and Omnivorous Diets," *Journal of the American College of Nutrition*, Vol. 14, No. 5, 1995, pp. 463-472.

[58] R. S. Gibson, *et al.*, "Are Changes in Food Consumption Patterns Associated with Lower Biochemical Zinc Status among Women from Dunedin, New Zealand?" *British Journal of Nutrition*, Vol. 86, No. 1, 2001, pp. 71-80.

[59] D. B. Milne and P. E. Johnson, "Assessment of Copper Status: Effect of Age and Gender on Reference Ranges in Healthy Adults," *Clinical Chemistry*, Vol. 39, No. 5, 1993, pp. 883-887.

[60] D. F. Cunha, *et al.*, "Is an Increased Body Mass Index Associated with a Risk of Cutaneous Leishmaniasis?" *Revista da Sociedade Brasileira de Medicina Tropical*, Vol. 42, No. 5, 2009, pp. 494-495.

[61] B. S. Kumari and R. K. Chandra, "Overnutrition and Immune-Responses," *Nutrition Research*, Vol. 13, No. 1, 1993, S3-S18.

[62] S. Samartin and R. K. Chandra, "Obesity, Overnutrition and the Immune System," *Nutrition Research*, Vol. 21, No. 1-2, 2001, pp. 243-262.

[63] O. I. Bermudez and K. L. Tucker, "Trends in Dietary Patterns of Latin American Populations," *Cadernos de Saúde Pública*, Vol. 19, No. 1, 2003, pp. S87-S99.

[64] M. N. Bassett, D. Romaguera and N. Samman, "Nutritional Status and Dietary Habits of the Population of the Calchaqui Valleys of Tucuman, Argentina," *Nutrition*, Vol. 27, No. 11-12, 2010, pp. 1130-1135.

[65] M. Gibbs, *et al.*, "The Adequacy of Micronutrient Concentrations in Manufactured Complementary Foods from Low-Income Countries," *Journal of Food Composition and Analysis*, Vol. 24, No. 3, 2011, pp. 418-426.

[66] IOM, "Prevention of Micronutrient Deficiencies: Tools for Policymakers and Public Health Workers," The National Academies Press, Washington DC, 1998.

[67] G. L. L. Machado-Coelho, *et al.*, "Risk Factors for Mucosal Manifestation of American Cutaneous Leishmaniasis," *Transactions of the Royal Society of Tropical Medicine and Hygiene*, Vol. 99, No. 1, 2005, pp. 55-61.

6

Microbial Load (Bacteria, Coliform and Mould Count/Flora) of Some Common Hot Smoked Freshwater Fish Species Using Different Packaging Materials

Olusegun Ayodele Oyelese, Jacob Oyeleye Oyedokun

Department of Aquaculture and Fisheries Management, Faculty of Agriculture and Forestry, University of Ibadan, Ibadan, Nigeria.

ABSTRACT

Three different packaging materials of (37 cm × 25 cm) size (Sealed Transparent Polythene Bag (STPB) Sealed Paper Bag (SPB) (Brown envelope), Open Mouth Polythene Bag (OMPB) (Black incolour)) were used for *Oreochromisniloticus* (O), *Clariasgariepinus* (C) and *Mormyrusrume* (M). Twenty fish samples per species (averaging 250 gm) were hot smoked dried whole for 36 hours at an average temperature of 100°C. Packaged hot at the rate of 6 fishes per package for each species (three packs for each packaging treatment *i.e.* 18 pieces were packed while the remaining 2 pieces were used for initial bacteria load and microbial load). Microbial load (Total Viable Count (TVC), Total Coliform Count (TCC) and Total Fungi Count (TFC)) for the fresh fish was initial hot smoked and finally at the end of 12 weeks was monitored. The TVC (bacterial load) of *O. niloticus* dropped from $(10.6 - 8.4) \times 10^4$ (fresh state-hot smoked) and *M. rume* $(9.8 - 7.0) \times 10^4$, while *C. gariepinus* slightly increased from $(12.4 - 12.6) \times 10^4$. After hot smoking, highest TVC of 8.6×10^4 (OMPBC), 8.3×10^4 (SPBC) and 8.2×10^4 (STPBC) was recorded in *C. gariepinus* among the 9 packaging at 12 weeks. However highest tendency for heavy TVC is in all OMPB with highest bacteria load in the OMPBC (8.6×10^4), 7.6×10^4 (OMPBO) and 6.6×10^4 (OMPBM). After 12 weeks highest ranged TFC of $(0.6 - 0.7) \times 10^4$ was recorded in *M. rume* as against 0.2×10^4 recorded in the initial smoked for all. TCC was highest in *C. gariepinus* $(4.0 - 4.3) \times 10^4$. Packaging did not limit the existence of micro-organisms. Six bacteria species (*Micrococcus* (*acidiophilus*, *luteus*), *Bacillus* (*subtilis*, *cereus*, *aureus*), *Staphylococcus aureus*, *Streptococcus lactis*, *Proteus* (*vulgaricus*, *morganii*), *Pseudomonas aureginosa*) and three fungi species (*Aspergillus* (*niger*, *tamari*), *Rhizopusnigricans*, *fusariumoxysporum*) were represented in all the packages. On the average five bacteria and two fungi species were represented, excepting for OMPBM and OMPBO with six bacteria species.

Keywords: Bacteria; Coliform; Mould Count/Flora; Freshwater Fish Species; Packaging Material

1. Introduction

Bacteria are unicellular microscopic organisms which occur almost everywhere in nature. Up to 1500 species of bacteria have been isolated since bacteria are living things; they acquire a source of food, moisture and suitable temperature to grow, when these conditions are adequate. Bacteria cause spoilage of improperly dried fish by multiplying inside the fish flesh thereby causing putrefaction. Once bacteria spoilage sets in there it is hard to remedy. The result of bacteria attack is off odour and flavor and when pathogenic bacteria are involved, it could result in illness to consumer [1].

The bacteria that most often involved in the spoilage of fish are part of the natural flora of the external slime of fishes and their intestinal content [2]. They lamented that the predominant kinds of bacteria causing spoilage vary with the temperatures at which the fish are held as follows:

- Chilling temperature
  Species of pseudomonas
  Achromebacter and
  Flavobacterium
- Higher temperature
  Genera micrococcus and

Bacillus
- Atmospheric temperature
Escherichia
Proteus
Serratia
Sarcina and
Clostridium

Bacteria are unicellular microscopic organisms which occur almost everywhere in nature [1]. Up to 1500 species of bacteria have been isolated since bacteria are living things. They acquire a source of food, moisture and suitable temperature to grow [3], when these conditions are adequate. Bacteria will grow by a process known as Binary Fission in which the cell divides into two new cells. Some bacteria causing fish spoilage might have a generation time of 20 minutes at 30°C [1]. In such a case, a single bacterium may give billions in 10 hours [4].

Whilst increase in the population of micro-organisms by geometric progression is theoretically possible, its practical application is limited by environment factors prevailing. These factors are:

1) Temperature

Table 1 below shows the ranges of temperature for the growth of micro-organism.

2) Water Content

Table 2 shows the minimum water activity for the growth of micro-organism.

3) Acidity or Alkalinity (pH)

Bacteria grow well over a wide range of hydrogen ion concentration pH ranging from 4.0 - 9.0. The optimum pH growth for most bacteria lies between pH 6.5 and 7.5 although some bacteria are capable of growing at the

Table 1. Temperature ranges for growth of micro-organism.

Types of micro-organism	Minimum (°C)	Optimum (°C)	Maximum (°C)
Psychrophiles	0	15 - 25	30
Mesophiles	10	37	43
Thermopliles	25	50 - 65.5	85

Table 2. Minimum water activity for growth of micro-organism [5].

Micro-organism	Minimum water activity
Normal bacteria	0.90
Normal yeast	0.88
Normal moulds	0.80
Halophilic bacteria	0.75
Dryness resistant moulds	0.05
Osmotic pressure resistant yeast	0.61

extremes of the pH ranges. Bacteria growth and toxin production are inhibited if the conditions are more lethal to micro-organisms than alkaline [6].

4) Nutrient Composition

Bacteria are living organisms and like other living things such as plants and animals, they require a source of energy to survive. Such energy can be obtained from sunlight or by breakdown of nutrients which are mainly carbohydrates, proteins, fats and oil, vitamins and other growth factors. The breakdown of each of these nutrients requires the possession of the appropriate enzymes by bacteria [1,6].

1.1. Bacterial in Smoked Fish

Smoked fish and shellfish products can be a source of microbial hazards including *listeria monocytogenes*, Salmonella species and *Clostridium botulinium*, *L. mnonocytogens* has been identified in several food borne outbreaks, in which pasteurized milk, coleslaw and soft cheese were implicated [7]. These organisms have also been isolated from a variety of fish and shellfish products [8].

1.2. Fungal Attacks in Smoked Fish

Insufficient dried fish (still containing approximately 40% moisture) especially at the processing location are prone to fungal infection, principally from the non-specific *Penicillium spp.*, *Aspergillus spp.* Substantial qualities of fish are usually discarded during drying due to fungal growth. Fungal spp. also associated with smoked fish include: *Aspergillus fumigates*, *Absidia spp.*, *Rhizopus spp.*, *Mucor spp.*, *Cladosporium spp.* [9-12]. It was observed that though smoking fish provides longer shelf life than other preservative methods, smoking will be effective if properly done (especially to reduce packaging).

Adebayo-Tayo *et al.* [9] identified 12 different fungi and aflatoxin B1 and G1 in three main markets in Nigeria on smoked dried fish with moisture content ranging from 22.7% - 27.6%. He said the level of infestation might be due to high percentage of moisture content of the smoked fish.

2. Materials and Methods

2.1. Collection of Samples\Packaging

Twenty pieces (sample) of each fish species of average weight 250 grams were collected for *Oreochromisniloticus* (O), *Clariasgariepinus* (C) and *Mormyrusrume* (M). Also fresh samples were collected for the initial proximate analysis while the remaining fresh fishes were transported to the processing unit for smoking. After which the initial proximate analysis of the hot smoked fish was also taken before packaging in the 37 cm × 25 cm packaging materials for each of the smoked fish species (using each of the three different packaging material for each fish species) at the rate of six (6) fish species per package and labeled e.g. for Oreochromis (STPBO—Sealed Transparent Polythene Bag Oreochromis, SPBO—Sealed paper Bag Oreochromis, OMPBO—Open Month Polythene Bag Oreochromis.

2.2. Hot Smoking of the Fish Species

The smoking kiln was locally improvised. Three broken blocks each of 0.3 m height was used to raise the wire gauze (on which the fish were laid) to avoid direct contact with fire. Big wire gauze of mesh size 2 cm was set on the fire when the fire was fully lit. The three species of the fish to be smoked were placed on the gauze. Big aluminum basin with a opening at the centre was used to cover the fish species in order to conserve the fire. It was through the opening that the temperature of the smoking kiln (chimney) was taken daily, until the three fish species were hot smoked dried. Hot smoking was done for 36 hours (this was achieved in three days at an average of 12 hours smoking per day) at an average temperature of 100°C.

Hot smoking was done with an exotic hard wood (Eucalyptus species), collected from the Forestry Department of the University of Ibadan. Turning of the fish species were done at the same time to maintain uniform drying\smoking at an interval of one hour (1.5 hr) thirty minutes for 3 days.

2.3. Packaging and Shelfing

After three days of intensive smoking, each species of the three freshwater fish species were packaged under three different packaging materials (Sealed Transparent Polythene Bag (STPB), Sealed Paper Bag (SPB) (Brown envelope), Open Mouth Polythene Bag (OMPB) (Black in colour)) under room ambient temperature range of 25°C - 32°C for 12 weeks. Mould growth: insect infestation was checked daily during this period for each of the fish species.

The three different materials used were:
A. Sealed Transparent Polythene Bag (STPB)
1. Tilapia (*Oreochromisniloticus*) (STPBO)
2. *Clariasgariepinus* (STPBC)
3. *Mormyrusrume* (STPBM)
B. Sealed Paper Bag (SPB)
1. Tilapia (*Oreochromisniloticus*) (SPBO)
2. *Clariasgariepinus* (STBC)
3. *Mormyrusrume* (STBM)
C. Open Mouth Polythene Bag (OMPB)

1. Tilapia (*Oreochromisniloticus*) (OMPBO)
2. *Clariasgariepinus* (OMPBC)
3. *Mormyrusrume* (OMPBM)
The fishes were packaged hot in the packaging bags and stored in the laboratory for 12 weeks.

2.4. Preparation of Media

All analytical procedures in this study are according to the A.O.A.C [13].

2.4.1. Nutrient Agar
Twenty eight (28) grams of powdered commercially prepared of nutrient agar was weighed on Analytical metller balance into a clean dry 1 litre conical flask and 1000 ml of distilled water placed inside a water bath set about 90°C, allow the agar to dissolve. Distribute them into MacCantney bottles and placed them inside autoclave and set the autoclave at 121°C for 15 mins.

2.4.2. Macconkey Agar (Mcca)
Fifty five (55) grams of macConkey Agar was weighted into a 1 litre capacity of conical flask and brings to boil to dissolve the agar. Distribute them into Mac Cartney bottles and autoclave as for Nutrient Agar.

2.4.3. Potato Dextrose Agar (PDA)
Thirty nine (39) grams of PDA was weighted into a 1 litre capacity of conical flask bring to boil and distributed them into Mac Cartney bottles and placed them inside an autoclave as for Nutrient Agar.

2.5. Pouring of Plates

After autoclaving the media were placed inside a water bath set at 45°C to maintain the media in a molten state.

1 g each of the sample was weighed into a test-tube containing 9 ml of sterile distilled water and serially dilute them until you reach your dilution factor (10-5) and plate out 1 ml of the last dilution factor into a sterile plates (sterilized by placing them in an over set at 160°C for an hour). Pour the media individually *i.e.* Nutrient Agar, Mac Conkey Agar and Potato Dextrose Agar into a separate plate *i.e.* each sample will have 3 plates and they were duplicated.

After solidifying the plates were incubated in an incubator set at 370°C for Nutrient Agar and Mac Conkey Agar while the potato Dextrose Agar was incubated at 280°C - 30°C. All the plates were incubated invertedly.

2.6. Microbial Count

The plate was counted at 48 hours for Nutrient Agar and Mac Conkey Agar while it was read for potato Detrose Agar t 72 hours.

Microbial Load (Bacteria, Coliform and Mould Count/Flora) of Some Common Hot Smoked Freshwater Fish
Species Using Different Packaging Materials

51

2.7. Lactic Acid Bacterial Count

Fifty five (55) grams of Man De Rogsa and shape medium (MRS) was weighed as for the above nutrient agar preparation procedures.

2.8. Statistical Analysis

Analysis of Variance (ANOVA) in completely randomized design was performed on the data obtained using SPSS (2006). Significant means were compared at 5% probability level using Duncan's New Multiple Range Test (DMRT) as provided in the same SPSS (2006).

3. Result

As shown in **Table 3** the microbial load varied significantly (P < 0.05) among the three species. In the fresh fish the highest TVC of 12.4×10^4 was recorded in *C. gariepinus*, this is followed by *O. niloticus* with 10.6×10^4 and lastly *M. rume* 9.8×10^4. However while TFC was zero in the fresh fish for the three fish species, highest TCC of 0.8×10^4, was recorded for *O. niloticus*, followed by *C. gariepinus* (0.4×10^4) and lastly *M. rume* with TCC of 0.3×10^4.

Table 4 and **Figure 1** show that the TVC of *O. niloticus* dropped from 10.6×10^4 (in the fresh state) to 8.4×10^4 in the initial hot smoked and *M. rume* dropped from 9.8×10^4 (fresh state) to 7.0×10^4 after hot smoking, while the TVC of *C. gariepinus* slightly increased from 12.4×10^4 in the fresh state to 12.6×10^4 after hot smoking. While TFC increased from zero to 0.2×10^4 for the three fish species; highest TCC of 9.8×10^4 was recorded in *C. gariepinus*, followed by 4.2×10^4 in *O. niloticus* and lastly 3.0×10^4 in *M. rume*.

Table 5 shows that the least bacteria load (TVC) was recorded in the SPBM and STPBM, both recording TVC 6.4×10^4 in each case. Generally highest TVC of 8.6×10^4 (OMPBC), 8.3×10^4 (SPBC) and 8.2×10^4 (STPBC) were recorded in all *C. gariepinus* among the nine packages at the end of 12 weeks storage/packaging. Next is *O. niloticus* packaging 7.6×10^4 (OMPBO), 7.4×10^4 (STPBO) and 7.2×10^4 (SPBO) and lastly *M. rume* $6.6 \times$

Table 3. Microbial load of fresh fish samples.

Fish species	Total Viable Count (TVC)	Total Coliform Count (TCC)	Total Fungi Count (TFC)
C. gariepinus	12.4×10^4	0.4×10^4	NIL
O. niloticus	10.6×10^4	0.8×10^4	NIL
M. rume	9.8×10^4	0.3×10^4	NIL

Table 4. Microbial load of initial hot smoked fish.

Fish species	Total Viable Count (TVC)	Total Coliform Count (TCC)	Total Fungi Count (TFC)
C. gariepinus	12.6×10^4	9.8×10^4	0.2×10^4
O. niloticus	8.4×10^4	4.2×10^4	0.2×10^4
M. rume	7.0×10^4	3.0×104	0.2×10^4

Table 5. Final microbial load at the end of twelve weeks storage/packaging of the three hot smoked freshwater fish species.

Fish species	Total Viable Count (TVC)	Total Coliform Count (TCC)	Total Fungi Count (TFC)
SPBC	8.3×10^4	4.0×10^4	0.6×10^4
OMPBC	8.6×10^4	4.3×10^4	0.6×10^4
STPBC	8.2×10^4	4.2×10^4	0.5×10^4
SPBO	7.2×10^4	3.8×10^4	0.4×10^4
OMPBO	7.6×10^4	3.4×10^4	0.3×10^4
STPBO	7.4×10^4	3.3×10^4	0.5×10^4
SPBM	6.4×10^4	3.2×10^4	0.6×10^4
OMPBM	6.6×10^4	3.1×10^4	0.7×10^4
STPBM	6.4×10^4	3.2×10^4	0.7×10^4

10^4 (OMPBO), 6.4×10^4 (SPBM) and 6.4×10^4 (STPBM) respectively. However all the OMPB packages (Open Mouth Polythene Bag)—OMPBC (8.6×10^4), OMPBO (7.6×10^4) and OMPBM (6.6×10^4) had the highest bacteria load in each of the 3 fish species. However the highest ranged fungi (TFC) of 0.6×10^4 - 0.7×10^4 was recorded in *M. rume*. This is followed by *C. gariepinus* with 0.5×10^4 - 0.6×10^4 while least TFC range of 0.3×10^4 - 0.5×10^4 was recorded in *O. niloticus*. TCC was highest in *C. gariepinus* ranging from 4.0×10^4 - 4.3×10^4 followed by *O. niloticus* (3.3×10^4 - 3.8×10^4) and lastly *M. rume* (3.1×10^4 - 3.2×10^4) respectively.

Generally *M. rume* was the best packaged in terms of bacteria load (TVC) with the least range of (6.4×10^4 - 6.6×10^4) followed by *O. niloticus* (7.2×10^4 - 7.6×10^4) and lastly *C. gariepinus* (8.2×10^4 - 8.6×10^4) which is the poorest in terms of bacteria loads. There were significant ($P < 0.05$) differences between and within the TVC (*i.e* bacteria load), TCC and TFC for the three species in this study.

TOTAL VIABLE COUNT (TVC)

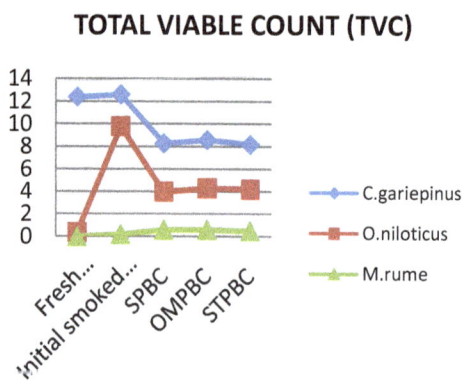

Figure 1. Total Viable Count (TVC) (Bacteria load) for the three fish species for the fresh fish, initial hot smoked and final/smoked packaged at the end of 12 weeks.

Table 6 shows that 6 bacteria species were identified in the fresh *O. niloticus*, while 5 species each were identified for *C. gariepinus* and *M. rume* in their fresh state. Also *Micrococcus acidiophilus* and *Proteus vulgaricus* were identified in the three fresh fish species under study, while *Streptococcus lactis* and *Staphylococcus aureus* were absent in *C. gariepinus*. However *Serraticmacescenes* was only present in the fresh *C. gariepinus*.

Table 7 shows that only *C. gariepinus* had only one fungi species (*Rhizopusnigrica*) represented in the initially hot smoked three (3) fish species. While 6 bacteria species were each represented in *C. gariepinus* and *M. rume*; *O. niloticus* had 5 bacteria species; also only *Staphylococcus aureus* was present throughout in the 3 initially hot smoked fish species.

Table 8 shows that only the OMPB for *M. rume* had 6 bacteria species and 2 fungi species, while the remaining 8 packages had 5 bacteria species and 2 fungi species. The prominent fungi species represented all the 9 packages are *Aspergillu ssp* (*niger*, *tamari*), *Rhizopusnigricans* (in SPBO, OMPBO, OMPBC, SPBM AND STPBM), WHILE *Fusariumoxysporum* is only represented in STPBO.

The prominent bacteria species represented in all the 9 packages are *Micrococcus sp* (*acidiophilus and luteus*), *Bacillus sp* (*aureus*, *cereus* and *luteus*). *Staphylococcus aureus* is present in 8 packages with the exception of OMPBO, *Streptococcus lactis* is also present in 8 packages excepting SPBO. *Proteus sp* (*vulgaricus* and *morgani*) were presented in 7 packages, excepting OMPBC and STPBM. Lastly, *Pseudomonas aureginosa* is present in only 3 packages (SPBO, OMPBO and OMPBC). Since micro-organisms are ubiquitous the type of packaging (as shown in the study) will not limit their existence.

Tables 6. Bacteria species identified from the fresh three fish species.

Fish species	Micro organism
C. gariepinus	*Micrococcus acidiophilus, Bacillus cereus, Serraticmacescenes, Bacillus subtilis, Proteus vulgaricus* (5 bacteria species)
O. niloticus	*Pseudomonas auregionosa, Streptococcus lactis, Micrococcus acidiophilus, Micrococcus luteus, Staphylococcus aureus, Proteusvulgaricus* (6 bacteria species)
M. rume	*Staphylococcus aureus, Bacillus subtilis, Micrococcus acidiophilus, Proteusvulgaricus, Streptococcus lactis* (5 bacteria species)

Table 7. Bacteria and fungi species identified from the initial hotsmoked three fish species.

Fish species	Micro organism
C. gariepinus	*Micrococcus luteus, Bacillus cereus, Staphylococcus aureus, Streptococcus lactis, Pseudomonas aureginosa, Proteus vulgaricus,* 6 bacteria species + 1 fungi (*Rhizopusnigrica*)
O. niloticus	*Streptococcus lactis, Micrococcus acidiophilus, Staphylococcus aureus, Bacillus subtilis, Micrococcus acidiophilus, Bacillus macerans.* 5 bacteria species + Nil (0) fungi species
M. rume	*Staphylococcus aureus, Bacillus subtilis, Micrococcus acidiophilus, Micrococcus luteus, Proteus morganii, Pseudomonas aureginosa* (6 bacteria species + Nil (0) fungi species)

Microbial Load (Bacteria, Coliform and Mould Count/Flora) of Some Common Hot Smoked Freshwater Fish
Species Using Different Packaging Materials

53

Table 8. Bacteria and fungi species identified from the smoked three fish species at 12 weeks of storage/packaging.

Packaging	Micro-organism (bacteria and fungi species)
SPBC	*Bacillus cereus, Streptococcus lactis, Staphylococcus aureus, Proteus vulgaricus, Micrococcus acidiophilus*, 5 bacteria + 2 fungi species *Aspergillustamari, Aspergillusniger*
OMPBC	*Staphylococcus aureus, Bacillus cereus, Micrococcus luteus, Pseudomonas aureginosa, Streptococcus lactis*, 5 bacteria + 2 fungi species *Rhizopusnigricans, Aspergillusniger*
STPBC	*Staphylococcus aureus, Bacillus cereus, Micrococcus acidiophilus, Proteus vulgaricus, Streptococcus lactis*, 5 bacteria + 2 fungi species *Aspergillustamari, Aspergillusniger*
SPBO	*Micrococcus luteus, Bacillus subtilis, Staphylococcus aureus, Proteus vulgaricus, Pseudomonas aureginosa*, 5 bacteria + 2 fungi species *Rhizopusnigricans, Aspergillusniger*
OMPBO	*Streptococcus lactis, Micrococcus acidiophilus, Bacillus cereus, Streptococcus lactis, Proteus vulgaricus, Pseudomonas aureginosa*, 6 bacteria + 2 fungi species *Rhizopusnigricans, Aspergillusniger*
STPBO	*Micrococcus acidiophilus, Streptococcus lactis, Proteus vulgaricus, Bacillus cereus, Staphylococcus aureus*, 5 bacteria + 2 fungi species *Aspergillusniger, Fusariumoxysporum*
SPBM	*Staphylococcus aureus, Micrococcus luteus, Bacillus macerans, Streptococcus lactis, Proteusmorganii*, 5 bacteria + 2 fungi species *Rhizopusnigricans, Aspergillustamari*
OMPBM	*Staphylococcus aureus, Bacillus cereus, Proteus vulgaricus, Bacillus subtilis, Streptococcus lactis, Micrococcus acidiophilus*, 6 bacteria + 2 fungi species *Aspergillustamari, Fusariumoxysporum*
STPBM	*Micrococcus acidiophilus, Micrococcus leteus, Streptococcus lactis, Staphylococcus aureus, Bacillus subtilis*, 5 bacteria + 2 fungi species *Rhizopusnigricans, Aspergillusniger*

4. Discussion

The highest bacteria load (TVC) of 12.4×10^4 was recorded in the fresh *C. gariepinus* followed by 10.6×10^4 in *O. niloticus* and lastly 9.8×10^4 in the fresh *M. rume*. However, initial hot smoked reduced the bacteria load of *O. niloticus* to 8.4×10^4 and *M. rume* to 7.0×10^4 while the initial hot smoked *C. gariepinus* TVC of 12.6×10^4 was not affected by hot-smoking since a slight increase of 0.2×10^4 was recorded after hot smoking.

Packaging had a significant (P < 0.05) effect at 12 weeks storage/packaging of smoked fish for *C. gariepinus* which reduced from 12.6×10^4 TVC to OMPBC (8.6×10^4)—SPBC (8.3×10^4)—STPBC (8.2×10^4) and *M. rume* with TVC reducing from 7.0×10^4 (in the initial hot smoked fish) to 6.6×10^4 (OMPBM)—6.4×10^4 (SPBM)—6.4×10^4 (STPBM). However all the OMPB packages (Open Mouth Polythene Bag)—OMPBC (8.6×10^4), OMPBO (7.6×10^4) and OMPBM (6.6×10^4) had the highest bacteria load in each of the 3 fish species. This is also revealed in **Figure 1**.

Total Coliform Count (TCC) generally increased from the fresh fish sample 0.4×10^4 to 9.8×10^4 (*C. gariepinus*) 0.8×10^4 to 4.2×10^4 (*O. niloticus*) and 0.3×10^4 to 3.0×10^4 in the initial smoked (*M. rume*). Total Coliform Count (TCC) dropped significantly (P < 0.05) from 9.8×10^4 in the initial hot smoked *C. gariepinus* to a range of 4.0×10^4 - 4.3×10^4 in all the 3 *C. gariepinus* packaging, while TCC virtually remained the same for the *M. rume* packaging and dropped from 4.2×10^4 to a range of 3.3×10^4 - 3.8×10^4 for *O. niloticus* at the end of 12 weeks. This is shown in **Figure 2**.

No Fungi count was recorded in the fresh fish sample

TOTAL COLIFORM COUNT (TCC)

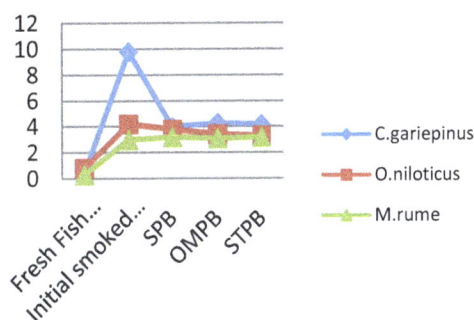

Figure 2. Total Coliform Count (TCC) for the three fish species for the fresh fish, initial hot smoked and final/smoked packaged at the end of 12 weeks.

for the 3 fish species. However a value of 0.2×10^4 fungi count was recorded for the 3 fish species after initial hot smoking. This value increased; highest for *M. rume* (0.6 $\times 10^4$ to 0.7×10^4) next is 0.4×10^4 to 0.6×10^4 in *C. gariepinus* and lastly 0.3×10^4 - 0.5×10^4 in *O. niloticus*. Since micro-organisms are ubiquitous the type of packaging (as shown in this study and **Figure 3**) will not limit their existence.

The bacterial load (TVC) count for all the three species of fish in the nine packages used for this study are below the maximum bacteria count of 5×10^5 cfu for good fish product according to the International Commission on Microbiology Safety for Food [14].

For *C. gariepinus* significant (P < 0.05) decreases were observed in the TCC 9.8×10^4 in the initial smoked fish which reduced to a range of 3.1×10^4 - 3.2×10^4 at the end of 12 weeks. This was in conformity with Wil-

TOTAL FUNGI COUNT (TFC)

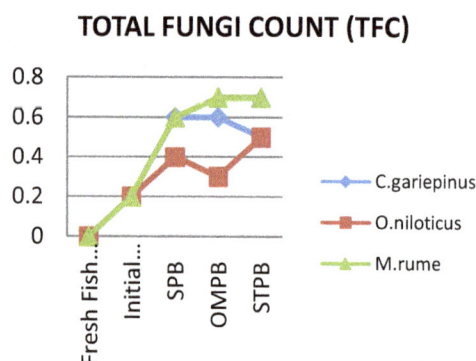

Figure 3. Total Fungi Count (TFC) for the three fish species for the fresh fish, initial hot smoked and final/smoked packaged at the end of 12 weeks.

liam, C.F and Dennis, C.W [15] who reported that the faecal coliforms count of fresh *C. gariepinus* fillets were similarly low after 8 days of cold storage.

Table 3 shows that there was absence of fungi in the fresh sample of the three fish species, while in **Table 4** only one species of fungi (*Rhizopusnigrica*) was present in the initially smoked *C. gariepinus*. At the end of the 12 weeks of storage/packaging three (3) more fungi species (*Aspergillusniger, Aspergillustamari* and *Fusarumoxysporum*) were represented at the rate of 2 fungi species per packaging. That is fungi species were represented in all the 9 packages. The results obtained were similar to those observed by Adebayo-Tayo *et al.* and Fafioye, O.O *et al.* [9,16]. During storage of smoked fish product there was significant (P < 0.05) increase in the fungi count with length of storage as seen in this study. This is in line with Oyebamiji, O. F *et al.* and Wogu, M.D *et al* [11,12] who worked on stored smoked fish products marketed in the open market. The presence of fungi may be due to the difference in the chemical composition of the fish species and to which different moulds react differently [16,17].

Only the OMPBM and OMPBO had 6 bacteria species represented while the remaining 7 packages had 5 bacteria species. The prominent bacteria species represented in all the nine (9) packages are *Micrococcus* sp (*acidiophilus and luteus*), *Bacillus* sp (*aureus, cereus and luteus*), *staphylococcus aureus* (is present in 8 packages) except in OMPBO. *Streptococcus lactis* also in 8 packages excepting SPBO. Others are *Proteus vulgaricus*, *P. morganii* and *Pseudomonas aureginosa*.

5. Conclusion

Highest Bacteria Count (TVC) was recorded in *C. gariepinus* packages among the nine packages at the end of 12 weeks. The 3 packaged fishes for *C. gariepinus* had the highest bacteria load with OMPBC (Open Mouth Polythene Bag Being the Highest). Highest tendency for

heavy bacteria load (TVC) is in the Open Mouth Polythene Bag which has been confirmed in the OMPB for all the 3 fish species. Highest ranged Total Fungi Count (TFC) was recorded in *M. rume* followed by *O. niloticus*. Total Coliform Count (TCC) was highest in *C. gariepinus* followed by *O. niloticus*. Packaging did not limit the existence of micro-organisms. There were 5 bacteria species and 2 fungi species represented in each of the packages (with the exception of OMPBM and OMPBO with 6 bacteria species). The prominent fungi species represented in the 9 packages at the end of 12 weeks are *Aspergillus* species (*niger* and *tamari*), *Rhizopusnigricans* and *Fusariumoxysporum*. Prominent bacteria species represented in all 9 packages are *Micrococcus* species (*acidiophilus* and *luteus*), *Bacillus* species (*aureus, cereus* and *luteus*). *Staphylococcus aureus* is present in 8 packages (excepting OMPBO) and also *Streptococcus lactis* (excepting SPBO). *Proteus* species (*vulgaricus and morganii*) (in 7 packages excepting OMPBC and STPBM) and lastly *Pseudomonas aureginosa* are present in only 3 packages (SPBO, OMPBO and OMPBC).

REFERENCES

[1] A. A. Eyo, "Fish Processing Technology in the Tropic National Institute for Freshwater Fisheries Research (NIFFR)," New Bussa, 2001, pp. 10-170.

[2] H. I. Ibrahim, A. A. Kigbu and R. Muhammed, "Women's Experiences in Small Scale Fish Processing in Lake Fafenwa, Fishing Community, Nasarawa State," *Nigeria livestock Research for Rural Development*, Vol. 23, 2011, p. 3.

[3] G. Hobbs, "Fish Handling and Processing," Ministry Of Agric, Fisheries and Food, Torry Counter, Edinburg, London, 1965, pp. 20-23.

[4] H. H. Huss, "Fresh Fish Quantity and Quality Changes," F.A.O Fisheries Series, Danish Rome, 1988, pp. 15-29.

[5] J. M. Shewan, "Bacteriological Standards for fish and Fishery Products," *Chemical Industrial*, Vol. 2, 1970, pp. 299-302.

[6] J. J. Conell, "Control of Fish Quality," 4th Edition, Churchhill Livingstone, Edinburg, 1995.

[7] B. W. Fleming, S. L. Cochi, K. L. MacDonald, J. Brondum, P. S. Hayes, B. D. Phkaytes, M. B. Holmes, A. Audurier, C. V. Broome and A. L Reingold, "Pasteurized Milk as a Vehicle of Infestation in an Outbreak of Lusteriosis," *New England Journal of Medicine*, Vol. 312, 1985, pp. 404-407.

[8] P. K. Ben Embarek, "Presence, Detection and Growth of Lusteramonocytogenes in Seafoods: A Review," *International Journal of Food Microbiology*, Vol. 23, No. 1, 1994, pp. 17-34.

[9] B. O. Adebayo-Tayo, A. A. Onilude and U. G. Patrick, "Micro Floral of Smoked-Dried Fishes Sold in Uyo, East-

Microbial Load (Bacteria, Coliform and Mould Count/Flora) of Some Common Hot Smoked Freshwater Fish
Species Using Different Packaging Materials

55

ern Nigeria," *World Journal of Agriculture Science*, Vol. 43, 2008, pp. 346-350.

[10] O. J. Abolagba and E. C. Uwagbu, "A Comparative Analysis of the Microbial Load of Smoked-Dried Fishes (*Ethmalosafunbruata* and *Pseudotolithuselongatus*) Sold in Oba and Koko Markets in Edo and Delta States, Nigeria at Different Season," *Australian Journal of Basic and Applied Science*, Vol. 5, 2011, pp. 500-544.

[11] O. F. Oyebamiji, T. R. Fagbohun and O. O. Olubanjo, "Fungal Infestation and Nutrient Quality of Traditionally Smoke Dried Freshwater Fish," *Turkish Journal of Fisheries and Aquatic Science*, Vol. 8, 2008, pp. 7-13.

[12] M. D. Wogu and A. D. Iyayi, "Micro-Flora of Some Smoked Fish Varieties in Benin City, Nigeria," *Ethiopian Journal of Environmental Studies and Management*, Vol. 4, No. 1, 2011, pp. 36-38.

[13] AOAC, "Official Methods of Analysis of AOAC International," 17th Edition, Association of Analytical Communities, Gaithersburg, 2002.

[14] ICMSF, "International Commission on Microbiological Specifications for Foods. Micro-Organisms in Foods 7. Microbiological Testing in Food Safety Management," Kluwer Academic Plenum Publishers, New York, 2002, p. 199.

[15] C. F. William and C. W. Dennis, "Food Microbiology," 4th Edition, Food Science Series, Mac Grow-Hill Book Company, Singapore, 1998, pp. 243-252.

[16] O. O. Fafioye, M. O. Efuntoye and A. Osho, "Study on the Fungal Infestation of Five Traditionally Smoked Fish," 2002.

[17] W. Reed, J. Burchard, J. Hopson, A. J Jeness and I. Yaro, "Fish and Fisheries of Northern Nigeria," Gaskiya Corporation, Zaria, 1967, 226 p.

Abbreviations

TVC: Total Viable Count

TCC: Total Coliform Count

TFC: Total Fungi Count

SPBC: Sealed Paper Bag-*Clariasgariepinus*

OMPBC: Open Mouth Polythene Bag-*Clariasgariepinus*

STPBC: Sealed Transparent Polythene Bag-*Clariasgariepinus*

SPBO: Sealed Paper Bag-Tilapia (*Oreochromisniloticus*)

OMPBO: Open Mouth Polythene Bag-Tilapia (*Oreochromisniloticus*)

STPBO: Sealed Transparent Polythene Bag-Tilapia (*Oreochromisniloticus*)

SPBM: Sealed Paper Bag-*Mormyrusrume*

OMPBM: Open Mouth Polythene Bag-*Mormyrusrume*

STPBM: Sealed Transparent Polythene Bag-*Mormyrusrume*

7

Role of Lactic Acid Bacteria-Myeloperoxidase Synergy in Establishing and Maintaining the Normal Flora in Man

Robert C. Allen[1], Jackson T. Stephens Jr.[2]

[1]Department of Pathology, School of Medicine, Creighton University, Omaha, USA; [2]ExOxEmis, Inc., Little Rock, USA.

ABSTRACT

Lactic acid bacteria (LAB) are incapable of cytochrome synthesis and lack the heme electron transport mechanisms required for efficient oxygen-based metabolism. Consequently, LAB redox activity is flavoenzyme-based and metabolism is fermentative, producing lactic acid, and in many cases, hydrogen peroxide (H_2O_2). Despite this seeming metabolic limitation, LAB dominate in the normal flora of the mouth, vagina and lower gastrointestinal tract in man. Myeloperoxidase (MPO) is produced by the neutrophil leukocytes and monocytes that provide the innate phagocyte defense against infecting pathogens. MPO is unique in its ability to catalyze the H_2O_2-dependent oxidation of chloride (Cl^-) to hypochlorite (OCl^-). In turn, this OCl^- directly reacts with a second H_2O_2 to produce singlet molecular oxygen ($^1O_2^*$), a metastable electronic excitation state of oxygen with a microsecond lifetime that restricts its combustive reactivity within a submicron radius of its point of generation. Each day a healthy human adult produces about a hundred billion neutrophils containing about 4 femtograms MPO per neutrophil. Inflammatory states and G-CSF treatment increase both neutrophil production and the quantity of MPO per neutrophil. After a short circulating lifetime, neutrophils leave the blood and migrate into body spaces including the mouth, vagina, urinary tract, and gastrointestinal tract. Greater than a hundred thousand neutrophils are lavaged from the mouths of healthy humans; the quantity lavaged is proportional to the blood neutrophil count. MPO selectively and avidly binds to most Gram-positive and all Gram-negative bacteria tested, but LAB do not show significant MPO binding. Neutrophils migrating to normal flora sites release MPO into the LAB-conditioned milieu containing adequate acidity and H_2O_2 to support extra-phagocyte MPO microbicidal action. In combination, LAB plus MPO exert a potent synergistic microbicidal action against high MPO-binding microbes. This LAB-MPO synergy provides a mechanism for the establishment and maintenance of LAB in the normal flora of man.

Keywords: Myeloperoxidase; Lactic Acid Bacteria; Hydrogen Peroxide; Hypochlorite; Singlet Oxygen; Selective Binding; Selective Microbicidal Action; Normal Flora

1. Introduction

In 1892 Döderlein suggested that fermentative lactic acid producing vaginal bacteria protected the host from pathogenic microbes [1]. The role of Döderlein's bacillus, *i.e.*, *Lactobacillus acidiphilus*, in healthy vaginal flora, was extended to include the normal intestinal bacterial flora of infants by Moro in 1900 [2]. At about the same time Tissier described the normal flora of breast-fed infants, establishing the importance of *Bifidobacteria* (aka, Tissier's bacillus) [3]. In 1908 the health benefits of the lactic acid bacteria that constitute healthy human flora were popularized by Metchnikoff's book, *Prolongation*

of Life [4].

The lactic acid bacteria (LAB) are Gram-positive, non-spore forming microbes incapable of cytochrome synthesis and lacking the heme electron transport mechanisms required for efficient respiration. As such, LAB metabolism is fermentative. The LAB are acid-tolerant, aerotolerant and lack heme catalase activity [5]. The absence of cytochromes restricts LAB redox metabolism to flavoenzymes. The resultant fermentative metabolism produces acids, and in many cases, hydrogen peroxide (H_2O_2) [6]. LAB fermentative metabolism is essential to food preservation. In their role as the dominant microbes of the normal flora, LAB serve the innate host defense against pathogenic microbes.

Many LAB produce H_2O_2 in the presence of oxygen [7,8]. The viridans streptococci that comprise the normal mouth flora [9] produce sufficient H_2O_2 to cause the characteristic green, *i.e.*, viridans, hemolysis seen on blood agar [10]. *Streptococcus sanguinis* and *Streptococcus mitis* are reported to produce H_2O_2 in the range of about 50 nanomoles (nmol)/min/mg of dry weight [11]. The rates of H_2O_2 production by *Streptococcus oralis* and *S. sanguinis* are described to be several nanograms/min $/10^6$ CFU [12].

2. Phagocyte Leukocyte Microbicidal Action

Neutrophil leukocytes and monocytes provide the innate phagocyte defense against pathogenic infection and are the microbicidal effector cells of the acute inflammatory response. Large quantities of neutrophils and monocytes are produced by the myelopoietic bone marrow and released into the circulating blood daily. Cytokines and chemotactic molecules direct phagocyte migration from the circulating blood to sites of infection where contact with opsonified microbes results in phagocytosis and formation of a phagosome. Fusion of the phagosome with lysosomal azurophilic granules containing cationic enzymes produces the phagolysosome. Both neutrophils and monocytes contain relatively large quantities of a cationic haloperoxidase, *i.e.*, myeloperoxidase (MPO) [13], capable of oxidizing chloride (Cl$^-$) to hypochlorite (OCl$^-$) in an acid milieu containing H_2O_2 [14].

As depicted in the schematic of **Figure 1**, phagocytosis is linked to activation of NADPH oxidase. The activated oxidase catalyzes the NADPH-dependent reduction of molecular oxygen (O_2) liberating the $NADP^+$ required for increased hexose monophosphate shunt (aka, pentose pathway) dehydrogenase activity. This increased glucose metabolism is proportional to increased non-mitochondrial oxygen consumption. These metabolic activities are

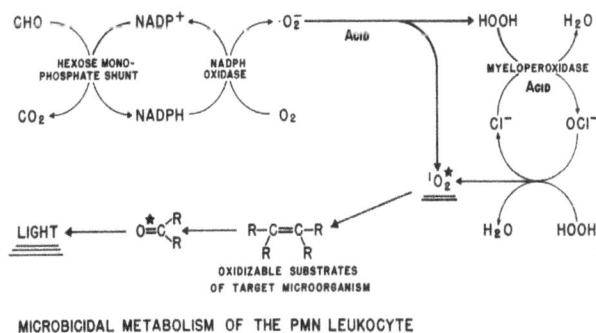

MICROBICIDAL METABOLISM OF THE PMN LEUKOCYTE

Figure 1. Schematic depiction of neutrophil production of the H_2O_2 that drives MPO oxidation of Cl$^-$ to OCl$^-$, followed by reaction of the OCl$^-$ with an additional H_2O_2 to produce $^1O_2^*$. The resulting microbicidal dioxygenation reactions yield excited carbonyl products that relax by photon emission, *i.e.*, chemiluminescence [15].

necessary for phagolysosomal acidification and H_2O_2 production, and provide an optimal milieu for H_2O_2-dependent MPO oxidization of Cl$^-$ to OCl$^-$ [15-18]. The reaction of OCl$^-$ with a second H_2O_2 yields electronically excited singlet molecular oxygen $\left(^1O_2^*\right)$, a highly reactive but metastable electrophilic oxygenating agent with a microsecond lifetime. Both OCl$^-$ and $\left(^1O_2^*\right)^*$ are potent microbicidal agents.

2.1. Phagocyte Production and Myeloperoxidase

Each day the bone marrow of a healthy human adult releases about a hundred billion neutrophils and monocytes into the circulating blood [19]. Approximately 5% of the dry weight of each neutrophil is MPO [20], a 145 kDa dimeric alpha-heme haloperoxidase carried in the neutrophil's azurophilic granules [13]. The average cell volume of a neutrophil is about 450 femtoliter (fL); its specific gravity is 1.1, and its cell water content is about 84% (16% dry weight). As such, each neutrophil contains about 4 femtograms (fg) MPO. Thus, the total MPO released per day is about 0.4 mg. Nanomolar concentrations (0.1 - 1.0 μg/mL) of MPO are highly microbicidal when presented with physiologically available chloride and sub-millimolar concentrations of H_2O_2 [21].

2.2. Tissue Neutrophils and MPO Availability

The circulating lifetime of neutrophils in the blood is less that twelve hours, and is followed by a less well characterized tissue phase that can last for a few days [19]. Although neutrophilia, *i.e.*, influx of neutrophils, is a characteristic feature of acute inflammation, small numbers of neutrophils are known to routinely migrate into the alimentary tract spaces, bladder and body cavities in the absence of inflammation. Continuous migration of neutrophils into the mouth [22] and vagina [23] has been well documented. Relatively large numbers of neutrophils (10^5) can be lavaged from the mouths of healthy humans. The quantity of neutrophils present in fluid lavaged from the mouth is proportional to the subject's circulating blood neutrophil counts [22]. Inflammation and G-CSF treatment increase neutrophil production as well as the quantity of MPO per neutrophil [24,25].

Migrating tissue phase neutrophils transport sufficient MPO to body flora spaces to have an effect on the microbial flora. The presence of MPO might be expected to exert microbicidal action against the resident H_2O_2-producing LAB [26,27], but the opposite is observed.

3. MPO Selectivity Binds and Kills Microbes

Direct MPO binding to bacteria can be visually demonstrated by contacting bacteria in suspension with MPO, and then pelleting the bacteria by centrifugation. The degree of bacterial pellet staining is proportional to the

MPO bound [21]. MPO was found to bind to all of the Gram-negative, and most of the Gram-positive bacteria tested, with the exception of viridans streptococcus, *i.e.*, *S. sanguinis*. The degree of MPO binding was quantified by reduced-minus-oxidized difference spectral analysis and by chemiluminescence-based Scatchard analysis. Both methods confirmed and expanded the visual observation of selective MPO binding. With the exception of the fermentative LAB, all bacteria tested showed high degrees of MPO binding [28-30].

Furthermore, MPO binding correlates with MPO killing. Bacteria showing strong MPO binding were rapidly and effectively killed when MPO was present in nanomolar quantities with about 100 micromolar (μM) H_2O_2 (*i.e.*, a greater than thousandfold dilution of 3% pharmacy H_2O_2). Streptococci and lactobacilli show relatively weak MPO-binding, and as such, these LAB are relatively protected from MPO microbicidal action.

3.1. Hypochlorite and Microbicidal Action

In the acidic milieu of the neutrophil's phagolysosome, or in acidic body spaces populated by LAB flora, MPO catalyzes the oxidation of chloride (Cl^-) to hypochlorous acid (HOCl). Hypochlorite, the conjugate base of hypochlorous acid, is the active ingredient in household bleach, and has been employed as a disinfectant and deodorizing agent since Claude Louis Berthollet introduced it as Eau de Javel in 1789 [31-33]. Hypochlorous acid is a weak acid with a pKa of 7.5, and as such, the acid predominates in phagolysosomal and fermentation milieux.

The chloronium (Cl^+) character of HOCl allows it to participate in a variety of reactions resulting in dehydrogenations and chloramine formation. The bactericidal action of hypochlorite is broad and complete at a concentration of 6 μM. However, a thousandfold higher hypochlorite concentration, *i.e.*, 6 mM, is required for the same level of microbicidal action when human erythrocytes are added at a ratio of about five erythrocytes per bacterium [21]. Although hypochlorite shows potent microbicidal action, it lacks specificity, confirming Alexander Fleming's comment that "leukocytes are more sensitive to the action of chemical antiseptics than are the bacteria, and, in view of this, it is unlikely that any of these antiseptics have the power of penetrating into the tissues and destroying the bacteria without first killing the tissues themselves" [34].

3.2. Reaction Radius of Singlet Oxygen

The critical importance of microbe-specific MPO binding is best understood relative to the reactive lifetime of $^1O_2^*$ and its restricted radius of reactivity [35,36]. Like chlorine gas (Cl_2), $^1O_2^*$ has singlet spin multiplicity, is a potent electrophilic reactant, and participates in highly exergonic reactions [18,37]. Unlike Cl_2, $^1O_2^*$ is a metastable electronically excited state with a relatively short lifetime. Singlet spin multiplicity is critically important for reactivity. Understanding the role of multiplicity in $^1O_2^*$ reactivity is best approached by considering the Wigner spin conservation rules [38,39]. In essence, the neutrophil leukocyte changes the spin quantum number of oxygen, thus removing the barrier to spin-allowed reaction with singlet multiplicity biological molecules. The wet combustive oxygenations that follow produce electronically excited oxygenation products. As shown in **Figure 1**, light emission or chemiluminescence (CL) is an energy product of $\pi^* \rightarrow n$ electron relaxation of the excited carbonyl functions generated.

The short lifetime of $^1O_2^*$ restricts reactivity to within a space of less than a micron from its point of generation [28,40,41]. MPO microbicidal effectiveness is limited by its proximity to the target microbe. For successful microbicidal action, primary production of HOCl, and especially, secondary production of $^1O_2^*$, must occur sufficiently close to the target microbe for adequate oxygenation activity. The concentration of HOCl decreases with the distance from the MPO production site. Once HOCl reacts with a second H_2O_2 molecule to generate $^1O_2^*$, the reactive radius is restricted to within about 0.2 μm. Chloramines produced by direct reaction of HOCl with amine components of the microbe, also react with additional H_2O_2 to produce $^1O_2^*$, and such reactions are facilitated by an acidic milieu.

The results of **Figure 2** illustrate the relationship of selective binding to selective killing. In this experiment, suspensions of *Escherichia coli* (~10^6) and *S. sanguinis* (~10^7) plus human erythrocytes (~10^7) were added together to yield a final ratio of about one *E. coli* to ten *S. sanguinis* and to five erythrocytes. The top plate shows the results in the absence of MPO. Note the presence of large colonies of *E. coli* and the absence of small *S. sanguinis* colonies. The middle plate shows the consequence of including 1.9 nM MPO. This small concentration of MPO significantly decreased the colonies of *E. coli* and also allowed the emergence of small colonies of *S. sanguinis*. The bottom plate shows the effects of 5.6 nM MPO. Only a single colony of *E. coli* developed, however, several hundred small colonies of *S. sanguinis* are now clearly visible.

In this experiment *S. sanguinis* metabolism is the only source of H_2O_2. *E. coli* are catalase positive and relatively well protected from low concentrations of H_2O_2. Erythrocytes show no significant MPO binding and contain abundant catalase. Introducing 10^7 erythrocytes with their several magnitudes greater mass than *E. coli* also provides competitive substrate for reaction with available oxidants. The absence of measurable hemolysis con

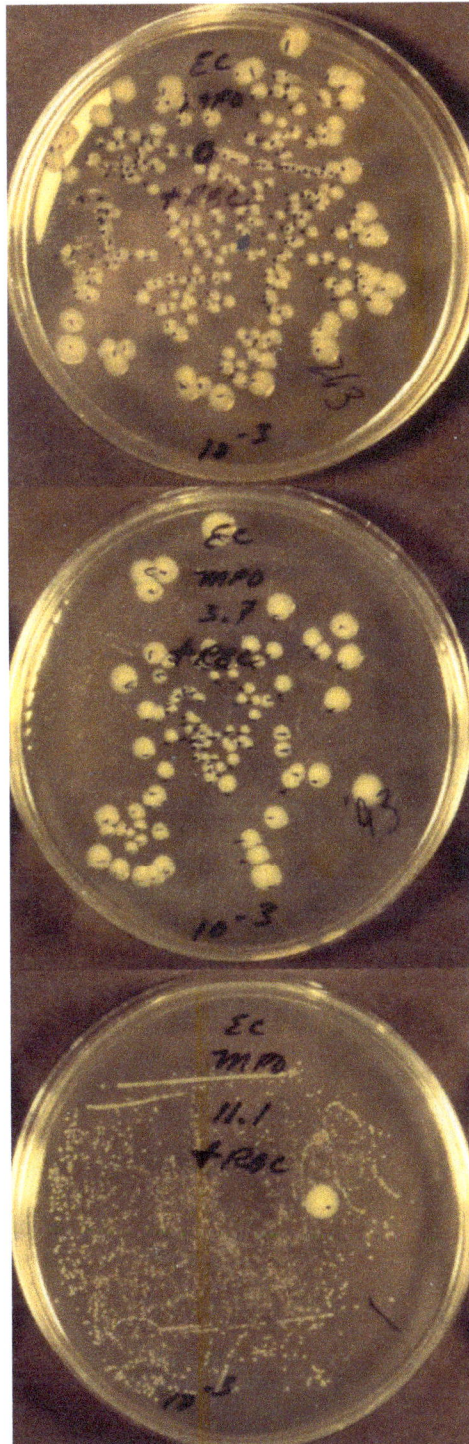

Figure 2. Effects of MPO on a co-suspension of live *E. coli* and *S. sanguinis* (about one *E. coli* per ten *S. sanguinis*) after thirty minutes exposure at 23°C. The 10^{-3} CFU dilution Petri plates are shown. Human erythrocytes were present at a ratio of about five erythrocytes per *E. coli*. No H_2O_2 was added. The suspension of the top plate contained no MPO, the suspension of the middle plate contained 1.9 nM MPO, and the suspension of the bottom plate contained 5.6 nM MPO. The plates were allowed to incubate for more than a day to allow growth of the smaller *S. sanguinis* colonies [21].

firmed the absence of erythrocyte damage during the course of the experiment [21].

LAB-MPO microbicidal action in the presence of erythrocytes demonstrates the highly selective nature of MPO-mediated oxidative attack. Only the MPO-binding *E. coli* were killed. The absence of significant MPO-binding to *S. sanguinis* and erythrocytes, and the proximity requirement imposed by the lifetime of $^1O_2^*$ allow LAB-MPO synergistic microbicidal action against *E. coli*. In the presence of relatively low H_2O_2 concentrations and low MPO concentrations, selectivity of MPO binding results in selectivity of killing.

4. LAB Fermentative Metabolism and MPO

LAB are incapable of cytochrome synthesis, and conesquently, metabolism is fermentative. Nonetheless, these LAB play dominant roles in the flora of the mouth, vagina and lower gastrointestinal tract. In contradistinction to other Gram-positive and to all Gram-negative bacteria, LAB show very low MPO binding. LAB fermentative metabolism provides the acidic milieu and sufficient H_2O_2 for MPO microbicidal action, as demonstrated by the results shown in **Figure 2**, and illustrated in the schematic of **Figure 3**.

LAB show poor MPO binding. Thus, the proximity requirement necessary for effective MPO microbicidal action spares LAB. At low MPO concentrations LAB are protected from their metabolic product H_2O_2. Instead, LAB-MPO synergistic microbicidal action is focused on competing MPO-binding microbes. The presence MPO in milieux containing LAB favors the killing of microbes showing significant MPO binding. As such, LAB-MPO synergy provides a mechanism for establishing and maintaining LAB as the dominant microbes of the normal flora of man.

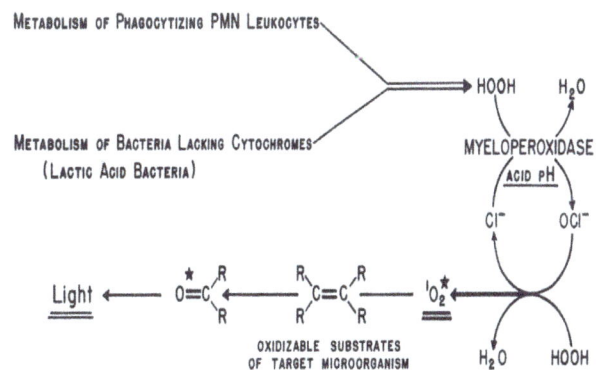

Figure 3. Similar roles of neutrophil metabolism and LAB metabolism in driving MPO combustive microbicidal action with chemiluminescence [17].

5. Acknowledgements

The support of ExOxEmis, Inc. is gratefully acknowledged.

REFERENCES

[1] "Das Scheidensekret und seine Bedeutung für das Puerperalfieber," *Centralblatt für Bacteriologie*, Vol. 11, 1892, pp. 699-700.

[2] E. Moro, "Über den *Bacillus acidophilus* n. spec. Ein Beitrag zur Kenntnis der normalen Darmbacterien des Säuglings," *Jahrbuch für Kinderheilkunde*, Vol. 52, 1900, pp. 38-55.

[3] H. Tissier, "Recherchers sur la Flora Intestinale Normale et Pathologique du Nourisson," Thesis, University of Paris, Paris, 1900.

[4] E. Metchnikoff, "Prolongation of Life," Putnam, New York, 1908.

[5] W. H. Holzapfel, P. Haberer, R. Geisen, J. Björkroth and U. Schillinger, "Taxonomy and Important Features of Probiotic Microorganisms in Food and Nutrition," *The American Journal of Clinical Nutrition*, Vol. 73, 2001, pp. 365S-373S.

[6] J. W. McLeod and J. Gordon, "Production of Hydrogen Peroxide by Bacteria," *Biochemical Journal*, Vol. 16, 1922, pp. 499-504.

[7] R. Whittenbury, "Hydrogen Peroxide Formation and Catalase Activity in Lactic Acid Bacteria," *Journal of General Microbiology*, Vol. 35, No. 1, 1964, pp. 13-26.

[8] C. Marty-Teysset, F. de la Torre and J. Garel, "Increased Production of Hydrogen Peroxide by *Lactobacillus delbrueckii* subsp. *bulgaricus* upon Aeration: Involvement of an NADH Oxidase in Oxidative Stress," *Applied and Environmental Microbiology*, Vol. 66, 2000, pp. 262-267.

[9] D. A. Johnston and G. P. Bodey, "Semiquantitative Oropharyngeal Culture Technique," *Journal of Applied Microbiology*, Vol. 20, 1970, pp. 218-223.

[10] J. P. Barnard and M. W. Stinson, "The Alpha Hemolysin of *Streptococcus gordonii* is Hydrogen Peroxide," *Infection and Immunity*, Vol. 64, 1996, pp. 3853-3857.

[11] J. Carlsson, Y. Iwami and T. Yamada, "Hydrogen Peroxide Excretion by Oral Streptococci and Effect of Lactoperoxidase-Thiocyanate-Hydrogen Peroxide," *Infection and Immunity*, Vol. 40, 1983, pp. 70-80.

[12] A. Garcia-Mendoza, J. Liebana, A. M. Castillo, A. de la Higuera and G. Piedrola, "Evaluation of the Capacity of Oral Streptococci to Produce Hydrogen Peroxide," *Journal of Medical Microbiology*, Vol. 39, No. 6, 1993, pp. 434-439.

[13] K. Agner, "Crystalline Myeloperoxidase," *Acta Chemica Scandinavica*, Vol. 12, 1958, pp. 89-94.

[14] R. C. Allen, "Studies on the Generation of Electronic Excitation States in Human Polymorphonuclear Leukocytes and their Participation in Microbicidal Action," Ph.D. Dissertation, Tulane University, New Orleans, 1973.

[15] R. C. Allen, S. J. Yevich, R. W. Orth and R. H. Steele, "The Superoxide Anion and Singlet Molecular Oxygen: Their Role in the Microbicidal Activity of the Polymorphonuclear Leukocyte," *Biochemical and Biophysical Research Communications*, Vol. 60, No. 3, 1974, pp. 909-917.

[16] R. C. Allen, "Halide Dependence of the Myeloperoxidase-Mediated Antimicrobial System of the Polymorphonuclear Leukocyte in the Phenomenon of Electronic Excitation," *Biochemical and Biophysical Research Communications*, Vol. 63, No. 3, 1975, pp. 675-683.

[17] R. C. Allen, "The Role of pH in the Chemiluminescent Response of the Myeloperoxidase-Halide-HOOH Antimicrobial System," *Biochemical and Biophysical Research Communications*, Vol. 63, No. 3, 1975, pp. 684-691.

[18] R. C. Allen, "Reduced, Radical, and Electronically Excited Oxygen in Leukocyte Microbicidal Metabolism," In: J. T. Dingle and P. J. Jacques, Eds., *Lysosomes in Biology and Pathology*, Vol. 6; In: A. Neuberger and E. L. Tatum, Eds., *Frontiers in Biology*, Vol. 48, North Holland, Amsterdam, 1979, pp. 197-233.

[19] D. F. Bainton, "Developmental Biology of Neutrophils and Eosinophils," In: J. I. Gallin and R. S. Snyderman, Eds., *Inflammation, Basic Principles and Clinical Correlates*, 3rd Edition, Lippincott Williams & Wilkins, Philadelphia, 1999, pp. 13-34.

[20] J. Schultz and K. Kaminker, "Myeloperoxidase of the Leukocyte of Normal Human Blood. I. Content and Localization," *Archives of Biochemistry and Biophysics*, Vol. 96, No. 3, 1962, pp. 465-467.

[21] R. C. Allen and J. T. Stephens Jr., "Myeloperoxidase Selectively Binds and Selectively Kills Microbes," *Infection and Immunity*, Vol. 79, No. 1, 2011, pp. 474-485.

[22] D. G. Wright, A. I. Meierovics and J. M. Foxley, "Assessing the Delivery of Neutrophils to Tissues in Neutropenia," *Blood*, Vol. 67, 1986, pp. 1023-1030.

[23] S. Cauci, S. Guaschino, D. de Aloysio, S. Driussi, D. De Santo, P. Penacchioni and F. Quadrifoglio. "Interrelationships of Interleukin-8 with Interleukin-1β and Neutrophils in Vaginal Fluid of Healthy and Bacterial Vaginosis Positive Women," *Molecular Human Reproduction*, Vol. 9, No. 1, 2003, pp. 53-58.

[24] R. C. Allen, P. R. Stevens, T. H. Price, G. S. Chatta and D. C. Dale. "*In Vivo* Effects of Recombinant Human Granulocyte Colony-Stimulating Factor on Neutrophil Oxidative Functions in Normal Human Volunteers," *The Journal of Infectious Diseases*, Vol. 175, No. 5, 1997, pp. 1184-1192.

[25] R. C. Allen, D. C. Dale and F. B. Taylor Jr., "Blood Phagocyte Luminescence: Gauging Systemic Immune Activation," *Methods in Enzymology*, Vol. 305, 2000, pp. 591-629.

[26] S. J. Klebanoff, "Myeloperoxidase-Halide-Hydrogen Peroxide Antibacterial System," *Journal of Bacteriology*, Vol. 95, 1968, pp. 2131-2138.

[27] R. C. Allen, E. L. Mills, T. R. McNitt and P. G. Quie, "Role of Myeloperoxidase and Bacterial Metabolism in Chemiluminescence of Granulocytes from Patients with Chronic Granulomatous Disease," *The Journal of Infectious Diseases*, Vol. 144, No. 4, 1981, pp. 344-348.

[28] R. C. Allen, "Method for Selectively Inhibiting the Growth of Microbes Using a Haloperoxidase-Halide-Peroxide System," US Patent No. 5,888,505, 1999.

[29] R. C. Allen and J. T. Stephens Jr., "Reduced-Oxidized Difference Spectral Analysis and Chemiluminescence-Based Scatchard Analysis Demonstrate Selective Binding of Myeloperoxidase to Microbes," *Luminescence*, Vol. 26, No. 3, 2010, pp. 208-213.

[30] R. C. Allen and J. T. Stephens Jr., "Phagocyte and Extra-Phagocyte Myeloperoxidase-Mediated Microbicidal Action," In: A. Mendez-Vilas, Ed., *Science Against Microbial Pathogens: Communicating Current Research and Technical Advances*, Formatex Res. Ctr., Badajoz, 2011, pp. 613-621.

[31] H. D. Dakin, "The Antiseptic action of Hypochlorites," *British Medical Journal*, Vol. 2, 1915, pp. 809-810.

[32] C. T. Butterfield, E. Wattie, S. Megregian and C. W. Chambers, "Influence of pH and Temperature on the Survival of Coliforms and Enteric Pathogens When Exposed to Chlorine," *Public Health Reports*, Vol. 58, No. 51, 1943, pp. 1837-1866.

[33] L. Friberg and E. Hammarström, "The Action of Free Available Chlorine on Bacteria and Bacterial Viruses," *Acta Pathologica Microbiologica Scandinavica*, Vol. 38, No. 2, 1956, pp. 127-134.

[34] A. Fleming, "The Action of Chemical and Physiological Antiseptics in a Septic Wound," *British Journal of Surgery*, Vol. 7, No. 25, 1919, pp. 99-129.

[35] M. Kasha and A. U. Khan, "The Physics, Chemistry, and Biology of Singlet Molecular Oxygen," *Annals of the New York Academy of Sciences*, Vol. 171, 1970, pp. 5-23.

[36] A. M. Held, D. J. Halko and J. K. Hurst, "Mechanism of Chlorine Oxidation by Hydrogen Peroxide," *Journal of the American Chemical Society*, Vol. 100, No. 18, 1978, pp. 5732-5740.

[37] R. C. Allen, "Role of Oxygen in Phagocyte Microbicidal Action," *Environmental Health Perspectives*, Vol. 102, No. S10, 1994, pp. 201-208.

[38] R. C. Allen, "Oxygen-Dependent Microbe Killing by Phagocytic Leukocytes: Spin Conservation and Reaction Rate," In: W. Ando and Y. Moro-oka, Eds., *Studies in Organic Chemistry*, Vol. 33, *The Role of Oxygen in Chemistry and Biochemistry*, Elsevier Press, Amsterdam 1988, pp. 425-434.

[39] R. C. Allen, "Molecular Oxygen (O_2): Reactivity and Luminescence," In: L. J. Kricka and P. E. Stanley, Eds., *Bioluminescence and Chemiluminescence* 2002, World Scientific, Singapore, 2002, pp. 223-232.

[40] R. W. Redmond and I. E. Kochevar, "Spatially Resolved Cellular Responses to Singlet Oxygen," *Photochemistry and Photobiology*, Vol. 82, No. 5, 2006, pp. 1172-1186.

[41] E. Skovsen, J. W. Snyder, J. D. C. Lambert and P. R. Ogilby, "Lifetime and Diffusion of Singlet Oxygen in a Cell," *The Journal of Physical Chemistry B*, Vol. 109, No. 18, 2005, 8570-8573.

Element Concentration in the Prepared and Commercial Feed as Well Their Status in the Breast Muscle of Chicken after Prolonged Feeding

Jayanta Kumar Goswami[1], Satya Sarmah[2], Dhirendra K. Sharma[1*]

[1]Bioinformatics Centre, Department of Zoology, Gauhati University, Guwahati, India; [2]Department of Biochemistry, College of Veterinary Sciences, Guwahati, India.

ABSTRACT

Quality poultry meat depends upon the feed and as such there are many commercially available feeds. However, their composition and standard by and large throughout the year may not remain same due to obvious reasons. Moreover, there is no mention of locally produced feed particularly in the north eastern part of India. The major objective of this study was to prepare mesh feed E1 with the available ingredients as well as their effect were compared with that of the two commercially available feed Amrit and Godrej (E2 and E3) in terms of Crude protein, fats and element composition. The findings showed that the protein content (240 g/kg) and fats (105 g/kg) in the breast muscle of female was higher in the E3 received against the broiler chicken received local feed. Element analysis of the E1, E2 and E3 depicted significantly higher value of Ca, K, Cu Zn and Se against the commercial feed. Other elements like Mg, Na, Fe, P, and Mn showed no variation while compared E1, E2 and E3 together. Thus the present findings suggest that the local feed E1 could be accepted at per with that of the commercial feed for poultry.

Keywords: Mesh Food; Elements; Comparison; Chicken Breast Muscle

1. Introduction

Poultry is a rapidly growing industry, and being a rich source of protein and micro nutrient, contributes a useful component to human diet. The modern broiler has been generally selected for rapid gains and efficient utilization of nutrients. Broilers are capable of thriving on widely varied types of diets, but do best on diets composed of low-fiber grains and highly digestible protein sources. Nutritional quality of broiler's meet always depends on proteins, free amino acids, minerals and vitamins. Lazar [1] was of the view that the ideal quantum of certain minerals in poultry meet should be for K (0.4%), P (0.2%), Na (0.09%) etc.

Significantly, the quality of the meat mostly depends upon the types of feed. Generally proteins are the major constituents of the poultry meat [2] along with the mineral constituents. In broiler nutrition, the concept of ideal protein has been widely accepted with growth rate and breast meat yield increases with balanced protein intake. It has also consistently been shown that if an adequate quantity of essential nutrients is maintained in relation to metabolic energy (ME), increasing concentrations of energy for broilers result in a more rapid weight gain and an improvement in feed conversion [3].

On the other hand differences in the quantum of the nutrients in terms of male and female, particularly in the breast muscle of female, which is finer and more tender [4] attracts significance. In addition, many of the stress, including the dietary stress stimuli could act in the homeostasis which in turn may cause loss of macro and microelements, deficiency of which in meat may affect the quality. Good broiler growth rates will be achieved if the daily nutritional requirement of the bird is met. The ability of the bird to achieve its daily nutritional requirement will, in part, depend upon the nutrient composition of the diet and what the bird actually responds to feed is as nutrient intake. For a good broiler growth and efficient nutrient utilization it is, therefore, vital to ensure that a

Element Concentration in the Prepared and Commercial Feed as Well Their Status
in the Breast Muscle of Chicken after Prolonged Feeding

63

good feed intake is achieved. Feed intake can be significantly affected by feed form. A poor feed form will inhibit feed intake and have a negative impact on growth rate [5].

The cost, easy availability and nutritional quality of feed coupled with genetically potent feed efficient stocks to produce safe food from poultry are the need of time [6]. The relationship between feed ingredient and animal product output is both direct and obvious, and the conventional feed stuffs are very expensive and scarce [7]. The commonly used feedstuffs in poultry diet *i.e.* maize, soybean meal, groundnut meal, fish meal etc. which mainly depends on their nutritive value as well as absence of any incriminating factor.

Moreover, the available quantum of macro and microelements either in commercial feed or in indigenously prepared feed is data deficient. Hence, an attempt has been made to determine the contents of Ca, Mg, P, Na, K, Fe, Cu, Zn, Mn, in the commercially available feed, indigenously prepared feed as well as their accumulation in the breast muscle of broiler chicken.

2. Materials and Method

2.1. Experimental Birds

300 VENCOBB commercial broiler chicks of uniform weight (day-old) were procured from a commercial hatchery. Chicks were weighed individually and wing bended. Chicks were randomly allotted according to nearest body weight to three experimental treatments groups. Each group was subdivided into two replicates of 50 chicks.

2.2. Experimental Ration

Three experimental Rations were considered to evaluate three types of rations, viz. prepared ration with conventional ingredients (E1) and two rations procured from the market Amrit (E2) an Godrej (E3) respectively.

Eighty numbers of samples were collected from sales and display centre of feed mills mainly from Jorhat, Golaghat, Sonitpur and Kamrup (urban) districts of Assam, India. These compounded feeds available in Assam were collected randomly for chemical analysis. While collecting the samples, at least three samples from each company were collected and pooled sample was analyzed. These two compounded feeds were in the form of crumbles and pellets.

The control diet (Ration-1) was formulated for pre-starter (0 - 1 week), starter (1 - 4 weeks) and finisher (4 - 6 weeks) separately with conventional ingredients. Pre-starter, starter and finisher rations were calculated for crude proteins (CP) and metabolic energy (ME) values. All the ingredients were obtained from local market.

2.3. Physical and Chemical Composition of the Prepared Rations of Broiler

As per standard protocol the control (Ration 1) was prepared as Pre-starter, Starter and broiler Finisher for the experimental birds [8]. The prepared and their analysis are presented in the **Tables 1, 2** and **3**.

Table 1. Physical and calculated chemical composition of pre-starter feed. ME: Metabolic energy, CP: Crude proteins.

Ingredients	Parts /100kg	ME/kg	MEKcal/kg	CP%	CP
Maize (energy source)	50	3430	1715	8.7	4.4
Groundnut cake (protein source)	15	2790	419	39.5	5.9
Soya bean (protein source)	26	3510	913	41.7	10.8
Fish Meal (protein source)	5	2100	105	40	2.0
Oyster shell grit	1	0	0	0	0
Dicalcium Phosphate	1.1	0	0	0	0
Methionine	0.25	0	0	0	0
Lysine	0.2	0	0	0	0
Trace Mineral	0.5	0	0	0	0
Salt	0.25	0	0	0	0
Probiotic	0.01	0	0	0	0
Choline chloride	0.1	0	0	0	0
Livol (liver tonic)	0.5	0	0	0	0
Toxin binder (antifungal)	0.1	0	0	0	0
Juricox (coccidiostate)	0.05	0	0	0	0
			3151.1		23.1

Table 2. Physical and calculated chemical composition of starter feed. ME: Metabolic energy, CP: Crude proteins.

Ingredients	Parts/100kg	ME/kg	ME (Kcal/kg)	CP%	CP
Maize	51	3430	1749	8.7	4.4
Groundnut cake (protein source)	16	2790	446	39.5	6.32
Soya bean (protein source)	24	3510	842	41.7	10
Fish Meal (protein source)	5.0	2100	105	40	2.0
Oyster shell grit	1.0	0	0	0	0
Dicalcium Phosphate	1.1	0	0	0	0
Methionine	0.25	0	0	0	0
Lysine	0.2	0	0	0	0
Trace Mineral	0.5	0	0	0	0
Salt	0.25	0	0	0	0
Probiotic	0.01	0	0	0	0
Choline chloride	0.1	0	0	0	0
Livol (liver tonic)	0.5	0	0	0	0
Toxin binder (antifungal)	0.1	0	0	0	0
Juricox (coccidiostate)	0.05	0	0	0	0
			3142		22.72

Table 3. Physical and calculated chemical composition of finisher feed. ME: Metabolic energy, CP: Crude proteins.

Ingredients	Parts/100kg	ME/kg	ME Kcal/kg	CP%	CP (total)
Maize	55	3430	1887	8.7	4.8
Groundnut cake	15	2790	419	39.5	5.9
Soya bean	21	3510	737	41.7	8.8
DORB	5	2100	105	12.9	0.6
OSG	1	0	0	0	0
DCP	1.1	0	0	0	0
Salt	0.5	0	0	0	0
Trace mineral	0.25	0	0	0	0
Methionine	0.25	0	0	0	0
Lysine	0.2	0	0	0	0
Choline chloride	0.1	0	0	0	0
Livol	0.5	0	0	0	0
Toxin binder	0.05	0	0	0	0
Juricox	0.05	0	0	0	0
			3147		20

Feed toxicity mainly for aflatoxin for three rations of different types tested by ELISA method. In the present trial three experimental rations and feed ingredients taken for preparation of experimental ration-1 were found aflatoxin negative and below the toxic level.

2.4. Housing and Management

The chicks were housed in a clean well ventilated room, previously disinfected with formalin. Each group was presented by two replicate of 50 birds. The experimental house was divided into six pens of equal size by using bamboo materials and wire netting. The doors, windows, wire netting etc. of the house was painted before starting the experiment. Fresh dried rice husk litter was spread on the floor of the pens at a depth of about 0.04 meter before placing the chicks in the pens to maintain brooding temperature. The room was provided with electric heaters to adjust the environmental temperature and provide the light as per requirement. Each pen was equipped with two 100 W electric bulb suspended 0.4 meter above the litter. The feeder and drinker were fixed such a way that the birds could eat and drink conveniently The chicks were vaccinated against new castle disease vaccine and infectious bursal disease and provided with required veterinary care. Each groups of chicks received F-strain RD virus vaccine intra-ocularly at the dose rate of 0.05 ml/bird on 5th day and IBD vaccine intra-ocularly at the dose rate of 0.05 ml/bird at 14 days of age. Booster dose of RD vaccine F-strain was given on 21 days of age. The room was provided with electric heaters to adjust the environmental temperature and provide the light as per requirement.

Feeding of Birds: Chicks were reared from 1 day old to 42 days on the experimental diets and were allowed *ad libitum* access to feed and water throughout the study. On the first day the chicks were provided with only crushed maize and then given pre-starter diet along with drinking water. They were fed with pre-starter, starter, finisher ration on the day of assigning the treatment and the beginning of the starter period the required amount of broiler starter feed were procured. The same procedure was followed in the procurement of broiler grower and broiler finisher diets.

Meat samples of breast muscle were taken from carcass of each group with scissor and sharp knife and the samples were wrapped in a polythene bag and kept in deep fridge for proximate analysis of meat as outlined by AOAC [9].

3. Methods

3.1. Crude Protein

Nitrogen in the sample was determined by the Kjeldahl method and was multiplied by factors 6.25 to determine the crude protein content of the feed.

The representative sample of feed was first weighed and digested with concentrated H_2SO_4 in presence of 10 g anhydrous sodium sulphate and 0.5 g copper sulphate. Digested materials were dissolved in distilled water and collected in a 250 ml volumetric flask. Then, known volume of aliquot was distilled after being made alkaline with 45% sodium hydroxide solution and the liberated ammonia was trapped in 2% boric acid solution. The same was titrated against N/10 H_2SO_4 (standard). Percentage of nitrogen was calculated by the following formula.

% of Nitrogen

$$= \frac{\text{Vol. of N}/10\ H_2SO_4\text{used of} \times 0.0014 \times 250 \times 100}{\text{Vol. of aliquot taken (ml)} \times \text{gm. of substance taken}}$$

% of CP = % of $N_2 \times 6.25$

3.2. Experimental Design

Groups	Provided with	Divided into	Provided ad libitum Feed
Group E1	Ration 1 prepared with conventional ingredients	A (50 broiler chick) A2 (50 broiler chick)	a-pre starter (0 - 7 d) b-starter (8 - 28 d) c-finisher (29 - 42 d)
Group E2	Ration 2 procured from market	A1 (50 broiler chick) A2 (50 broiler chick)	a-pre starter (0 - 7 d) b-starter (8 - 28 d) c-finisher (29 - 42 d)
Group E3	Ration 3 procured from market	A1 (50 broiler chick) A2 (50 broiler chick)	a-pre-starter (0 - 7 d) b-starter (8 - 28 d) c-finisher (29 - 42 d)

Element Concentration in the Prepared and Commercial Feed as Well Their Status
in the Breast Muscle of Chicken after Prolonged Feeding

65

Feeding trial was conducted for a period of 42 days, A1 and A2 were simply the replicas of E1, E2 and E3 respectively.

Crude Fats: The wet tissue was homogenized with a mixture of chloroform and methanol (3:1 v/v) in such proportions that a miscible system was formed with the water in the tissue. Dilution with chloroform and water separates the homogenate into two layers, the chloroform layer containing all the lipids and the methanolic layer containing all the non-lipids. A purified lipid extract was obtained merely by isolating the chloroform layer [10] and weighed.

Element Analysis: Samples from the finisher feed E1 (Indigenous feed), E2 (Amrit) and E3 (Godrej) as well as from the breast muscle of the 42 days old chickens were considered for crude protein (CP) and elemental assessment. Breast muscles obtained from the chickens were kept in 10% formalin for 24 h and thereafter 4 - 5 g of them were subjected to acid digestion [11]. In brief, the weighed samples were added with HNO_3-$HCLO_4$ (3:1) acid digestion until a clear precipitation was obtained. The sample was diluted to 50 ml with deionized water and subjected for element analysis in Atomic absorption spectrophotometer (Perkin-Elmer 3110) at SAIF, NEHU, Shillong, India. The content of Na and K was determined by the emission technique (acetylene air flame). The P was determined by colorimetric method [12]. The data collected were subjected to ANOVA followed by Fischer's test of significance.

4. Results and Discussion

The results of the present investigation have been depicted in the **Table 4** and **5**. Administration of feed as shown exerts clear impact on the breast muscle of 42 days broiler presenting distinct variation among the E1, E2 and E3. The source of protein for the indigenously prepared feed was the maize, groundnut and soyaben against the unknown composition of the two commercially available feed.

Table 4. Element (mg/kg) present in the indigenous feed (E1) and two commercially available feeds E2 (Amrit) and E3 (Godrej).

	E1	E2	E3
Ca	5.64 ± 0.92^a	7.38 ± 0.83	8.42 ± 0.72
Mg	1.75 ± 0.06	1.22 ± 0.42	1.39 ± 0.54
P	5.24 ± 0.39	5.80 ± 0.45	6.34 ± 0.24
Na	1.05 ± 0.03	1.24 ± 0.09	0.94 ± 0.33
K	8.44 ± 0.42^a	9.12 ± 0.36^a	5.42 ± 0.11
Fe	13.54 ± 1.45	9.39 ± 2.45^a	1.9 ± 2.34
Cu	0.89 ± 0.01^a	0.70 ± 0.01	0.92 ± 0.01
Zn	1.54 ± 0.03^a	1.34 ± 0.02	0.82 ± 0.01
Mn	1.51 ± 0.02^a	1.49 ± 0.02^a	0.54 ± 0.01
Se	0.11 ± 0.01^a	0.42 ± 0.01	0.32 ± 0.02

Mean ± SD, Superscript [a] denotes significant difference.

Table 5. Evaluation of protein (g/kg), fats (g/kg) and elements in the breast muscle of male (M) and female (F) Broiler chicken fed with indigenous feed (E1), and two commercially available feed E2 (Amrit) and E3 (Godrej) for a period of 42 days.

Para-meters	E1		E2		E3	
	M	F	M	F	M	F
Protein	221.64 ± 5.50	210.0 ± 3.9	225 ± 4.5	231.2 ± 3.98	215.64 ± 5.2	240.0 ± 4.2^a
Fats	58.6 ± 8.34	80.0 ± 9.3	73.4 ± 8.2	70.1 ± 6.42	78.39 ± 6.34	105.5 ± 6.9^a
Ca	33.24 ± 7.2	43.64 ± 8.2	42.30 ± 5.9	38.84 ± 4.92	32.64 ± 6.23	33.89 ± 4.21
Mg	310.6 ± 6.5	354.0 ± 7.2	331.0 ± 5.0	350.3 ± 3.9	400.1 ± 8.4	450.4 ± 5.4^a
P	266.22 ± 6.8	411.4 ± 6.2^a	299.2 ± 4.5	302.6 ± 5.4	322.2 ± 5.6	311.7 ± 7.3
Mn	11.2 ± 4.9	11.3 ± 2.4	11.0 ± 1.5	11.6 ± 3.4	14.2 ± 4.11	11.82 ± 3.42
Cu	10.90 ± 3.3	13.2 ± 3.4	11.5 ± 1.61	8.2 ± 2.11	16.5 ± 2.21	7.2 ± 3.62
Fe	2.15 ± 0.12	4.7 ± 0.21^a	1.8 ± 0.3	3.9 ± 0.11	3.9 ± 0.42	3.2 ± 0.21
Zn	1.4 ± 0.11	1.5 ± 0.4	1.3 ± 0.31	1.5 ± 0.31	1.4 ± 0.3	2.9 ± 0.41^a
Se	1.1 ± 0.07	1.1 ± 0.12	1.2 ± 0.20	1.1 ± 0.34	1.5 ± 0.31	1.2 ± 0.21
Na	115.4 ± 5.92	116.4 ± 4.2	118.4 ± 3.9	111.2 ± 3.52	115.7 ± 2.7	116.2 ± 3.98
K	450.2 ± 8.2^a	438.1 ± 5.7^a	388.2 ± 3.9	411.11 ± 4.2	380.4 ± 4.3	375.2 ± 3.84

Mean ± SD, superscript (a) denotes significant difference.

Elements contents present in the feed (E1) and commercially available feed E2 and E3 have been found to be consistent with the chicken growths. Indigenously prepared feed E1 was found to contain significantly lower Ca (5.64 mg/g) and Se (0.11 mg/g) as well as higher Cu (0.89 mg/g), Zn (1.54 mg/g) compared to the commercial feeds E2 and E3. As well, higher quantum of K, Fe and Mn and relatively higher Se in E2 and E3 were assayed. Analysis of two other categories of feed viz pre-starter and starter feed either showed identical or no significant variation, hence not shown in the text. The higher quantity of Ca in the feed lowers the palatability and interfere with the utilization of P, Zn, Mg and Mn, though its precise requirement is not known [13] The maximum requirement up to the level of finished meat has been stated as Mn, Zn, and Fe at 50 mg [13] against the availability of these minerals at 1.51, 1.54 and 13.54 mg/kg of finisher food of this investigation (**Table 4**).

Analysis of breast muscle after 42 day of feeding and that too with finisher demonstrated significantly higher value of crude protein (240 g/kg) and fats(105.5 mg/kg) in the E3 feed female breast muscle against E2 (231.2 and 70.1 g/kg proteins and fats respectively) and E1 (210.0 and 80.0 g/kg proteins and fats respectively) feed. The quality of feature of the meat depends upon the protein and fats. The variation recorded in this investigation might be an attribution of the feed content and there has been insignificant variation of protein between E1 (Indigenous) and E2 (*Amrit*). However, less amount of fat accumulation could be noticed in male under the feeding of E1 while it is higher in E3 female, which might be related to the egg production. The higher quantum of protein (22.26%) in the breast muscle of male and female and higher value of fats in female (2.78%) in broiler chicken have already been highlighted [14]. Further a higher fat content in female broiler [15] and its difference is associated with the metabolic differences, higher competitiveness, and variation in fat deposition capabilities, nutritional requirement and higher hormonal influences [16]. The muscles are finer and tender in female and the female accumulate more fats compared to male [4]. Earlier, it was suggested low fat quantity in male abdomen [17]. In fact, breast muscle contain approximately 24% crude protein against 22% in broiler chicken [18], while it was recorded as 219 g/kg of protein and 16.7 g/kg of fats respectively in male and female broiler chicken [19].

The present finding is very much suggestive of the negative correlation between the fats and protein content and in conformity with the work of various workers [19,20]. However, it is true that the fat content in meat largely depends upon the factors like animal species, breed, gender and anatomic origin of the muscles. The important feature of broiler chicken meat from dietary aspects might be from dietetic aspects and excessive accumulation of fat lies in an imbalance between feed intake and consumption of energy [21]. The present feeding suggested higher quantum of protein in breast muscle in the finished product and it could reasonably be argued that the indigenously prepared feed might be considered for quality meat production.

Evaluation of elements analysis showed presence of higher Ca (43.64 mg/g) in the E1 fed female broiler, significantly different from that of E2 and E3 fed chicken breast muscles. However, the present findings could not support the data of some workers for Calcium [19]. This might be due to the difference in breed or strain, however, demands further characterization.

The Mg quantum in the breast muscle was evaluated and ranged between 310 to 450 mg/g of breast tissue and significantly higher Mg was noted in male and female E3 fed feed. The variation might, probably be due to the feed composition and notably the higher Mg is related to the cardiovascular disease [22]. Moreover, higher Mg in the diet has been linked to a 22% lower risk of Ischemic heart disease [23]. Yet the breast muscle of the E1 and E2 group has had higher range of Mg against 130 mg/g in Ross 308 broiler chicken [19]

The macro element P was in higher direction in the range of 266 to 411 mg/g, while the P in E1 fed female breast muscle showed significantly higher quantity (P < 0.01; 411 mg/g). The possible reason might be due to the variation of feed intake and therefore, it is suggested that the male and female should distinctly be segregated. On the other hand, the importance of P in various activities have already been suggested and it was stated that P influences on the release of energy from protein, carbohydrate and fats while the body was exposed to stress bearing factors [24].

Trance quantities of minerals present in the tissue serve a variety of function in their bodies. The present findings for Mn and Cu showed their identical presence amongst the E1, E2 and E3 feed categories, while the Fe quantum of female fed with E1 feed depicted significantly higher values (4.7 µg/g) and the Se and Zn in E3 feed breast muscle exhibited higher values (2.9 µg/g). Significantly higher quantity of K was noted in male, fed with E1 feed. Presence of identical quantity of Na and Se was recorded amongst the three groups (**Table 2**). As mentioned elsewhere, Na, K, Zn, Cu, Mn, Se, Fe occurring tissue affects the osmotic balance in the body [25]. The median value of potassium in BBQ breast meat has been estimated at 284 mg/100g. The average amount of potassium in 100 g of chicken breast is 273.77 mg against the 450.0 µg/g in the breast tissue fed with local

Element Concentration in the Prepared and Commercial Feed as Well Their Status
in the Breast Muscle of Chicken after Prolonged Feeding

67

feed E1 in this investigation. It is evident that Cu regulates cholesterol biosynthesis and the distribution of fatty acids in poultry [25]. The concentration of selenium (Se) in chicken breast meat in Scandinavia is, however, only about 0.01 mg/100g. Thigh meat from broilers raised on a feed supplemented with 40 g rapeseed oil, 10 g linseed oil per kg diet and 0.27, 0.44, 0.78 or 1.16 mg Se per kg diet could be described as a functional food. This broiler meat is a good contribution to a better strategy for increasing the food content of Se and very long chain *omega*-3 fatty acids [26]. Presence of optional quantum of elements in feed mixtures enable the proper functioning of an organism and good production performance [27] and thus the prepared feed E1 could be of ideal with that of the commercially available feeds.

REFERENCES

[1] V. Lazar, "Poultry-Raising," University of Agriculture in Brono, 1990, p. 210.

[2] Ingr, "Meat Technology," Brono, 1996, p. 290.

[3] J. Michard, "Latest Trends in Feeding Modern Broilers," *International Poultry Production*, Vol. 18, No. 3, 2010, pp. 7-9.

[4] I. Ingr, "Evaluation of Animal Products," Brono, 1993, p. 180.

[5] Arbor Acres, "Broiler Management Guide," Aviagen, 2009, p. 18.

[6] T. Jayalakshmi, R. Kumararaj, T. Sivakumar and T. T. Vanan, "Carcass Characteristics of Commercial Broilers Reared under Varying Stocking," *Tamilnadu Journal of Veterinary and Animal Sciences*, Vol. 5, No. 4, 2009, pp. 132-135.

[7] B. O. Esonu, U. D. Ogbonna, G. A. Anyanwu, O. O. Emelanom, M. C. Uchegbu, E. B. Etuk and A. B. I. Udedibe, "Evaluation of Performance, Organ Characteristics and Economic Analysis of Broiler Finisher Fed Dried Rumen Digesta," *International Journal of Poultry Science*, Vol. 5, No. 12, 2006, pp. 1116-1118.

[8] D. V. Reddy, "Applied Nutrition," Oxford & IBH Publishing Co. Pvt. Ltd, New Delhi, 2008, pp. 132-167.

[9] AOAC, "Official Methods of Analysis," 16th Edition, Association of Analytical Chemists, Washington DC, 2000.

[10] E. G. Bligh and W. J. Dyer, "A Rapid Method of Total Lipid Extraction and Purification," *Canadian Journal of Biochemistry and Physiology*, Vol. 37, No. 8, 1959, pp. 911-917.

[11] M. J. Khanke, "Analysis of Fish and Seafood Wet Digestion in Analytical Methods for Atomic Absorption Spectroscopy," 1966, p. 197.

[12] D. W. Bolin and O. E. Stamberg, "Rapid Digestion Method for Determination of Phosphorus," *Industrial and Engineering Chemistry, Analytical Edition*, Vol. 16, No.

5, 1944, p. 345.

[13] M. L. Scott, M. C. Nesheim and R. J. Young, "Nutrition of the Chicken," 2nd Edition, M.L. Scot & Associates, Ithaca, 1975, p. 555.

[14] S. Bogosavljevic-Boskovic, S. Mitrovic, R. Djokovic, V. Doskovicand and V. Djermanovic, "Chemical Composition of Chicken Meat Produced in Extensive Indoor and Free Range Rearing Systems," *African Journal of Biotechnology*, Vol. 9, No. 5, 2010, pp. 9069-9075.

[15] M. Snaz, A. Flore, P. Perez de Ayala and C. J. Lopezborte, "Effect of Fatty Acid Saturation in Broiler Diets on Abdominal Fat and Breast Muscle Fatty Acid Composition and Susceptibility Lipid Oxidation," *Poultry Science*, Vol. 78, 1999, pp. 378-382.

[16] E. Tumova and A. Teimouri, "Fat Deposition in the Broiler Chicken: A Review," *Scientia Agriculturae Bohemica*, Vol. 41, No. 2, 2010, pp. 121-128.

[17] W. A. Backer, J. V. Spencer, L. W. Minoch and J. A. Vastrate, "Prediction of Fat and Fat Free Live Weight in Broiler Chicken Using, Backskin Fat, Abdominal Fat and Live Body Weight," *Poultry Science*, Vol. 50, 1979, pp. 835-845.

[18] J. Simeonovova, "Technology of Poultry, Eggs and Other Minor Animal Products," MZLU Brno, 1999, p. 247.

[19] P. Suchy, P. Jelinek, E. Stakova and J. Hucl, "Chemical Composition of Muscles of Hybrid Broiler Chickens during Prolonged Feeding Czech," *Journal of Animal Sciences*, Vol. 47, 2002, pp. 511-518.

[20] E. Matusovicova, "Technology of Poultry Production," *Priorda, Bratislava*, 1986, p. 393.

[21] M. Skrivan, "Poultry Raising Agrospoj," *Praha*, 2000, p. 203.

[22] J. Y. Shin, P. Xun, Y. Nakamura and K. He, "Egg Consumption in Relation to Risk of Cardiovascular Disease and Diabetes: A Systematic Review and Meta-Analysis," *American Journal of Clinical Nutrition*, Vol. 98, No. 1, 2013, pp. 146-159.

[23] W. Zhang, H. Iso, T. Ohira, C. Date and A. Tamakoshi, "Association of Dietary Magnesium Intake with Mortality from Cardiovascular Disease," *Journal of Atherosclerosis*, Vol. 221, No. 2, 2012, pp. 587-595.

[24] A. Wojeik, T. Mituniewicz, K. Iwanczuk-Czernik, J. Sowinska, D. Witkowska and L. L. Chorazy, "Contents of Macro and Micro Elements in Blood Serum and Breast Muscle of Broiler Chickens Subjected to Different Variants of Pre-Slaughter Handling Czech," *Animal Sciences*, Vol. 54, No. 4, 2009, pp. 175-181.

[25] R. I. Bakalli, G. M. Pesti, W. L. Regland and V. Konjufea, "Dietary Copper in Excess of Nutritional Requirement Reduces Plasma and Breast Muscle Cholesterol of Chickens," *Poultry Science*, Vol. 74, No. 2, 1995, pp. 360-365.

[26] A. Haug, O. A. Christophersen and T. Sogn, "Chicken Meat Rich in Selenium and *Omega*-3 Fatty Acids," *The Open Agriculture Journal*, Vol. 5, 2011, pp. 30-36.

[27] A. Gergely, M. Kontraszt, A. Herman, T. Acs, J. Gundel, T. Palfy, S. Mihok and A. Lugasi, "Microelements in Muscle Tissues of Turkeys Kept on Intensive and Extensive Farming Systems," *Proceedings on Trace elements in the Food Chain*, Budapest, 2006, pp. 436-440.

9

Innovative Food Safety Strategies in a Pioneering Hotel

Su-Ling Wu

Department of Leisure Management, Yu Da University, Miaoli County, Chinese Taipei.

ABSTRACT

A case study approach was adopted to identify the innovative food safety strategies in place at one pioneering hotel that had voluntarily implemented food safety control. An investigation of food safety strategies and the reasons for their implementation in the hotel foodservice were carried out using multiple sources of data, including interviews with key decision makers in the hotel, observations of the business environment, and a review of documentation. The findings suggest that not only food control strategies but also marketing and corporate strategies are important when addressing the problems of food safety. The findings of this study also demonstrate the relationship between motivating factors and food safety strategies. Analysis of the interviews indicates that a hotel's food safety strategies depend significantly on the attitude of senior management, the firm's capability, and corporate image.

Keywords: Food Safety Strategies; Food Safety Control System (FSCS); Hotel

1. Introduction

Despite the advances in technology and hygiene standards in the past three decades, the outbreaks of fatal food born illness continued to increase globally, such as outbreaks of Salmonella serotype Enteritidis (SE), Listeria monocytogenes (Lm) and Bovine Spongiform Encephalopathy (BSE) in Europe and Northern America; Escherichia coli O157:H7 in Japan; cholera in Asia, Africa, and Latin America [1]. Life style changes and rapid global economic growth contributed to this trend [1]. These two factors lead to great numbers of people eating processed or imported food, and consuming food in foodservice establishments; and the increase of global movement in food supply and people also increase the exposure rate of contacting food born illness due to the unawareness of food born hazard on imported food or on food consumption while traveling aboard [1]. To help curb the rate of food born illness generated by the food industry, government agencies in Taiwan and other countries continue to pass new regulations, such as mandatory or voluntary enforcements of Good Manufacturing Practices (GMP), Good Hygienic Practices (GHP), Hazard Analysis Critical Control Points (HACCP) and ISO 9000 and 22000 series. Moreover, other stakeholders, such as food safety advocacy groups, customers, the press and media, and others, are pushing for firms to take food safety strategies from farm-to-fork, such as risk assessments throughout the external supply chains and the internal production process proactively into their managerial decision making process to minimize contamination which lead to loss of reputation, revenue, and even disruption of business. The incidence of food borne illness may damage the reputation of a country as a tourist destination [2]. The dining experiences of tourists were among the most important predictors of overall travel satisfaction [3,4]. Other researches indicated that the perception of food risks at their travel destinations also influenced their purchasing decisions for other products, especially when they were traveling to less developed countries [5,6]. Therefore, implementation of a certified food safety standard such as the GHP and HACCP standards would benefit especially less developed countries or regions, like Taiwan, by offering food quality assurance, strengthening the foodservices' competitiveness and ultimately enhance the country or region's reputation as tourist destination. In Taiwan approximately 40% of food borne illness outbreaks was linked to commercial food establishments (http://www.fda.gov.tw). Due to these repeated outbreaks of food borne illness, the Taiwanese Health Department in 2000 published a Food Safety Control System (FSCS), which includes principles from both HACCP and Good Hygienic Practice (GHP), in order to bring Taiwan in line with international standards. Beginning in September 2008, the Taiwanese Heath Department has been promoting the FSCS standards with stringent guidelines for the hotel industry. These standards have been gradually taking effect in outlets with central kitchens or kitchens in international tourist hotels. In order to promote the Taiwanese tourism industry, since 2009 the Taiwan Tourism Bureau has provided financial incentives to encourage the hotel industry to improve their facilities and service quality to meet international standards by obtaining local or internationally recognized certificates such as FSCS (http://admin.taiwan.net.tw).

Consumers react strongly to negative reports of food safety and their purchasing behaviors are negatively influenced [7]. Consumer food safety concerns have increased rapidly in recent years, as several high profile food incidents grabbed the head lines. Nevertheless serious incidents continue to occur. Food safety concerns have become an important factor for consumers in determining food selection and consumption. Chemical food additives, meats from diseased pigs and pesticide residuals are considered to be the most severe food safety problems by consumer in Taiwan, especially while they dine in commercial food outlets and consume processed food or meat products [8]. Similar results were also found in two studies which indicated that food safety issues influenced consumers' dining-out decisions [9,10]. Fears of being featured on the media and press were one of reasons for foodservice initiating food safety control program [11]. Therefore, the foodservice industry should try to build their reputation by offering less pre-packaged, processed food and more on healthy, nutritious fresh certified produces. Food Safety Control System (FSCS) specifies the requirements needed for food safety management to ensure the final food is safe for consumption. Therefore, the innovative food safety strategies currently adopted and the strategies being carried out in a pioneering hotel are the focus of this study.

Previous hospitality food safety research has focused only on the food safety control strategy. However, that the levels and types of food safety control strategy include not only a functional strategy (control and marketing strategy), but also a business (competitive advantage) and a corporate strategy (a link with quality and other corporate objectives) [12,13]. The main aim of this study was to understand the food safety strategies (activities) in a hotel food service, while government bureaus have been gradually putting requirements and standards in place, and encouraging tourist hotels to gain FSCS certification. This study explores experiences of the implementation and operation of FSCS in a pioneering international tourist hotel in Taiwan, and thus identifies the key motivating factors and the challenges to implementing food safety strategies.

2. Method

An investigation of food safety strategies and the reasons food safety strategies are or are not implemented in hotel food services was carried out using multiple sources of data, including interviews, observation of environmental conditions, and the review of documentation to increase the construct validity and reliability of this study [14]. The method was designed to yield results that would assist the prediction of future drivers of food safety strategies and their uptake in the hotel food service industry. A franchised international tourist hotel that had voluntarily implemented the Food Safety Control System (FSCS) was identified as a suitable participant for the purposes of the study.

Semi-structured interviews were conducted with senior management in the hotel, such as managers, chefs, and food safety supervisors, who were in management teams that had received special training in food safety control system principles [13]. The interviewer (the researcher) specifically directed open-ended questions to the interviewee in relation to food safety strategies. A set of questions was as follows:

1) What are the food safety practices in your company? How are they carried out in your firm? Why?

2) Is it convenient for you to practice food safety strategies? If not, how could it be made convenient for you?

3) What is your biggest food safety management difficulty?

4) Is there anything that would assist you in your food safety management efforts?

With the consent of the participants, interviews were taped and later transcribed. A pattern matching analytical procedure was used to increase the internal validity of a study [14]. It was anticipated that this analysis would lead to a set of food safety strategies being identified and their motivators.

The case study was conducted in a hotel that was a locally owned franchise of an international hotel chain. The chain's first hotel opened in the early 1990s in Taiwan and there are now 9 hotels in the chain. The company has been striving to reinforce food safety strategies voluntary since early 1999 in all its hotels. It is company policy that all hotels in the chain adopt a food safety control system (FSCS). The hotel that participated in this study obtained FSCS certification in 2011.

3. Results and Discussion

Seven interviews with staff from the chain hotel were conducted for the case study. The interviewees included managers in the food and beverage departments, executive chefs, food safety supervisors, personnel managers and general managers. Most had been in the hotel industry for more than 10 years. All of them had received training in food safety and received FSCS certification. The average time per interview was 35 minutes. Observations of environmental conditions and examination of documentation such as annual reports, policy statements, employee manuals, product advertisements, and corporate advertisements were also collected in order to obtain a greater understanding of operational systems. Many interesting issues were found during analysis of the interviews. The interview data revealed that the study hotel engaged in many food safety strategies, such as safety control, marketing, and corporate organization.

3.1. Safety Control Strategies

The study hotel's capacity to provide sufficient personnel, a training program, time, and resources was found to be the main barrier to the consistent use of a food safety control system [15,16]. As far as possible, however, the principles of the FSCS were adhered to by the various departments of the hotel, as well as the hotel's suppliers. Production equipment; cleaning and sanitation; personal hygiene; training; chemical control; receiving, storage, and shipping; traceability and recall; and pest control (http://food.fda.gov.tw) were all subject to provisions outlined in the FSCS. The interviewees indicated that the hotel's Food Safety Control System (FSCS) could be successfully carried out with proper layout and equipment and an effective training and assessment program.

3.1.1. Layout and Equipment

Analysis of the data from the interviews revealed that proper layout and equipment in the kitchen promoted food safety and supported food safety systems. With a more centralized kitchen, the interviewees indicated that less cost, better food quality, and less food borne risk could be achieved [13].

"Our hotel has a centralized kitchen in the basement of the building where food ingredients are pre-processed and supplied to all the kitchens in the hotel. Cost can be reduced through easier control in inventory, better deal in purchasing with large quantities, and less labor and equipment needed in each kitchen. Food quality is easier to control, not only in food freshness, texture, and portion size, but also in food safety. Different areas for different kinds of food in process and storage are designed to eliminate cross contamination. Examples of physical separation include the space of food preparation and refrigeration separated according to the food type such as seafood, meat, cold process food, hot process food, and bakery goods." (Food and Beverage Manager, 20/07/2010, interview at the hotel office).

3.1.2. In-House Training and Assessment

Effective food safety training should be designed to closely match employees' expectations and the organization's goals and be accompanied by inducements that motivate staff in order to optimize training performance

[13,17,18].

"In Taiwan, kitchen staffs need to have eight hours food safety seminar attendance each year or 32 hours in four years to keep chef certification valid. FSCS certified kitchen needs to have 80% of kitchen staffs with chef certifications. In addition, at least three staffs need to have additional 32 hours FSCS training course and pass qualify exam at the end of the course in order to form a FSCS team." (Executive chef, 20/07/2010, interview at the hotel office).

For the food safety seminar, overall, the interviewees agreed that effectiveness of training is an important factor in gaining the employees' commitment to the company's goals in terms of the FSCS. However, all the interviewees agreed that the current training courses and certification awarded by the HACCP associations should offer a balance of practical skills development and knowledge requirements [13]. Proper food safety handling skill should also be included in training courses and assessed by both theory and a practical exam. Overall, the interviewees indicated that hotel staff who attended the HACCP courses all felt that the information sessions were overlong long and that the content was boring.

"The idea that the HACCP course focuses on both GHP and HACCP plan development seems ok theoretically. However, it is hard for our staffs to sit and listen eight hours a day. An interactive short course should be developed by incorporating more case discussion after showing real examples of HACCP operation in pictures or video rather than too much written words theory. In addition, practical skill development on food preparation should also be included." (Personnel Manager, 20/07/2010, interview at the hotel office).

An effective training system should not be without effective food safety management and compliance assessment. The interviewees indicated that hotel food safety was supervised and audited by an assistant manager in the Stewarding department who worked closely with the executive chef, all restaurants and kitchens, and was fully empowered by the General Manager to communicate across departments on the issue of food safety. The General Manager of the hotel was very conscious of the food safety issue, and had considerable experience in the operation of hotel kitchens and FSCS implementation. He always checked and made corrections on the food safety operation while walking around the hotel, and often walked with the food safety supervisor. A video recording system was also used to assist the assessment of activities in the hotel.

"Our hotel has video camera around the food process area, and the food safety auditor will review the video to see any defaults during operation. The auditor will also walk around the kitchen floor, take pictures, and make correction. The audit report and video will be discussed with the kitchen staff during weekly meeting in which the staff will be able to identify the problem from practical examples. It is also hope to increase the staff involvement through the weekly discussion. The practical examples can also be a good material for food safety training." (Food safety auditor, 20/07/2010, interview at the hotel office).

The Area Manager from the franchisor's head office also conducted spot-checks of food safety standards implementation. In addition, every restaurant in the hotel formed an audit team, which then assessed the other restaurants on a monthly rotation, and reported to the executive chef. The Executive Chef in an interview indicated that temperature measuring and recording were especially important for FSCS implementation, but that not all the chefs were able to do the measuring and recording work. However, the senior management ensured that they received assistance.

"Many experienced chefs, especially in Chinese cuisine, usually have lower levels of education. Some of them have difficulty to read and write and do not have hospitality relevant degrees. Therefore, now we have kitchen stewards to assist the chef on measuring and record keeping to overcome this problem." (Executive chef, 20/07/2010, interview at the hotel office).

As for environmental issues, many interviewees indicated that wearing disposable gloves was not necessary because, unlike the circumstances in a factory, most of their products were ready to eat after a short processing time. They noted that the bacteria on hands after proper washing was not enough to cause sickness, while the disposable gloves not only caused environmental problems, but also allowed employees to forget to actually wash their hands.

3.2. Marketing Strategies

Many studies [9,10,19,20] have shown that awareness of food safety or having the experience of food poisoning did influence consumers' dine out decisions.

3.2.1. Transparent/Open Kitchen
The interviewees indicated that the customers' trust concerning food safety and quality maintained or improved the company's brand image [13]. The presentation of transparent and open kitchens staffed with employees well trained in food safety was a sign to customers that food was being prepared in a safe way and that there was nothing to hide.

3.2.2. Advertising
The hotel advertised their food safety efforts by posting where their food came from and when they received FSCS certification on both their hotel's webpage and restaurant entrances.

3.2.3. In-House Training for the Public
In one interview, a General Manager indicated that the hotel held in-house FSCS training courses sponsored by one of the HACCP associations, and the training was not limited to their hotel staff, but that others could participate.

"At the end of the course, the trainee will take turn to observe and assess our Hotel FSCS system, then form a discussion after that. It is another way to advertising our hotel by having this activity." (General Manager, 20/07/2010, interview at the hotel office).

The interviewee indicated that they hoped the customers and the trainee could judge the positive nature of the food safety effort from the dining environment and staff presentation.

3.3. Corporate Strategies

3.3.1. Corporate Social Responsibility
In this study addressing food safety issues was found to be part of the company's policy and was documented on the mission and vision statements of the company. All the chain hotels were required to obtain food safety control system (FSCS) certification. The interviewees agreed that protecting the health and safety of the customers and employees by providing a healthy and safe environment was part of the hotel's corporate social responsibility (CSR). Many food safety efforts made by the company are outlined in annual reports and highlighted on the hotel webpage.

3.3.2. Management Team
A FSCS management team was led by the hotel General Manager and consisted of the executive committees and other heads of department, including the Food and Beverage Manager, Chief Engineer Controller, Executive Chef, Sales and Marketing Manager, Personnel Manager, Restaurant manager, Banquet Manager, Stewards, Bar Manager. The FSCS team met monthly to discuss data gathered during assessments of the hotel, and areas needing further improvement. The General Manager indicated that FSCS was a task that needed all departments to work together in order to effectively achieve required standards of food safety.

3.4. Food Safety Strategies and Their Motivators

Analysis of the interviews indicates that the commitment of senior management to food safety, the firm's capacity to act, and ideas of competitive advantage have the most positive effect on overall food safety strategies. **Table 1** shows extracts of significant food safety strategies and primary motivators from the interview data.

The commitment of the management of a hotel to standards of food safety is the key driving factor for making

Table 1. Thematic conceptual matrix-food safety strategies and motivator.

	Food safety Strategies	Motivators
Safety Control Strategies	Layout & equipment In-house training/assessment	Senior management, firm's capability, Senior management, firm's capability
Marketing Strategies	Transparent/open kitchen Advertisement In-house training for public	Corp image, product differentiation, Corp image, Corp image
Corporate Strategies	CSR management team	Corp image, Senior management

the resources available for the implementation of a food safety control system [13]. In Asia, most corporations still operate on a traditional hierarchical basis in which decisions are made from the top [21]; therefore, commitment on the part of senior management is especially important. The implementation of a food safety control system in the hotel was mainly influenced by the senior management, who realized that FSCS needed the cooperation of all departments. The interviewees also indicated that a firm's willingness to support FSCS relied on the proper provision of facilities and an in-house food safety training program with incentives to retain and motivate employees were significant motivators in promoting food safety program.

The objectives of competitive advantage are cost reduction and differentiation [22], such as developing sound safe food products to improve reputation and brand loyalty. As mentioned earlier, cost saving can be easily achieved, as well as food quality and safety, with a centralized kitchen. However, interviewees indicated that the overall cost saving was minimal, compared to the investment cost. The interviewees stated that the senior management all realized the international trend toward food safety management and were willing to invest.

The franchising company's policy was to have all the hotels in the chain obtain food safety control system (FSCS) certification in order to create a brand image which customers could trust in terms of food safety and quality. Although the system was mostly carried out behind the centralized kitchen, transparent and open kitchen facilities in the dining areas were built for customers to see the hotel's product differentiation and efforts to achieve food safety and quality. Advertisement on the hotel webpage and in-house training for the public on food safety issues were also used to improve the hotel's image. One interviewee indicated that it was senior management's intention to place food safety issues in the company's strategic report, which can maintain or improve a company's image. It was also the company's policy to have a management team led by top manage-

ment to run the food safety strategy more efficiently.

4. Conclusions

The main aim of this study was to understand the food safety strategies (activities) in a pioneering hotel, and the associated relationships between the motivating antecedents and food safety strategies and to determine the key motivating factors in order to provide a source of new hypotheses for further study.

This study has provided details on "what" and "how" with respect to carrying out the food safety strategies and on "how" top management commends each motivating factor. The effect of perceptions of corporate image, a firm's capability, and senior management's commitment to a food safety strategy in a large international chain hotel provided insight into the issues faced when attempting to encourage or mandate stronger safe food control systems. The results of this study suggest the most effective food safety practices and information for helping the hotel industry to further improve food safety strategies or practices. These findings can also be employed as a basis for further review by policy makers and educational institutions to ensure that the policy is appropriate for the hotel industry.

This study identified three key motivating factors—corporate image, firm's capability, and senior management's commitment—which can be developed as new hypotheses for further testing and refinement. This reduction process of data by identifying key factors is significant for forming future research.

Moreover, the result found in this study representing the relationship between the motivating factors and food safety strategies can be applied to other sectors of the food service industry or other countries confronted by similar situations and problems. The intention of this study was not to draw conclusions from the limited sample, but to provide direction for future large scale research. Therefore, any findings from the analysis of the interviewees' responses have not been used to generalise.

REFERENCES

[1] World Health Organization, "International Travel and Health," 2002. http://www.who.int/ith/ITH2002.pdf

[2] World Health Organization, "Food Safety," 1999. http://www.who.int/foodsafety/publications/general/brochure_1999/en/

[3] C. G. Q. Chi and H. Qu, "Examining the Relationship between Tourists' Attribute Satisfaction and Overall Satisfaction," *Journal of Hospitality Marketing & Management*, Vol. 18, No. 1, 2009, pp. 4-25.

[4] B. Ozdemir, A. Aksu, R. Ehtiyar, B. Çizel, R. B. Çizel and E. T. İçigen, "Relationships among Tourist Profile, Satisfaction and Destination Loyalty: Examining Empiri-

cal Evidences in Antalya Region of Turkey," *Journal of Hospitality Marketing & Management*, Vol. 21, No. 5, 2012, pp. 506-540.

[5] G. Fuchs and A. Reichel, "Tourist Destination Risk Perception: The Case of Israel," *Journal of Hospitality & Leisure Marketing*, Vol. 14, No. 2, 2006, pp. 83-108.

[6] Y. Yilmaz, Y. Yilmaz, E. T. İçigen, Y. Ekin and B. D. Utku, "Destination Image: A Comparative Study on Pre and Post Trip Image Variations," *Journal of Hospitality Marketing & Management*, Vol. 18, No. 5, 2009, pp. 461-479.

[7] E. G. McKeown and W. B. Werner, "Content Analysis of Consumer Confidence in Food Service in Relation to Food Safety Laws, Publicity, and Sales," *Journal of Hospitality Marketing & Management*, Vol. 19, No. 1, 2009, pp. 72-81.

[8] M. F. Chen, "Consumer Trust in Food Safety—A Multidisciplinary Approach and Empirical Evidence from Taiwan," *Risk Analysis*, Vol. 6, 2008, pp. 1553-1569.

[9] A. Radam, M. L. Abu and M. R. Yacub, "Consumers' Perceptions and Attitudes towards Safety Beef Consumption," *The IUP Journal of Marketing Management*, Vol. IX, No. 4, 2010, pp. 29-55.

[10] A. Knight, M. R. Worosz and E. C. D. Todd, "Dining for Safety Consumer Perceptions of Food Safety and Eating Out," *Journal of Hospitality & Tourism Research*, Vol. 33, No. 4, 2009, pp. 471-486.

[11] J. Sneed and D. Henroid, "HACCP Implementation in School Foodservice: Perspectives of Foodservice Directors," *The Journal of Child Nutrition & Management*, 2003.
http://docs.schoolnutrition.org/newsroom/jonm/03spring/sneed/

[12] S. B. Banerjee, "Corporate Environmental Strategies and Actions," *Management Decision*, Vol. 39, No. 2, 2001, pp. 36-44.

[13] S. L. Wu, "Factors Influencing the Implementation of Food Safety Control Systems in Taiwanese International Tourist Hotels," *Food Control*, Vol. 28, No. 2, 2012, pp. 265-272.

[14] R. K. Yin, "Case Study Research: Design and Methods," 2nd Edition, Sage, Thousand Oaks, 1999.

[15] P. J. Panisello and P. C. Quantict, "Technical Barriers to Hazard Analysis Critical Point," *Food Control*, Vol. 12, 2001, pp. 165-173.

[16] L. A. Brannon, V. K. York, K. R. Robert, C. W. Shanklin and A. D. Howells, "Appreciation of Food Safety Practices Based on Level of Experience," *Journal of Foodservice Business Research*, Vol. 12, No. 2, 2009, pp. 1-32.

[17] G. C. Ogbeide, "A Case Study of Restaurant Training Motivations and Outcomes," *International Journal of Tourism and Hospitality Research*, Vol. 19, No. 1, 2008, pp. 72-177.

[18] J. Salazar, H. R. Ashraf, M. Tchebg and J. Antun, "Food Service Employee Satisfaction and Motivation and the Relationship with Learning Food Safety," *Journal of Culinary Science & Technology*, Vol. 4, No. 2, 2005, pp. 93-108.

[19] A. Knight, E. C. D. Todd and M. R. Worosz, "Dining for Safety Consumer Perceptions of Food Safety and Eating Out," *Journal of Hospitality & Tourism Research*, Vol. 33, No. 4, 2009, pp. 471-486.

[20] U. Z. A. Ungku Fatimaha, H. C. Boo, M. Sambasivan and R. Salleh, "Foodservice Hygiene Factors—The Consumer Perspective," *International Journal of Hospitality Management*, Vol. 30, No. 1, 2011, pp. 38-45.

[21] Y. Boshy, "Action Learning Worldwide," Palgrave, New York, 2002.

[22] R. Welford, "Corporate Environmental Management 1: Systems and Strategies," Earthscan, London, 1998.

10

Antioxidant Capacity in Vanilla Extracts Obtained by Applying Focused Microwaves

Adalith Rojas-López, María P. Cañizares-Macías[*]

Departamento de Química Analítica, Facultad de Química, Universidad Nacional Autónoma de México, México D.F., México.

ABSTRACT

ORAC method and a continuous flow injection method based on Folin-Ciocalteau reaction (FI-FC) were used for determining the antioxidant activity in extracts obtained by using focused microwaves. Analysis of the antioxidant capacity (AC) of the main compounds of vanilla (vanillin, p-hydroxybenzaldehyde, p-hydroxybenzoic acid and vanillic acid) was also carried out. Vanilla extracts obtained by using focused microwaves had a higher AC (between 72% and 117%) than the obtained by conventional methods. Vanillin had a linear correlation with the antioxidant capacity of the extracts and it is the most influential compound in the antioxidant power. The AC calculated by the ORAC method and the FI-FC method had a ratio 2:1 because of different kinetics and reaction mechanisms of the antioxidants with the reagents, so it is necessary more than one method to establish the antioxidant power in food. On base on the results of the present study microwaves energy can be used to obtain vanilla extracts to improve the AC of them.

Keywords: Focused Microwaves; Vanilla; Antioxidant Capacity; ORAC Method; Flow Injection; Folin-Ciocalteau

1. Introduction

The early news from vanilla are in the middle 15th century, when Aztecs conquered the Totonaca Empire who used vanilla to make a drink named "xocolatl" (chocolate). *Libellus de Medicinalibus Indorum Herbis*, Cruz Balbiano codex, were the first document where vanilla is named with its Nahuatl name: "tlilxochitl" (black flower) [1]. For three centuries Mexico was the only producer of vanilla in the world. In 1841, Edmond Albins, a farmer from Reunion Island had the idea of pollinating the flower with a bamboo stick. From then vanilla began to spread through Africa and Asia [2].

Vanilla is the only orchid that produces an important commercial fruit for its flavour and smell. There are more than 110 species of vanilla, but the Vanilla planifolia, also known as Vanilla fragans, is the most important source of the natural commercial vanilla [1,3,4].

Vanilla has more than 250 compounds but only 26 have concentrations higher than 1 mg·kg^{-1}. Among non-volatiles compounds of vanilla are: tannins, polyphenols, free amino acids and resins and among volatiles compounds are: carbonyl aromatic and aliphatic alcohols, aromatic acids, aromatic esters, phenols, lactones, aliphatic and aromatic hydrocarbons, terpenoids, etc. Vanillin (1% - 3%), vanillic acid (0.1% - 0.2%), p-hydroxy-benzaldehyde (0.1% - 0.2%) and p-hydroxybenzoic acid (0.01% - 0.02%) are the most important phenolics compounds produced during cured process [5-7].

Antioxidants have beneficial effects on the conservation of the food, avoiding their oxidation, besides having healthy effects; so industries of food, pharmaceutical and cosmetic have been used antioxidants to give better advantage over yours products. The main synthetic antioxidants used by the industry are: propyl gallate, butylated hydroxy anisole, hydroxyl toluenehydroxyanisole, tertiary butylhydroquinone, most of them are not very healthy, therefore natural antioxidants instead of the synthetic compounds have been more used [8-10]. Plants and spices are the main target to look for natural antioxidants. There are some studies to identify the compounds that should act as antioxidants [11]. Among phytochemicals best characterized in vegetable food polyphenols have been the most studied due to their high dairy ingest [12]. Moreover, polyphenols are able to fix other reactive compounds as $HO\bullet$, $NO_2\bullet$, N_2O_3, $ONOOH$ and $HOCl$ and also are able to fix metallic ions (specially iron and cop-

---

[*]Corresponding author.

per) to obtain poorly active compounds in the formation of reactive species [13].

Inhibition of the chain reaction, oxygen captation, inhibition of singlet oxygen, metals quelation and inhibition of oxidative enzymes are mechanism that makes very complex the antioxidant action in the food. Therefore, several analytical methods are necessary to get a precise evaluation of the antioxidant capacity (AC) [14]. Based on the chemical reaction involved, two categories of the main antioxidant capacity tests can be named:

a) Tests based on the hydrogen atom transfer reaction: these methods measure the competitive kinetic reaction and the quantification comes from the kinetic curves. Usually these are made from a free radicals synthetic generator, an oxidative molecular probe and an antioxidant. The Oxygen Radical Absorbance Capacity (ORAC) test is one of the more used [15-17].

b) Tests based on the electron transfer reaction: in these methods a redox reaction is carried out and the reduction of the oxidant marks the finish of the reaction. The reaction with Folin-Ciocalteau reagent is one of these. This method is principally used to determine total polyphenols [18,19].

In the last decade, there has been an increasing demand for new extraction techniques enabling automation, shortening extraction times and reduction of organic solvents consumption [20,21]. Advances in preparation of solid samples have brought a great number of new techniques such as focused microwave-assisted energy [22-24]. Microwaves have been used for different chemical process accelerating digestion [25] and organic and inorganic synthesis [26-28]. Focused microwaves have been also used successfully to extract vanillin from vanilla beans increasing the efficiency at 100%, so it has been possible to accelerate extraction procedures of different compounds without oxidation of them when the conditions of extraction are controlled [29].

Vanilla is one of the species more consumed, mainly in extracts, with a high antioxidant capacity (AC). Vanillin is its main compound and the majority phenol. In this paper the antioxidant capacity in ethanolic vanilla extracts obtained by applying of focused microwaves was measured. A maceration procedure [30] was also carried out and the results were compared. Vanillin concentration was determined and a study about the relation between vanillin and AC was carried out. AC was determined by two tests way: a) ORAC method, using fluorescein as probeand 2,2'-azobis(2-methylpropionamidine) dihydrochlorideas radicals' generator and b) a flow injection method using Folin-Ciocalteau reagent (FI-FC).

2. Material and Methods

2.1. Reagents and Solutions

All reagents were analytical grade and used water was distilled.

Fluorescein (Sigma-Aldrich) and 2,2'-azobis(2-methylpropionamidine) dihydrochloride, 97% (AAPH) from Sigma-Aldrich were used by the ORAC method. A fluorescein stock solution (10 $\mu g \cdot mL^{-1}$) was prepared in a 0.05 M pH 7 phosphates buffer. 0.03 $\mu g \cdot mL^{-1}$ fluorescein solutions to obtain the kinetics curves were prepared from the stock solution using the same buffer. A 0.22 M AAPH solution was also used. A 1:10 dilution from Folin-Ciocalteau reagent (Merck) and a 0.5 M NaOH (Baker) solution were used by the FI-FC method. Caffeic acid (Sigma-Aldrich) was used as reference to determine the AC by both methods.

Standard solutions obtained from a 1000 $\mu g \cdot mL^{-1}$ vanillin (Sigma-Aldrich) stock solution were used to build the vanillin calibration curve. Vanillic acid, p-hydroxybenzaldehyde and p-hydroxybenzoic acid were also purchased by Sigma-Aldrich.

A 70:30 ethanol (Baker): water mixture was also used to obtain the vanilla extracts.

2.2. Instruments

A fluorimeter PMT-FL Fialab and a thermostated bath Lab-line 1800 were used to determine the AC by the ORAC method. A peristaltic pump (Gilson minipuls), an injection valve (Rheodyne), Teflon and Tygon tubing, a Hellma 18 μL internal volume flow cell and a Cary UV-Vis spectrophotometer (Varian) as detector were used to build the FI manifold. A FI manifold reported by Valdéz and Cañizares [7] was used to determine vanillin.

A 300 W maximum power focused microwaves oven (Prolabo) was used to obtain the vanilla extracts. An ultrasonic bath (Branson Model 2510R), a centrifuge (HETTICH EBA 20) and a pHmeter (Oakton) were also used. A controlled temperature oven (Rios Rocha) was used for the moisture assay.

2.3. Samples

All vanilla samples were Mexican. Two different cured vanilla beans quality were analyzed: a) *gourmet* quality, from Papantla, Veracruz: 15 cm length, dark brown colour and glossy beans; b) second quality (named *zacatillo*) from Ayotoxco, Puebla, used to make handicrafts: from 11 cm to 17 cm length, light brown colour with clear stripes beans. Green beans from Papantla, Veracruz were also used.

Gourmet vanilla was purchased in a local market from Papantla. A Cooperative of Women from Ayotoxco, Puebla, who cultivate, harvest and cure vanilla, gave the Zacatillo vanilla beans.

For the analyses a stock of 500 g of each kind of vanilla beans was used. Beans were cut to a size of 2 mm × 2 mm and homogenized. They were stored in refrigeration until were used.

Vanilla green beans from Ayotoxco were also analyzed. The beans were collected and stored at 4°C until analysis.

2.4. Getting of the Extracts

Focused microwaves-assisted extraction (FMAE) [29]: 1 g of cut vanilla beans (2 × 2 mm) was poured into a 10 cm test tube of 3 cm of internal diameter, which was placed into a water bath (test tube of 4.5 cm of internal diameter). 25 ml of a 70% (v/v) ethanol-water solution were added to the sample. A refrigerant was adapted to the test tube to avoid loss of the extractant and a 150 W microwaves irradiation power was applied to the sample. Twenty cycles of 1 min irradiation each one with a delay time between them of 3 min were carried out. The total extraction time was 80 min. Once the extraction was over, solution was filtered and water was added until reaching a final volume of 100 mL.

Maceration extraction (ME) [29,30]: 1 g of cut vanilla beans (2 × 2 mm) was poured into a 10 ml volumetric flask covered with 2 ml of ethanol and 1 ml of water. The solution was macerated for 12 hours. 2 ml of ethanol were added mixing well all the content. Maceration continued for 3 days. The solution was drained funnel dry, packing solids firmly and percolating slowly with a 50% ethanol solution until reaching a final volume of 10 ml.

2.5. Determination of the Antioxidant Capacity (AC) in Extracts of Vanilla Beans

2.5.1. ORAC Method

It was used, with some modifications, for the determination the AC in the vanilla extracts obtained from the vanilla beans. 2 mL of the 0.03 $\mu g \cdot mL^{-1}$ fluorescein (FL) solution, 2 mL of blank (phosphates buffer (0.5 M, pH 7) and 1 mL of standard (caffeic acid) or sample (diluted 1:1000) were set into a 13 × 100 cm testing tube. During 15 min were incubated at 37°C. Later 1 mL of ΛΛPH was added and then the fluorescence intensity was measured setting this moment as cero time (t_0). After 5 min of a new incubation step the signal was measured again. This procedure was repeated until fluorescence intensity decayed at 20% from initial value. The fluorescence intensity was measured $\lambda ex = 486$ nm and $\lambda em = 500$ nm. The measurements were carried out in triplicate for each sample and standard.

A calibration curve was obtained by plotting the total area under the curve (AUC) against caffeic acid concentrations in the 1 $\mu g \cdot mL^{-1}$ - 6 $\mu g \cdot mL^{-1}$ range. The area under the kinetic curve (AUC) was calculated for each standard and from these values the total area under the kinetic curve (AUC$_{Total}$). In **Figure 1** the integration ranges and the fluorescence intensity decreasing curve

are shown. The trapezoid method according to Equation (1) was used to calculate the AUC;

$$AUC = \left(\frac{(y_0 + y_5)}{2} \times 5 \right)$$
$$+ \left(\frac{(y_5 + y_{10})}{2} \times 5 \right) + ... + \left(\frac{y_{n-5} + y_n}{2} \times 5 \right) \quad (1)$$

where y_0 is the initial fluorescence signal and y_5, y_{10}, y_{n-5}, $\cdot$, y_n are the fluorescence intensity values at different incubation time: 5, 10, $n - 5$, $\cdots$, n minutes.

AUC$_{Total}$ was solved from Equation (2):

$$AUC_{Total} = AUC_{ANTIOXIDANT} - AUC_{BLANK} \quad (2)$$

2.5.2. FI-FC Method

This method has been also used to determine AC en different kind of food. FI methods are faster than batch method and it has been proved that they are very precise. Folin-Ciocalteu reagent has been used for phenolic compounds analysis both by FI [31] and by batch methods obtained a very good precision [19,32]. In **Figure 2** the used FI manifold is shown. 100 μL of sample or caf-

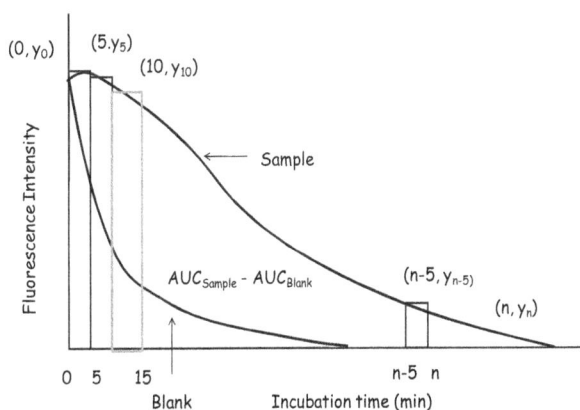

Figure 1. Trapezoid method (area under the curve (AUC)) to determine the antioxidant capacity (AC) using the ORAC method.

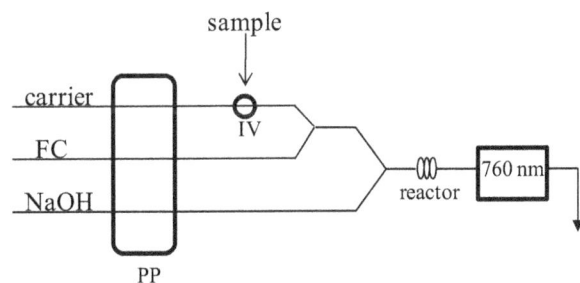

Figure 2. Flow injection manifold to determine the antioxidant capacity (AC) using the Folin-Ciocalteu reagent (FI-FC). PP: peristaltic pump; IV: injection valve; FC: Folin-Ciocalteu reagent (10%); carrier: distilled water; NaOH: 0.5 M; reactor: 100 cm.

feic acid standard were injected into the carrier (distilled water) to merge with Folin-Ciocalteau reagent and later with a 0.5 M NaOH stream. The formed complex was measured at 760 nm. Extracts were diluted 1:500 to measure the AC in the vanilla.

The precisions of the methods were calculated by determination of the antioxidant capacity of six independent aliquots of the sample for each evaluated method.

2.5.3. Determination of Vanillin
A flow injection manifold was used to quantify vanillin in accordance with Valdez and Cañizares [7]. 100 μl of sample were injected into a carrier of 0.01 N NaOH, which carries the sample to the reactor where the vanillin was hydrolyzed. Finally, the hydrolyzed plug passed through a flow cell located in the UV-VIS spectrophotometer, where the product was measured at 347 nm.

2.6. Determination of Moisture in Cured Vanilla Beans

The procedure was carried out in accordance with the established in the American Society for Testing and Material [33]. From one and five grams of the sample were weighed and placed into a previously dried container, which was covered and then weighed again. Later, the stopper was removed from the container and both were placed in the oven. The sample was dried for 4 h at 105°C ± 3°C. At the end of the specified period, the container was quickly covered and placed in a desiccator during 1 h. After this time the sample was weighed. The drying and weighing procedures were repeated until the mass loss between two successive weightings was not more than 0.005 g (or until the specimen showed a gain in mass). The moisture was calculated by using the following equation:

$$\text{Moisture content} = \left[\frac{(M-D)}{(M-T)}\right] \times 100$$

where:
M = original mass of the specimen and weighing bottle
D = oven-dry mass of the specimen
T = mass of the empty weighing bottle

3. Results and Discussion

3.1. Characteristics of the Quantification Methods

It has been widely accepted that the radical system used for the antioxidant evaluation may influence the experimental results [34]. There are many methods of analysis to determine the antioxidant capacity and radical scavenging activity in food (iron(II) chelating activity, total

antioxidant capacity, DPPH assays and lipid peroxidation) and a number of different reagents have been used. Each one of these methods has different values of AC depending on the kind of reaction and the analytical measurement (fluorescence, UV-Vis, electrochemical, etc.) [35]. Although Trolox is the most used reagent by ORAC method due to its high sensibility, caffeic acid has been also used because it has a high antioxidant power [36]. So, in this paper all results have been expressed as mg caffeic acid/g of vanilla bean.

3.1.1. ORAC Method
The calibration curve was built from the caffeic acid kinetics graphics. In **Figure 3** is showed some signals of three different concentrations of caffeic acid; some signals of vanilla extracts obtained by both extraction methods are also showed. For blank, initial fluorescence intensity falls down quickly because there is not an antioxidant and the peroxiles radicals generated by thermic decomposition of AAPH reacts with fluorescein obtaining a non-fluorescent product. If the reaction is carried out with an antioxidant the fall is not so quick therefore when the caffeic acid concentration increases the lag phase is larger, because caffeic acid fixes the peroxil radicals.

The linear range was from 1 μg·mL^{-1} to 6 μg·mL^{-1} with a lineal equation $AUC_{Total} = 3589.7 (\pm179.70)$ [caffeic acid] $+ 1809.8 (\pm727.74)$ and a regression coefficient of 0.9962. At concentration higher than 6 μg·mL^{-1} the decay of the fluorescence intensity in 20% was too long.

The detection limit (DL) was evaluated on the standard deviation of the response of the blank and the slope using the ratio 3.3 s/m, where s is the standard deviation of the response of the blank and m is the slope of the calibration curve of the analyte. The detection limit was 0.81 μg·mL^{-1} and the quantification limit using the ratio 10 s/m was 1.65 μg·mL^{-1}.

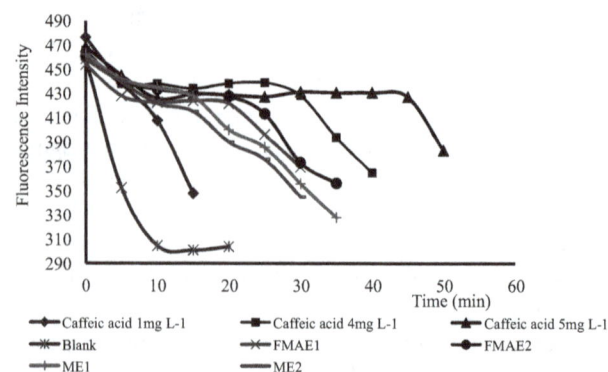

Figure 3. Plot of fluorescence intensity signal vs time for three caffeic acid standard solutions and for two vanilla extracts obtained by focused microwaves assisted extraction (FMAE$_1$ and FMAE$_2$) and by maceration extraction (ME$_1$ and ME$_2$).

3.1.2. FI-FC Method

The linear equation Abs = 0.011 (±0.0001) [caffeic acid] + 0.0085 (±0.0058) was obtained for a caffeic acid concentration range between 5 $\mu g \cdot mL^{-1}$ and 100 $\mu g \cdot mL^{-1}$. The regression coefficient was 0.9998, the detection limit of 3.55 $\mu g \cdot mL^{-1}$ and the quantification limit of 6.35 $\mu g \cdot mL^{-1}$.

3.2. Determination of Vanillin in Extracts

Extracts obtained by both methods were between light and medium brown color (depending of the kind of bean), with a vanilla characteristic odor. When focused microwaves were used the vanillin amount in the extracts obtained from both kinds of beans increased. Once again it was demonstrated that when focused microwaves are applied the vanillin extraction from vanilla beans increases: for zacatillo beans between 75% and 87% and for gourmet beans between 81% and 93%, comparing with the obtained results by the official method (ME). For green beans the vanillin concentration was too low therefore this was only determined when microwaves were used. The results are showed in **Table 1**. In this same table moisture of the beans is also showed. Moisture is a quality control parameter for vanilla beans and it has to be between 25% and 30% for gourmet vanilla and between 20 and 24 % for vanilla of second quality [30]. The evaluated beans were in these classifications. Although the vanilla concentration was higher when moisture in the vanilla beans was higher, it was not possible to conclude that moisture higher increases the vanillin concentration in the vanilla extracts. On the other hand, in a previous paper was demonstrated that when cured vanilla beans were dried at 135°C for 4 h, the vanilla concentration decreased between 30% and 40% [29]. The waves of microwave are non-ionizant radiations that cause the mobility of the molecules due to the migration of ions and the bipolar rotation. When the electric camp decreases the thermic disorder that produce energy as heat is restored. This law makes that the interaction between the analytes in a matrix and the extractant is better and the extraction is more efficient [22]. There are some examples of focused microwave extraction that improve the extraction of polar and non-polar compounds in different matrixes as the extraction of ketoprofen in pharmaceutical cream formulation [37], the extraction of heterocyclic amines in meat [38] or in order to clarify the possible hazard of acidic pharmaceuticals (ibuprofen, naproxen, ketoprofen, and diclofenac) in river water and sediments [39]. Using microwave in solid samples allows more efficient and faster sample treatment, so it is being more used in analysis laboratories. Microwaves are non-destructive energy, so can be applied in food under controlled radiation which has allowed to be used, for example to accelerate the oxidation process in olive oil [40].

On the other hand, several studies have demonstrated that vanillin helps to the stability in the beans avoiding oxidation, but it is necessary good conservation conditions.

3.3. Antioxidant Capacity (AC) in Vanilla Beans

ORAC method is the most used method to measure the AC in food because of its high sensitivity [15,41]. Nevertheless there are a number of methods to determine the AC where the found values are very different between methods. For this reason, usually more than one method is used to evaluate the AC in food [42,43].

Vanilla, specie, is mainly used as flavoring and it is used in a very low amount in our daily diet but it could help to get the daily recommended antioxidants. So, it is important to develop extraction methods of improving of the antioxidant power in the extracts, such as the focused microwaves assisted extraction method evaluated in this research. In **Figure 3** is showed the signals obtained by the ORAC method of different vanilla extracts obtained by the maceration method (ME) and by the FMAE method. The signals of vanilla extracts show a large lag phase which is indicative of a high AC but they are lower when the extracts are obtained by the ME. The extracts analyzed by the FI-FC method had values of AC lower than by the ORAC method although by both methods the AC increased when microwaves were applied. The results of the AC obtained by both methods are showed in **Table 2**. The AC for gourmet beans was higher than for zacatillo beans between a 55% and a 57%, using both

Table 1. Concentration of vanillin in extracts obtained by focused microwaves (FMAE) and by maceration (ME). It is also shown the moisture for each kind of bean.

Sample	Vanillin concentration (%, w/w)		Moisture (%)
	ME	FMAE	
Zacatillo			24.70 ± 0.20
1	0.62 ± 0.012	1.16 ± 0.03	
2	0.57 ± 0.025	1.08 ± 0.02	
3	0.65 ± 0.015	1.14 ± 0.03	
Gourmet			27.80 ± 0.24
1	1.07 ± 0.020	2.02 ± 0.02	
2	0.95 ± 0.015	1.81 ± 0.02	
3	1.05 ± 0.031	1.96 ± 0.03	
Green beans	nd	0.27 ± 0.01	-

nd: no detected.

Table 2. Values of the antioxidant capacity (AC) by the ORAC method and by the Flow Injection-Folin Ciocalteau (FI-FC) method for vanilla extracts obtained by focused microwave (FMAE) and by maceration (ME).

| Sample | AC (expressed as mg caffeic acid g^{-1} bean) | | | |
| | ORAC | | FI-FC | |
	FMAE	ME	FMAE	ME
Gourmet				
1	77.22 ± 1.03	37.86 ± 1.65	35.74 ± 0.30	20.32 ± 0.13
2	69.93 ± 1.10	33.69 ± 1.42	32.33 ± 0.95	18.37 ± 0.36
3	75.13 ± 0.85	37.17 ± 2.05	34.77 ± 0.65	19.99 ± 0.44
Zacatillo				
1	45.09 ± 0.87	24.74 ± 0.74	24.03 ± 0.16	12.18 ± 0.22
2	44.58 ± 0.78	20.52 ± 0.84	20.48 ± 0.56	12.20 ± 0.42
3	46.66 ± 1.00	23.27 ± 1.25	21.61 ± 0.34	12.52 ± 0.16
Green	6.70 ± 0.32	nd	nd	nd

nd. no detected.

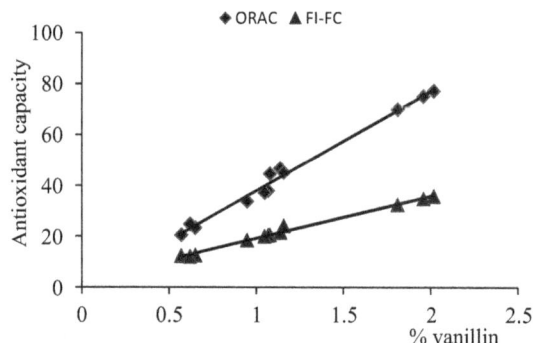

Figure 4. Correlation between the vanilla concentration and the ORAC method (♦) and the FI-FC method (▲). Antioxidant capacity expressed as mg caffeic acid g bean^{-1}.

analysis methods. Also, when the vanillin concentration increased the AC does, so the highest AC was for extracts obtained using focused microwaves because they had an amount of vanillin higher than when the maceration method was used. The AC:vanillin ratio by the ORAC method was between 35:1 and 41:1 and by the FI-FC method was between 17:1 and 21:1, with an excellent correlation coefficient between vanillin concentration and AC: of 0.994 for the ORAC method and of 0.996 for the FI-FC method (**Figure 4**). The AC by the ORAC method increased between 82% and 117% in the extracts obtained by focused microwaves in comparison with the obtained by the maceration method; by the FI-FC method the increased was between 72% and 97%. The results show that the AC value is higher when microwaves are applied independently of the analysis method used.

F test was calculated to compare the standard deviation from the two methods of extraction using the two analysis methods, ORAC and FI-FC. The F test showed that there was no difference in the variances of the methods because experimental value (1.56) was lower than critical value (4.026). So the precision was similar by both methods. So t test was calculated based on equal variance. The null hypothesis was: the AC is the same by both methods, FMAE and ME. The results showed that the values of the media were statistically different, so the AC is higher when FMAE was used to obtain vanilla extracts, t value was higher than the critical value ($t_9 =$ 2.26).

The vanillin concentration in green beans was very low and therefore the AC, so it is shown the importance of the vanillin in the in antioxidant capacity in cured vanilla beans. In green beans, vanillin is fixed to glucose as glucovanillin and its antioxidant power is not strong enough so green vanilla beans have a very poor AC.

On the other hand, the AC value is different between an analysis method and another one for a same compound, so when caffeic acid was used as standard, which is a strong antioxidant, the vanilla AC values by the ORAC method were approximately two times higher than by the FI-FC method (**Table 2**) but by both methods of analysis the AC increased or decreased in the same proportion for the same extracts. This difference is due to that vanillin is majority compound of vanilla and the analytical response is not the same for the two methods. The linear correlation between the ORAC and FI-FC methods was 0.99. Different electrochemical methods for the determination of antioxidant power in food have been compared with DPPH and ABTS radical cation assays by Buratti et al. [42,44]. The antioxidant power values are different but the linear correlation among them have been higher than 0.92.

So, a calibration curve using vanillin as standard to react with the Folin-Ciocalteau was evaluated. The calibration curve had a slope of 0.0081 mg·mL^{-1} (1.36 times lower than the slope for caffeic acid). For at same concentration (expressed as mmol·L^{-1}) the analytical signal of vanillin was lower than for caffeic acid.

The AC of vanillin was also measure using the ORAC method and was compared with others compounds of the vanilla beans.

3.4. Antioxidant Capacity of Vanillin, p-Hydroxybenzaldehyde (PHB), Vanillic Acid and p-Hydroxibenzoic Acid (HBA) by ORAC Method

Besides vanillin the vanilla beans have others compounds which are extracted at the same time of vanillin. The most important are p-HBA, vanillic acid and HBA. Vanillin is between 10 and 30 times more concentrated than PHB and vanillic acid and the ratio increases for HBA between 100 and 140 times.

Standards of PHB, HBA and vanillic acid to concentrations closed at the vanillin concentration in the diluted extracts were tested. The vanillin concentration in the extracts obtained by focused microwaves from zacatillo beans was of 467 mg·L^{-1} so it was diluted 1000 times and AC was determined, the vanillin concentration was 0.50 mg·L^{-1} approximately. The others compounds were analyzed at this same concentration. At 0.5 mg·L^{-1} of each compound the AC could be only measured for vanillin and for vanillic acid, obtaining 2.3 mg caffeic acid L^{-1} and 0.53 mg caffeic acid L^{-1}, respectability. The AC of vanillin is 4.3 times higher than vanillic acid. PHB and HBA did not show antioxidant capacity. This behavior is because the antioxidant activity of phenolics acids and their esters depends on the number of hydroxyl groups in the molecule that would be strengthened by steric hindrance [45]. The electron withdrawing properties of the carboxylate group in benzoic acids has a negative influence on the H-donating abilities of the hydroxyl benzoates. The monohydroxy benzoic acids show no antioxidant activity in the *ortho* and *para* positions in terms of hydrogen-donating capacity against radical generated in the aqueous phase but the *m*-hydroxy acid has antioxidant activity. This is consistent with the electron withdrawing potential of the single carboxyl functional group on the phenol ring affecting the *o*- and *p*-positions. The monohydroxy benzoates are, however, effective hydroxyl radical scavengers, due to their propensity to hydroxylation and the high reactivity of the hydroxyl radical. With a methylene group between the phenolic ring and the carboxylate group, as in the vanillin, the *o*- and *m*-hydroxy derivatives have high antioxidant activities. Carried out a study of AC of vanillin using different assays were carried out by Akiro Tai *et al.* [46]. They showed that vanillin had a stronger antioxidant capacity than ascorbic acid and Trolox by the ORAC method but it showed no activity in the DPPH radical and galvinoxyl radical scavenging assays. By the FI-FC method a 5 mg·L^{-1} concentration of each analyte was used. At this concentration only vanillin and vanillic acid showed AC where the vanillin concentration was 2.43 times higher than the vanillic acid concentration (4.94 mg caffeic acid L^{-1} vs 2.03 mg caffeic acid L^{-1}). The AC for 500 mg·L^{-1} PHB and HBA solutions was 4.53 mg caffeic acid L^{-1} and 35.7 mg caffeic acid L^{-1}, respectively. So, the analysis of AC using different method is suggested to assure the antioxidant properties from a compound.

4. Conclusions

The AC value was different among analysis methods so it is convenient to use more than one method to establish the antioxidant power of the food. In this paper is shown that the vanilla has a high AC independently of used method, but the obtained numeric values were different. So, when caffeic acid is used as standard, which is a strong antioxidant, the vanilla AC values by the ORAC method were approximately two times higher than by the FI-FC method but by both methods of analysis the AC increased or decreased in the same proportion for the same extracts. The linear correlation between the ORAC and FI-FC methods was 0.99. Also the antioxidant power values reported by Buratti *et al.* [42,44] were different but the linear correlation among them was higher than 0.92.

When microwaves were applied the AC in the extracts increased between 72% and 117% with a strong relationship with the vanillin concentration. The vanillin concentration in the obtained extracts by using microwaves increased between 75% and 93%. So, it has been showed that when microwaves are applied to obtain natural vanilla extracts the AC increased because the vanillin concentration is higher and it is the most influential compound in the antioxidant power.

Vanilla, specie, is mainly used as flavoring and it is used in a very low amount in our daily diet but it could help to get the daily recommended antioxidants. So, it is important to develop extraction methods of improving of the antioxidant power in the extracts, such as the focused microwaves assisted extraction method used. On the other hand, several studies have demonstrated that vanillin helps to the stability in the beans avoiding oxidation, but it is necessary good conservation conditions.

The results showed that the vanilla AC is given, principally, for the vanillin amount in the beans and in the extracts. They also showed that green vanilla beans have a very poor AC.

5. Acknowledgements

The Faculty of Chemistry and the Programa de Apoyo a Proyectos de Investigación e Innovación Tecnológica, PAPIIT (grant No. IT202812-3) of the Universidad NacionalAutónoma de México are gratefully acknowledged for financial support. The Ayotoxco vanilla beans were supplied by the Ayotoxco farmers with the support of the InstitutoPoblano de Desarrollo Rural A.C.

REFERENCES

[1] E. Hágsater, M. A. Soto-Arenas, G. A. Salazar-Chávez, R. Jiménez Machorro, M. A. López-Rosas and R. L. Dressler, "Las Orquídeas de México," Instituto Chinoín A.C., Redacta S.A. de C.V., Mexico City, 2005.

[2] S. R. Rao and G. A. Ravishankar, "Vanilla Flavour: Production by Conventional and Biotechnological Routes," *Journal of the Science of Food and Agriculture*, Vol. 80, No. 3,

2000, pp. 289-304.

[3] A. K. Sinha, U. K. Sharma and N. Sharma, "A Comprehensive Review on Vanilla Flavor: Extraction, Isolation and Quantification of Vanillin and Others Constituents," *International Journal Food Science and Nutrition*, Vol. 59, No. 4, 2008, pp. 299-326.

[4] M. J. Dignum, J. Kerler and R. Verpoorte, "Vanilla Production: Technological, Chemical and Biosynthetic Aspects," *Food Reviews International*, Vol. 17, No. 2, 2001, pp. 199-219.

[5] A. K. Sinha, S. C. Verma and N. Sharma, "Development and Validation of an RP-HPLC Method for Quantitative Determination of Vanillin and Related Phenolic Compounds in Vanilla Planifolia," *Journal of Separation Science*, Vol. 30, No. 1, 2007, pp. 15-20.

[6] E. Odoux, J. Escoute and J. Verdell, "The Relation between Glucovanillin, β-D-Lucosidase Activity and Cellular Compartmentation during the Senescence, Freezing and Traditional Curing of Vanilla Beans," *Annals Applied Biology*, Vol. 149, No. 1, 2006, pp. 43-52.

[7] C. Valdez-Flores and M. P. Cañizares-Macías, "On-Line Dilution and Detection of Vanillin in Vanilla Extracts Obtained by Ultrasound," *Food Chemistry*, Vol. 105, No. 3, 2007, pp. 1201-1208.

[8] S. P. Wong, L. P. Leong and J. Koh, "Antioxidant Activities of Aqueous Extracts of Selected Plants," *Food Chemistry*, Vol. 99, No. 4, 2006, pp. 775-783.

[9] N. V. Yanishlieva, E. Marinova and J. Pokorny, "Natural Antioxidant from Herbs and Spices," *European Journal of Lipid Science and Technology*, Vol. 108, No. 9, 2006, pp. 776-793.

[10] Pastene, M. Gómez, H. Speisky and L. Nuñez-Vergara, "Un Sistema Para la Detección de Antioxidantes Volátiles Comúnmente Emitidos Desde Especias y Hierbas Medicinales," *Quimi Nova*, Vol. 32, No. 2, 2009, pp. 482-487.

[11] J. C. Espín, M. T. García-Conesa and F. A. Tomás-Barberán, "Nutraceuticals: Facts and Fiction," *Phytochemical*, Vol. 68, No. 22, 2007, pp. 2986-3008.

[12] M. E. Drago-Serrano, M. López-López and T. R. Sainz-Espuñes, "Componentes Bioactivos de Alimentos Funcionales de Origen Vegetal," *Revista Mexicana de Ciencias Farmacéuticas*, Vol. 37, No. 1, 2006, pp. 58-68.

[13] B. Halliwell and J. M. Gutteridge, "Free Radicals in Biology and Medicine," 3rd Edition, Oxford University Press, Oxford, 1999.

[14] E. N. Frankel and A. S. Meyer, "The Problems of Using One-Dimensional Methods to Evaluate Multifunctional Food and Biological Antioxidants," *Journal of Science Food Agriculture*, Vol. 80, No. 13, 2000, pp. 1925-1941.

[15] R. H. Bisby, R. Brooke and S. Navaratnam, "Effect of Antioxidant Oxidation Potential in the Oxygen Radical Absorption Capacity (ORAC) Assay," *Food Chemistry*, Vol. 108, No. 3, 2008, pp. 1002-1007.

[16] B. Ou, M. Hampsch-Woodill and R. L. Prior, "Development and Validation of an Improved Oxygen Radical Absorbance Capacity Using Fluorescein as the Fluorescent Probe," *Journal of Agriculture and Food Chemistry*, Vol. 49, No. 10, 2001, pp. 4619-4626.

[17] A. Zulueta, M. J. Esteve and A. Frígola, "ORAC and TEAC Assays Comparison to Measure the Antioxidant Capacity of Food Products," *Food Chemistry*, Vol. 114, No. 1, 2009, pp. 310-316.

[18] M. Celeste, C. Tomás, A. Cladera, J. M. Estela and V. Cerdà, "Enhanced Automatic Flow Injection Determination of the Total Polyphenol Number of Wines Using the Folin-Ciocalteau Reagent," *Analytica Chimica Acta*, Vol. 269, No. 1, 1992, pp. 21-28.

[19] V. L. Singleton, R. Orthofer and R. M. Lamuela-Raventós, "Analysis of Total Phenols and Other Oxidation Substrates and Antioxidants by Means of Folin-Ciocalteu Reagent," *Methods in Enzymology*, Vol. 299, No. 14, 1999, pp. 152-178.

[20] D. Sterbova, D. Matejícek, J. Vlcek and V. Kubán, "Combined Microwave-Assisted Isolation and Solid-Phase Purification Procedures Prior to Chromatographic Determination of Phenolic Compounds in Plant Materials," *Analytica Chimica Acta*, Vol. 513, No. 2, 2004, pp. 435-444.

[21] Y. Zuo, L. Zhang, J. Wu, J. W. Fritz, S. Medeiros and C. Rego, "Ultrasonic Extraction and Capillary Gas Chromatography Determination of Nicotine in Pharmaceutical Formulations," *Analytica Chimica Acta*, Vol. 526, No. 1, 2004, pp. 35-39.

[22] J. L. Luque-García and M. D. Luque de Castro, "Where Is Microwave-Based Analytical Equipment for Solid Simple Pre-Treatment Going?" *Trends Analytical Chemistry*, Vol. 22, No. 2, 2003, pp. 90-98.

[23] J. L. Luque-García, J. Velasco, M. C. Dobarganes and M. D. Luque de Castro, "Fast Quality Monitoring of Oil from Prefried and Fried Foods by Focused Microwave-Assisted Soxhlet Extraction," *Food Chemistry*, Vol. 76, No. 2, 2002, pp. 241-248.

[24] E. E. Stashenko, B. E. Jaramillo and J. R. Martínez, "Comparison of Different Extraction Methods for the Analysis of Volatile Secondary Metabolites of Lippia Alba (Mill.) N. E. Brown, Grown in Colombia and Evaluation of Its *in Vitro* Antioxidant Activity," *Journal of Chromatography A*, Vol. 1025, No. 1, 2004, pp. 93-103.

[25] K. J. Lamble and S. J. Hill, "Microwave Digestion Procedures for Environmental Matrices," *Analyst*, Vol. 123, No. 7, 1998, pp. 103R-133R.

[26] S. Vadahanambi, S. J. H. Jung and I. K. Oh, "Microwave Syntheses of Graphene and Graphene Decorated with Me-

tal Nanoparticles," *Carbon*, Vol. 49, No. 13, 2011, pp. 4449-4457.

[27] S. Mishra, G. Sen, G. U. Rani and S. Sinha, "Microwave Assisted Synthesis of Polyacrylamide Grafted Agar (Ag-g-PAM) and Its Application as Flocculant for Wastewater Treatment," *International Journal of Biological Macromolecules*, Vol. 49, No. 4, 2011, pp. 591-598.

[28] S. V. Jadhav, E. Suresh and H. C. Bajaj, "Microwave-Assisted Solvent-Free Synthesis of α,α'-Bis(substituted benzylidine)cycloalkanones Catalyzed by SO_4 2-/ZrO_2 and B_2O_3/ZrO_2," *Green Chemistry Letters Review*, Vol. 4, No. 3, 2011, pp. 249-256.

[29] A. Longares-Patrón and M. P. Cañizares-Macías, "Focused Microwaves-Assisted Extraction and Simultaneous Spectrophotometric Determination of Vanillin and p-Hydroxybenzaldehyde from Vanilla Fragans," *Talanta*, Vol. 69, No. 4, 2006, pp. 882-887.

[30] Mexican Norm: NMX-FF-074-SCFI-2009, "Productos No Industrializados Para Uso Humano-Vainilla (*Vanilla fragrans* (Salisbury) Ames Especificaciones y Métodos de Prueba. Non Industrialized Food Products for Human Consumption-Vanilla (*Vanilla fragrans* (Salisbury) Ames). Specifications and Test Methods," Secretaría de Comercio y Finanzas, Mexico City, 2009.

[31] M. P. Cañizares, M. T. Tena and M. D. Luque de Castro, "On Line Coupling of a Liquid-Liquid Extraction Flow-Reversal System to a Spectrophotometric Flow-Through Sensor for the Determination of Polyphenols in Olive Oil," *Analytica Chimica Acta*, Vol. 323, No. 1-3, 1996, pp. 55-62.

[32] T. Ercetin, F. S. Senol, I. Erdogan, I. Orhan and G. Toker, "Comparative Assessment of Antioxidant and Cholinesterase Inhibitory Properties of the Marigold Extracts from *Calendula arvensis* L. and *Calendula officinalis* L.," *Industrial Crops Products*, Vol. 36, No. 1, 2012, pp. 203-208.

[33] American Society for Testing and Material, "Annual Book of ASTM," ASTM D 1348-94, Vol. 6.03, 2003, pp. 287-292.

[34] L. Yu, J. Perret, B. Davy, J. Wilson and C. L. Melby, "Free Radical Scavenging Properties of Wheat Extracts," *Journal of Food Science*, Vol. 67, No. 22, 2002, pp. 2600-2603.

[35] E. A. Abdelilah, K. Hajar and H. Abdellatif, "Phenolic Profile and Antioxidant Activities of Olive Mill Wastewater," *Food Chemistry*, Vol. 132, No. 1, 2012, pp. 406-412.

[36] M. F. Barroso, C. Delerue-Matos and M. B. P. P. Oliveira, "Electrochemical Evaluation of Total Antioxidant Capacity of Beverages Using a Purine-Biosensor," *Food Chemistry*, Vol. 132, No. 2, 2012, pp. 1055-1062.

[37] S. Labbozzetta, L. Valvo, P. Bertochi, S. Alimunti, M. C. Gaudiano and L. Manna, "Focused Microwave-Assisted Extraction and LC Determination of Ketoprofen in the Presence of Preservatives in a Pharmaceutical Cream Formulation," *Chromatographia*, Vol. 69, No. 3, 2009, pp. 365-368.

[38] A. Martín-Calero, V. Pino, J. H. Ayala, V. González and A. M. Alfonso, "Ionic Liquids as Mobile Phase Additives in High-Performance Liquid Chromatography with Electrochemical Detection: Application to the Determination of Heterocyclic Aromatic Amines in Meat-Based Infant Foods," *Talanta*, Vol. 79, No. 3, 2009, pp. 590-597.

[39] M. Varga, J. Dobor, A. Helenkár, L. Jurecska, J. Yao and G. Záray, "Investigation of Acidic Pharmaceuticals in River Water and Sediment by Microwave-Assisted Extraction and Gas Chromatography—Mass Spectrometry," *Microchemica Journal*, Vol. 95, No. 2, 2010, pp. 353-358.

[40] M. P. Cañizares-Macías, J. A. García-Mesa and M. D. Luque de Castro, "Determination of Oxidative Stability on Olive Oil Using Focused-Microwave Energy to Accelerate the Oxidation Process," *Analytical and Bioanalytical Chemistry*, Vol. 378, No. 2, 2004, pp. 479-483.

[41] D. P. Singh, J. Beloy, J. K. McInerney and L. Day, "Impact of Boron, Calcium and Genetic Factors on Vitamin C, Carotenoids, Phenolic Acids, Anthocyanins and Antioxidant Capacity of Carrots (*Daucus carota*)," *Food Chemistry*, Vol. 132, No. 3, 2012, pp. 1161-1170.

[42] S. Buratii, N. Pelligrini, O. Brenna and S. Mannino, "Rapid Electrochemical Method for the Evaluation of the Antioxidant Power of Some Lipophilic Food Extracts," *Journal of Agriculture and Food Chemistry*, Vol. 49, No. 11, 2001, pp. 5136-5141.

[43] D. Lettieri-Barbato, D. Villaño, B. Beheydt, F. Guadagni, I. Trogh and M. Serafini, "Effect of Ingestion of Dark Chocolates with Similar Lipid Composition and Different Cocoa Content on Antioxidant and Lipid Status in Healthy Humans," *Food Chemistry*, Vol. 132, No. 3, 2012, pp. 1305-1310.

[44] S. Buratti, S. Benedetti and M. S. Cosio, "Evaluation of the Antioxidant Power of Honey, Propolis and Royal Jelly by Amperometric Flow Injection Analysis," *Talanta*, Vol. 71, No. 3, 2007, pp. 1387-1392.

[45] C. A. Rice-Evans, N. J. Miller and G. Paganga, "Structure Antioxidant Activity Relationships of Flavonoids and Phenolic Acids," *Free Radical Biology and Medicine*, Vol. 20, No. 7, 1996, pp. 933-956.

[46] A. Tai, T. Sawano, S. F. Yasama and H. Ito, "Evaluation of Antioxidant Activity of Vanillin by Using Multiple Antioxidant Assays," *Biochemica and Biophysica Acta*, Vol. 1810, No. 2, 2011, pp. 170-177.

Abreviations

AC: Antioxidant Capacity

AAPH: 2,2'-azobis(2-methylpropionamidine) dihydrochloride

AUC: Area under Curve

FI-FC: Flow Injection-Folin Ciocalteau method

FMAE: Focused Microwaves-Assisted Extraction

HBA: p-hydroxybenzoic acid

ME: Maceration Extraction

ORAC: Oxygen Radical Absorbance Capacity

PHB: p-hydroxybenzaldehyde

11

The Antioxidant and Free Radical Scavenging Activities of Chlorophylls and Pheophytins

Ching-Yun Hsu[1], Pi-Yu Chao[2], Shene-Pin Hu[3], Chi-Ming Yang[4*]

[1]Department of Nutrition and Health Sciences, Chang Gung University of Science and Technology, Taoyuan, Taiwan; [2]Department of Food and Nutrition, Chinese Culture University, Taipei, Taiwan; [3]School of Nutrition and Health Science, Taipei Medical University, Taipei, Taiwan; [4]Biodiversity Research Center, Academia Sinica, Taipei, Taiwan.

ABSTRACT

Chlorophylls are important antioxidants found in foods. We explored the mechanisms through which the a and b forms of chlorophyll and of pheophytin (the Mg-chelated form of chlorophyll) reduce oxidation: we used comet assay to measure prevention of H_2O_2 DNA damage; we tested for quenching of 1,1-diphenyl-2-picrylhydrazyl (DPPH); we measured the ability to chelate Fe(II); and, we tested their ability to prevent formation of thiobarbituric acid reactive substances (TBARS) during Cu-mediated peroxidation of low density lipoprotein (LDL) in a chemical assay. All chlorophylls and pheophytins showed significant dose-dependent activity in the assays, with the pheophytins being the strongest antioxidants. Thus, these chemicals can prevent oxidative DNA damage and lipid peroxidation both by reducing reactive oxygen species, such as DPPH, and by chelation of metal ions, such as Fe(II), which can form reactive oxygen species.

Keywords: Chlorophyll; Pheophytin; Comet Assay; Antioxidation

1. Introduction

Epidemiological studies have connected diets high in vegetables with reduced cancer risks [1]. Chlorophylls and chlorophyll related compounds are among the best candidates for the chemicals responsible for the general protection afforded by vegetables. Chlorophyll compounds are highly abundant in green vegetables, and they have been shown to prevent carcinogenesis through multiple chemical mechanisms. Biological studies using purified chlorophyll compounds have been able to replicate some of the dietary benefits of eating green vegetables [2].

Chlorophyll is a photosensitive light harvesting pigment with special electronic properties. The general structure of chlorophyll (**Figure 1**) consists of a porphyrin ring chelating a Mg atom. A 20-carbon phytol tail at C17 makes the chlorophyll highly hydrophobic and allows chlorophyll to incorporate into biological lipid membranes. In the naturally occurring chlorophylls the functional group of C7, a -CH_3 or a -CHO group, define the a and b forms of chlorophyll, respectively.

Naturally occurring a and b derivatives of chlorophyll, such as pheophytin, chlorophyllide, and pheophorbide, are present in plants as breakdown products [3], and in animals as the products of chlorophyll digestion [4]. The simplest derivative, pheophytin (**Figure 1**), has had the Mg atom dechelated from its porphyrin ring. A semisynthetically prepared chlorophyll derivative called chlorophyllin is commercially available. In commercial chlorophyllin, the Mg atom is replaced with a Na or Cu atom to form a water soluble, hydrophilic salt complex.

All the chlorophyll compounds act through similar anti-carcinogenic mechanisms, with varying degrees of strength. One proposed mechanism is that the chlorophyll porphyrin ring acts as an interceptor molecule, or desmutagen, by directly binding to other planar cyclic molecules through a shared pi-cloud [5-7]. Carcinogenic molecules with planar structures have been shown to be bound [7] and prevented from acting as carcinogens in biological systems [2,7-10]. However, other carcinogenic molecules are not directly bound by chlorophyll compounds, yet are still less effective as mutagens in the presence of chlorophyll compounds [7]. Possible additional anti-carcinogenic mechanisms include antioxidation [5,11,12], chelation of pro-oxidant ions such as Fe(II)

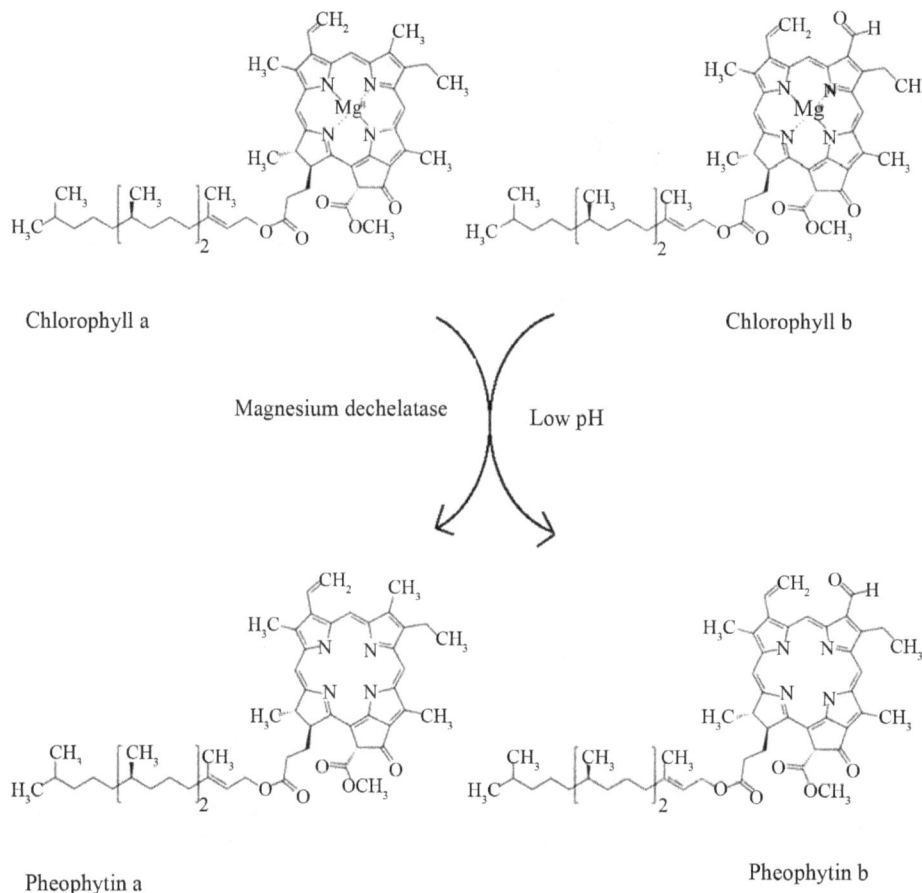

Figure 1. Structures of chlorophylls a and b and pheophytins a and b.

[13,14], or the stimulation of cellular defenses [15].

This paper presents data on the anti-oxidant and chelation properties of chlorophyll a and b and pheophytin a and b in relation to biological systems. Previous studies have shown that chlorophyll and pheophytin act as lipid antioxidants in stored edible oils [5], and can reduce free radicals in standard assays [5,11]. And, general protection from carcinogens by chlorophyllin has been observed in whole animal studies [2]. However, there is a gap between the mechanistically informative chemical studies and the biologically relevant assays regarding the nutritional antioxidant properties of chlorophylls.

Using chlorophyllins, pheophorbides, and chlorophyllides, we recently observed the reduction of free radicals and the protection of cultured human lymphocytes against oxidative DNA damage [12]. Here we study chlorophylls and pheophytins with respect to the protection of lymphocytes against oxidative DNA damage by H_2O_2, and explore whether they can act as direct reducers of free radicals or as chelators of Fe(II). We further test if the natural chlorophylls can prevent lipid peroxidation of low-density lipoprotein (LDL), as does the semisynthetic chlorophyllin [13,14,16].

2. Materials and Methods

2.1. Preparation of Chlorophyll and Derivatives

Chlorophyll derivatives were prepared from spinach purchased in a local market in Taipei, Taiwan, as previously described [17]. Briefly, to extract chlorophylls a and b, spinach was washed with cold water, quickly freeze-dried, powdered in a mortar filled with liquid nitrogen, and stored at $-70°C$ until extraction. Total pigment was extracted from the powdered spinach by grinding in 80% acetone. The crude extract was centrifuged (1500 g, 5 min). From the supernatant, chlorophyll a and b were purified by liquid chromatography using a combination of ion-exchange and size exclusion chromatography with a CM-Sepharose CL-6B column. Analyses of chromatography fractions were performed by measuring the absorbance at 663.6 and 646.6 nm, which are the major absorption peaks of chlorophyll a and b. Chlorophyll a and b were further Mg-chelated to form pheophytin a and b by acidification with acetic acid.

2.2. LDL Preparation

Blood was drawn from the veins of healthy subjects and

collected into sterile glass tubes containing 1.5 mg/mL EDTA-K3. Plasma was isolated immediately for preparation of LDL. A two-step LDL fraction (1.006 > d > 1.063 g/mL) was isolated by two-step sequential flotation ultracentrifugation [18], using a Hitachi CP85β ultracentrifuge (4°C, P70AT2-376 rotor, 44,000 rpm) for 16 h (d < 1.006 g/mL) to remove VLDL and for 20 h (d < 1.063 g/mL) to collect LDL. The isolated LDL fraction from each study subject was dialyzed separately at 4°C against 0.15 M NaCl 50 mM in phosphate buffer, pH 7.4, 22 h before use in the oxidative susceptibility of LDL assay.

Comet Assay. Blood samples (10 mL) were obtained from healthy donors, and lymphocytes were isolated using a separation solution kit (Ficoll-Paque Plus lymphocyte isolation sterile solution; Pharmacia Biotech, Sweden). For the experimental procedure, cells were harvested within one day of blood samples having been taken, and cultured in AIM V® medium (serum-free lymphocyte medium; Invitrogen, Carlsbad, CA) in 5% CO_2/ 95% air, humidified, at 37°C for 24 h.

Subsequent to culture, lymphocytes were exposed to one of the four different chlorophyll compounds at various concentrations for 30 min at 37°C. All test substances were dissolved in dimethyl-sulfoxide (DMSO); the solvent concentration in the incubation medium never exceeded 1%. Control incubations contained the same concentration of DMSO. Then, DNA damage was induced by adding 10 μM H_2O_2 for 5 min on ice. Treatment on ice minimizes the possibility of cellular DNA repair subsequent to H_2O_2 injury. Cells were centrifuged (100 g, 10 min), washed, and resuspended in the same medium for the comet assay.

The comet assay [19] measures DNA single-strand break (ssbs) damage by fixing cells in soft agar on a microscope slide, subjecting them to electrophoresis, staining the DNA, and observing under a microscope. A tail of stained DNA is formed as it migrates out of the nucleus; the length of the tail of DNA is relative to the amount of DNA ssbs damage. We calculated tail moment, which increases as DNA damage increases. The assay was conducted as previously described [12]. Control tests with each of the components (H_2O_2, chlorophyll compounds, DMSO) of the test alone were performed, and the tail moment by comet assay and viability by MTS assay were measured.

2.3. Measurement of DPPH Radical Scavenging

The ability of chlorophylls or pheophytins to scavenge free radicals was explored by their ability to reduce 1,1-diphenyl-2-picrylhydrazyl (DPPH) as previously described [12]. The DPPH scavenging capacity of chlorophyll or pheophytin was expressed as the percentage of inhibition by the following formula:

$$\% \text{ inhibition} = \left[\left(A_{control} - A_{sample} \right) / A_{control} \right] \times 100$$

where $A_{control}$ is the absorbance of the sample at 0 min, and A_{sample} is the absorbance of the sample at 30 minutes.

2.4. Chelation of Fe(II) Cation

It has been proposed that most of the hydroxyl radicals in living organisms are due to iron ion-dependent generation through the Fenton reaction (ref). Following Dinis et al. [20], the binding of ferrous ions to chlorophylls or pheophytins was estimated by the decrease in the peak absorbance of the Fe(II)-ferrozine complex. Briefly, 0 to 100 mM of chlorophyll or pheophytin was incubated with 20 μM Fe(II) (ammonium ferrous sulfate) in 5% ammonium acetate, pH 6.9. The reaction was initiated by the addition of 100 μM ferrozine. After the mixture had reached equilibrium (10 min), the absorbance at 562 nm was measured. The chelating effect was calculated as:

% chelating effect

$$= \left[1 - \frac{A_{562} \text{ in the presence of sample}}{A_{562} \text{ nm in the absence of sample}} \right] \times 100$$

2.5. Antioxidant Activity in Human Low Density Lipoprotein (LDL) Oxidation System

LDL oxidation was determined by measuring the amount of thiobarbituric acid reactive substances (TBARS). For assay of TBARS, the dialyzed LDL was diluted in saline to 0.9 mg LDL-C/mL. LDL (50 mL in each assay tube) was incubated with 25 mM $CuSO_4$ in the atmosphere at 37°C for 5 h to induce lipid peroxidation. The TBARS assay involved determining malonaldehyde (MDA) formation after peroxidation of LDL at intervals of 0, 30, 60, 70, 80, 90, 100, 120, 150, 180, 240 and 300 min. The incubation was terminated by adding 100 mL TCA (15.2%, w/v), and the mixture was cooled to 4°C. A precipitate formed at the bottom of the tube, and a 100 mL aliquot of supernatant was taken for MDA assay [21]. The supernatant was added to 1 mL TBA solution (0.6%, w/v) and boiled at 100°C for at least 30 min. Finally, 1, 1, 3, 3-tetramethoxy propane was used as the source of MDA in the calibration of the TBARS assay. The assay was performed according to the procedure of Yagi [22] with modification. Pheophytin a and b and chlorophyll a and b were tested in this system. The control had no treatment. In addition, a known antioxidant, trolox, was measured with the assay. The TBA value was measured by its absorption at 532 nm wavelength. The percent inhibition of oxidation was calculated as:

$$\% \text{ inhibition} = \left[1 - \frac{\text{TBA value of treated sample}}{\text{TBA value of control}} \right] \times 100$$

2.6. Statistical Analyses

Data are reported as the mean ± SD of triplicate determinations. Statistical analyses were performed using a Student's t-test to compare differences between control and pretreated groups. One-way ANOVA was used to test for differences amongst the chlorophyll compounds. Post hoc comparison of means was performed by Duncan's multiple comparison, and $p < 0.05$ was considered to represent a statistically significant difference between test populations.

3. Results and Discussion

In order to see if chlorophyll compounds act as antioxidants in a biological setting, we employed a comet assay to observe oxidative DNA damage caused by H_2O_2. As a preliminary, we check to see that chlorophyll, pheophytin or DMSO was non-toxic to the cells. The lymphocyte viability and DNA damage (tail moment) of cells treated with H_2O_2 alone, with chlorophylls a or b alone, with pheophytins a or b alone, with DMSO alone, or with nothing is shown in **Table 1**. We then tested to see if chlorophyll and pheophytins could act as antioxidants versus H_2O_2 in this system.

As expected, both chlorophyll compounds reduced the comet tail moment caused by H_2O_2 in the experiment, showing that they act as dose-dependent antioxidants

(**Figure 2**). We had previously used the same DPPH measurement to detect radical scavenging (Hsu 2005) and had shown the relationship of antioxidant strengths among chlorophyll compounds (50 μM chlorophyll compounds versus 10 μM H_2O_2) were following the sequences as pheophorbide > chlorophyllin > chlorophyllide = pheophytin > chlorophyll (Hsu 2005). For each given natural chlorophyll compound there was no significant difference in antioxidant effect between the a and b forms.

Table 1. Effect of chlorophyll-related compounds upon human lymphocyte viability and DNA damage[1].

	Viability (%)[2]	Comet assay (TM)[3]
Chlorophyll a (50 μM)	96.4 ± 3.2	103 ± 22
Chlorophyll b (50 μM)	97.2 ± 2.3	102 ± 25
Pheophytin a (50 μM)	98.3 ± 2.4	104 ± 36
Pheophytin b (50 μM)	98.4 ± 2.1	102 ± 26
DMSO (as solvent)	96.8 ± 2.5	109 ± 25
H_2O_2 10 μM	99.4 ± 2.4	6386 ± 803
Control	100	103 ± 21

[1]Mean ± SD; [2]Viability (measured by the MTS assay) was determined both prior to (100%) and following chlorophyll-related compounds pretreatment. [3]Mean Tail Moment (TM) was calculated by means of the comet assay.

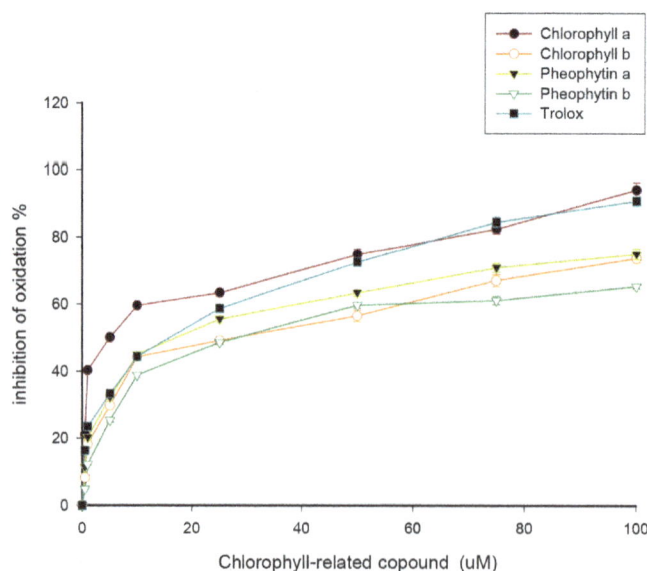

Oxidative inhibition of chlorophyll-related compounds extracts from spinach on TBARS assay

Figure 2. Prevention of oxidative DNA damage by chlorophylls and pheophytins. The results of comet assays following oxidation by HO₂ are plotted as tail moments versus the concentrations of chlorophylls or pheophytins tested as antioxidants. The dose-dependent reduction in tail moment shows that fewer single stranded DNA breaks were caused by hyrdoxyl radicals when chlorophyll compounds were present. At all concentrations of chlorophylls and pheophytins tested, a significantly reduced level of DNA single stranded breaks was formed following H_2O_2 exposure ($p < 0.05$). It may also be seen from this figure that at 5 and 20 μM concentrations all compounds tested had equal antioxidant capacities.

In the Fenton reaction, Fe(II) catalyzes the decomposition of H_2O_2 to generate hydroxyl radicals, which can then damage DNA. Previously, we showed that the DNA was being hydroxylated during the comet assay, and that chlorophyll compounds prevent both the hydroxylation of DNA and the formation of single stand breaks in the DNA [12]. Two mechanisms could hypothetically account for the antioxidant effects of the chlorophyll compounds. The chlorophyll compounds could directly scavenge the hydroxyl free radicals generated from H_2O_2, or the chlorophyll compounds could chelate the Fe(II), preventing the Fenton reaction.

In order to test for free radical scavenging, chlorophyll and pheophytin were tested in a DPPH assay. Chlorophyll and pheophytin were able to reduce DPPH in a dose-dependent manner, showing that they can act as free radical scavengers (**Figure 3**); there was little difference between the activities of the a and b forms. The scavenging capacities of pheophytin did appear to be more pronounced than that of chlorophyll ($p < 0.05$). The 50% inhibitory concentrations (IC$_{50}$) were about 161 μM (pheophytin a), 198 μM (pheophytin b), and greater than 200 μM (chlorophylls a and b). Combined with our previous data [12], the relationship of free radical scavenging strengths (IC$_{50}$ values) among chlorophyll compounds is chlorophyllide > chlorophyllin > chlorophyll > pheophytin > pheophorbide. In general, this agrees with previous chlorophyll compound DPPH assay data [5,11], although their conditions and concentrations used were different.

We then tested the ability of different concentrations of chlorophylls or pheophytins to chelate Fe(II) from solution. As the concentration of the chlorophylls or pheophytins was increased, the ability of the Fe(II) to complex with ferrozine was decreased to more than half its original amount (**Figure 4**). We interpret this as evidence that the chlorophyll compounds chelated the Fe(II) atoms. In the case of the chlorophylls, this would require substitution of the Fe(II) for the Mg atom. This may explain why pheophytin was more active than chlorophyll in the assay. There was little difference in activity between the a and b forms.

We are not aware of previous reports exploring chlorophylls or pheophytins as antioxidant Fe(II) chelators. However, Arimoto-Kobayashi *et al.* [23] created an "Fe-chlorophyllin", a chlorophyll with Fe replacing the Mg, structurally the same as our putative Fe-chelated chlorophyll or Fe-chelated pheophytin. Arimoto-Kobayashi *et al.* showed that their Fe-chlorophyllin could bind to a planar carcinogenic molecule and enhance its degradation rate. They speculate that the Fe atom makes the Fe-chlorophyllin into an oxidative molecule that can both desmutagenically bind to planar carcinogen molecules and detoxify them by oxidization. This suggests that if dietary chlorophyll or pheophytin is spontaneously converted into Fe-chlorophyllin, not only would Fenton reactions be avoided, but also the resulting Fe-chlorophyllin would be a superior anti-carcinogen.

Recently, Nelson and Ferruzzi [24] synthesized "FePhe" out of pheophytin and Fe(II), which is essentially the same idea as Fe-chlorophyllin. Their purpose was to create an alternate bioavailable source of iron, although no dietary tests have been reported yet. It has been suggested [9] that in meals including red meat and

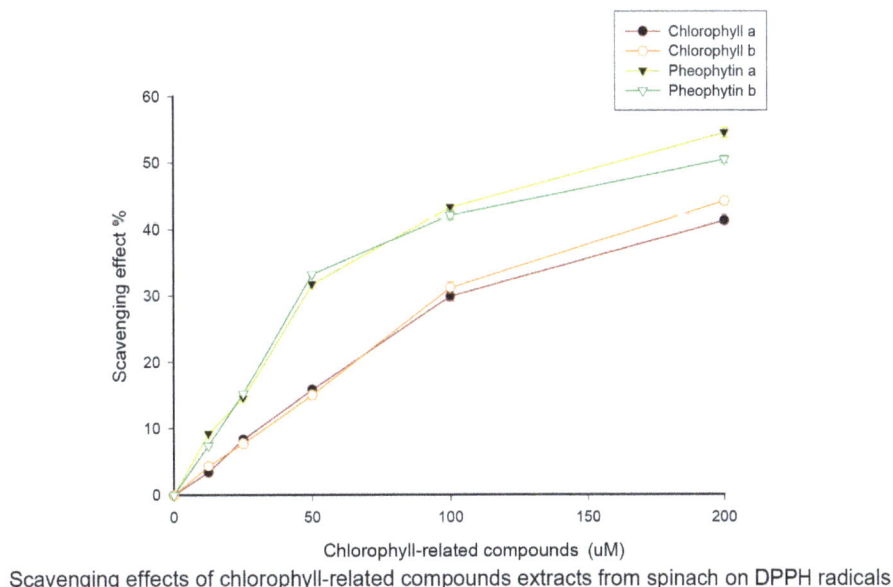

Scavenging effects of chlorophyll-related compounds extracts from spinach on DPPH radicals

Figure 3. Chlorophylls and pheophytins scavenge the DPPH radical. The relative scavenging effect is plotted for different concentrations of test substances. All chlorophylls and pheophytins were able to scavenge the DPPH radical in a dose-dependent manner.

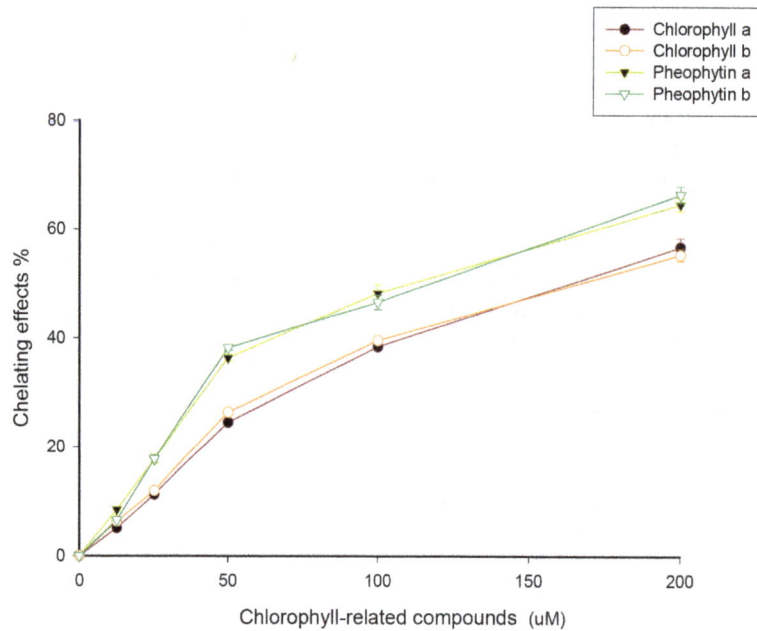

Chelating effects of chlorophyll-related compounds extracts from spinach on Fe^{2+} ion

Figure 4. Inhibition of the Fe(II)-ferrozine reaction as a measure of Fe-chelation. The Fe(II) was pre-incubated with chlorophyll compounds before reacting with ferrozine. Data is presented as the proportional reduction in Fe(II)-ferrozine versus a control reaction (% chelating effect). Pheophytins had significantly higher % chelating effects than the chlorophylls ($p < 0.05$). The approximate I_{50} values were: pheophytin a, 119 µM; pheophytin b, 123 µM; chlorophyll a, 171 µM; chlorophyll b 175 µM.

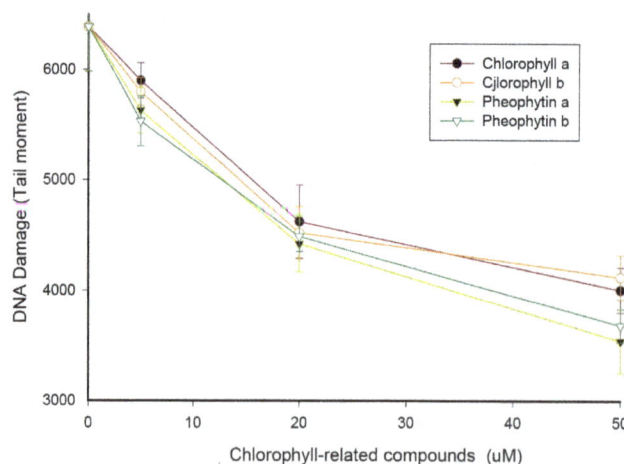

Effect of Chlorophyll-related compounds pretreatment on 10 uM hydrogen peroxide-induced DNA damage in isolated human lymphocytes.

Figure 5. Chlorophylls and pheophytins inhibit lipid peroxidation. Peroxidation of low-density lipoprotein was catalyzed by Cu. Addition of various concentrations of chlorophyll a, chlorophyll b, pheophytin a, or pheophytin b showed a dose-dependent antioxidant effect. The antioxidant trolox is included as a positive control.

green vegetables, the chlorophyll from the vegetables neutralizes the iron-containing heme from the meat. It is possible that chlorophyll compounds not only would absorb reactive Fe species from heme, but also would later transmit the Fe atoms as nutrients in a more benign fashion, in the same way as is intended for the engineered FePhe of Nelson and Ferruzzi. Clearly, further research

into the interactions of Fe and chlorophyll compounds is warranted.

Following the evidence that chlorophyll and pheophytin act both as free radical scavengers and as Fe(II) chelators, we hypothesized that they might also be able to prevent the catalysis of lipid peroxidization. We performed a test of the ability of chlorophyll or pheophytin to prevent lipid peroxidation in an assay using a biological lipid (LDL) and Cu, another atom which catalyzes Fenton reactions. The antioxidative effects of the chlorophyll compounds were determined by measuring the creation of thiobarbituric acid reactive substances (TBARS). **Figure 5** shows that both chlorophyll and pheophytin inhibited copper-mediated LDL oxidative activity in a dose-dependent manner. Trolox, a synthetic water-soluble derivative of the antioxidant Vitamin E, was used as a positive control. The 50% inhibition (I_{50}) value, where a lower value of I_{50} indicates a higher antioxidant activity, for each compound was: trolox, 16 µM; chlorophyll a, 5 µM; chlorophyll b, 26 µM; pheophytin a, 17 µM; pheophytin b, 26 µM.

That these lipid peroxidation results are biologically relevant is supported by the research of de Vogel's *et al.* [9]. When rats were fed a diet containing heme, increased amounts of TBARS were found in their fecal water; however, when chlorophyll was added to this diet, the TBARS excretion returned to normal [9]. Based on our research, the mechanisms responsible for antioxidant effects on lipid peroxidation may involve both chelating of Fe or Cu atoms, and the scavenging of hydroxyl radicals, as well as possibly scavenging peroxy radicals, as suggested by Endo *et al.* [5].

The results of our research confirm that chlorophyll compounds are important health promoting dietary factors which can protect the body through multiple chemical mechanisms. In particular, we have shown that chlorophylls and pheophytins act as antioxidants to prevent oxidative DNA damage and lipid peroxidation both by chelating reactive ions and by scavenging free radicals.

REFERENCES

[1]   G. Block, B. Patterson and A. Subar, "Fruit, Vegetables, and Cancer Prevention: A Review of the Epidemiological Evidence," *Nutrition and Cancer*, Vol. 18, No. 1, 1992, pp. 1-29.

[2]   U. Harttig and G. S. Bailey, "Chemoprotection by Natural Chlorophylls *in Vivo*: Inhibition of Dibenzo[a,l]pyrene-DNA Adducts in Rainbow Trout Liver," *Carcinogenesis*, Vol. 19, No. 7, 1998, pp. 1323-1326.

[3]   K. I. Takamiya, T. Tsuchiya and H. Ohta, "Degradation Pathway(s) of Chlorophyll: What Has Gene Cloning Revealed?" *Trends Plant Science*, Vol. 5, No. 10, 2000, pp. 426-431.

[4]   M. G. Ferruzzi, M. L. Failla and S. J. Schwartz, "Assessment of Degradation and Intestinal Cell Uptake of Carotenoids and Chlorophyll Derivatives from Spinach Puree Using an *in Vitro* Digestion and Caco-2 Human Cell Model," *Journal of Agricultural and Food Chemistry*, Vol. 49, No. 4, 2001, pp. 2082-2089.

[5]   Y. Endo, R. Usuki and T. Kaneda, "Antioxidant Effects of Chlorophyll and Pheophytin on the Autoxidation of Oil

in the Dark II. The Mechanism of Antioxidative Action of Chlorophyll," *Journal of the American Oil Chemists' Society*, Vol. 62, No. 9, 1985, pp. 1387-1390.

[6]   N. Tachino, D. Guo, W. M. Dashwood, S. Yamane, R. Larsen and R. Dashwood, "Mechanisms of the *in Vitro* Antimutagenic Action of Chlorophyllin against Benzo[a]-pyrene: Studies of Enzyme Inhibition, Molecular Complex Formation and Degradation of the Ultimate Carcinogen," *Mutation Research*, Vol. 308, No. 2, 1994, pp. 191-203.

[7]   S. Chernomorsky, A. Segelman and R. D. Poretz, "Effect of Dietary Chlorophyll Derivatives on Mutagenesis and Tumor Cell Growth," *Teratogenesis, Carcinogenesis, and Mutagenesis*, Vol. 19, No. 5, 1999, pp. 313-322.

[8]   P. A. Egner, J. B. Wang, Y. R. Zhu, B. C. Zhang, Y. Wu, Q. N. Zhang, G. S. Qian, S. Y. Kuang, S. J. Gange, L. P. Jacobson, K. J. Helzlsouer, G. S. Bailey, J. D. Groopman and T. W. Kensler, "Chlorophyllin Intervention Reduces Aflatoxin-DNA Adducts in Individuals at High Risk for Liver Cancer," *Proceedings of the National Academy of Sciences*, Vol. 98, No. 25, 2001, pp. 14601-14606.

[9]   J. de Vogel, D. S. Jonker-Termont, M. B. Katan and R. van der Meer, "Natural Chlorophyll But Not Chlorophyllin Prevents Heme-Induced Cytotoxic and Hyperproliferative Effects in Rat Colon," *Journal of Nutrition*, Vol. 135, No. 8, 2005, pp. 1995-2000.

[10]   M. T. Simonich, P. A. Egner, B. D. Roebuck, G. A. Orner, C. Jubert, C. Pereira, J. D. Groopman, T. W. Kensler, R. H. Dashwood, D. E. Williams and G. S. Bailey, "Natural Chlorophyll Inhibits Aflatoxin B1-Induced Multi-Organ Carcinogenesis in the Rat," *Carcinogenesis*, Vol. 28, No. 6, 2007, pp. 1294-1302.

[11]   U. Lanfer-Marquez, R. Barros and P. Sinnecker, "Antioxidant Activity of Chlorophylls and Their Derivatives," *Food Research International*, Vol. 38, No. 8, 2005, pp. 885-891.

[12]   C. Y. Hsu, C. M. Yang, C. M Chen., P. Y. Chao and S. P. Hu, "Effects of Chlorophyll-Related Compounds on Hydrogen Peroxide Induced DNA Damage within Human Lymphocytes," *Journal of Agricultural and Food Chemistry*, Vol. 53, No. 7, 2005, pp. 2746-2750.

[13]   M. Sato, K. Imai and T. Murata, "Effect of Sodium Copper Chlorophyllin on Lipid Peroxidation: The Antioxidative Activities of the Commercial Preparations of Sodium Copper Chlorophyllin," *Yakugaku Zasshi*, Vol. 100, No. 5, 1980, pp. 580-584.

[14]   J. P. Kamat, K. K. Boloor and T. P. Devasagayam, "Chlorophyllin as an Effective Antioxidant against Membrane Damage *in Vitro* and *ex Vivo*," *Biochim Biophys Acta*, Vol. 1487, No. 2-3, 2000, pp. 113-127.

[15]   J. W. Fahey, K. K. Stephenson, A. T. Dinkova-Kostova, P. A. Egner, T. W. Kensler and P. Talalay, "Chlorophyll, Chlorophyllin and Related Tetrapyrroles Are Significant

Inducers of Mammalian Phase 2 Cytoprotective Genes," *Carcinogenesis*, Vol. 26, No. 7, 2005, pp. 1247-1255.

[16] S. Kapiotis, M. Hermann, M. Exner, H. Laggner and B. M. Gmeiner, "Copper- and Magnesium Protoporphyrin Complexes Inhibit Oxidative Modification of LDL Induced by Hemin, Transition Metal Ions and Tyrosyl Radicals," *Free Radical Research*, Vol. 39, No. 11, 2005, pp. 1193-1202.

[17] C. M. Yang, K. W. Chang, M. H. Yin and H. M. Huang, "Methods for the Determination of the Chlorophylls and Their Derivatives," *Taiwania*, Vol. 43, No. 2, 1998, pp. 116-122.

[18] V. Schumasker and D. Puppione, "Sequential Flotation Ultracentrifugation," In: J. Segrest and J. Albers, Eds., *Methods in Enzymology*, Academic Press, San Diego, Vol. 128, 1986, pp. 155-170.

[19] N. P. Singh, M. T. McCoy, R. R. Tice and E. L. Schneider, "A Simple Technique for Quantitation of Low Levels of DNA Damage in Individual Cells," *Experimental Cell Research*, Vol. 175, No. 1, 1988, pp. 184-191.

[20] T. C. Dinis, V. M. Maderia and L. M. Almeida, "Action of Phenolic Derivatives (Acetaminophen, Salicylate, and 5-Aminosalicylate) as Inhibitors of Membrane Lipid Peroxidation and as Peroxyl Radical Scavengers," *Arch Biochem Biophys*, Vol. 315, No. 1, 1994, pp. 161-169.

[21] B. Wallin, B. Rosengren, H. G. Shertzer and G. Camejo, "Lipoprotein Oxidation and Measurement of Thiobarbituric Acid Reacting Substances Formation in a Single Microtiter Plate: Its Use for Evaluation of Antioxidants," *Analytical Biochemistry*, Vol. 208, No. 1, 1993, pp. 10-15.

[22] K. Yagi, "Assay for Blood or Plasma," In: L. Packer, Ed., *Methods in Enzymology, Oxygen Radical in Biological System*, Academic Press, New York, 1984, pp 328-331.

[23] S. Arimoto-Kobayashi, N. Inada, H. Nakano, H. Rai and H. Hayatsu, "Iron-Chlorophyllin-Mediated Conversion of 3-Hydroxyamino-1-methyl-5H-pyrido[4,3-b]indole (Trp-P-2(NHOH)) into Its Nitroso Derivative," *Mutation Research*, Vol. 400, No. 1-2, 1998, pp. 259-269.

[24] R. E. Nelson and M. G. Ferruzzi, "Synthesis and Bioaccessibility of Fe-Pheophytin Derivatives from Crude Spinach Extract," *Journal of Food Science*, Vol. 73, No. 5, 2008, pp. H86-H91.

12

Short Communication: Effect of Timing of Introduction to Pasture Post Calving and Supplementation with *Saccharomyces cerevisiae* on Milk Fatty Acid Profiles in Early Lactation Dairy Cows

Karina Mary Pierce[1], Radwan M. Alibrahim[2], Rafael Alejandro Palladino[3], Stephen Joseph Whelan[1], Finbar John Mulligan[4]

[1]School of Agriculture and Food Science, University College Dublin, Dublin, Ireland; [2]Arasco Feeds, Riyadh, KSA; [3]Department of Animal Production, University of Buenos Aires, Buenos Aires City, Argentina; [4]School of Veterinary Medicine, University College Dublin, Dublin, Ireland.

ABSTRACT

There is increased public awareness of the effect of dietary fatty acid (FA) profile on human health. Therefore, when devising nutritional management strategies for dairy cows it is important to evaluate the effects of said strategies on the FA profile of the milk. This experiment investigates the effects of two early PP nutritional management strategies (NM); abrupt introduction to pasture (AP) or a total mixed ration for 21 d followed by a gradual introduction to pasture over 7 d (GP), with (Y) or without (C) live yeast (YS) on milk fatty acid (FA) profile. Forty multiparous dairy cows were assigned to one of four dietary treatments in a two (AP vs. GP) by two (Y vs. C) factorial, randomized block design. The experiment was conducted from d 1 to 70 PP. Pasture, TMR and concentrate samples were taken weekly to assess the chemical and FA composition. Milk yield was recorded daily and individual milk samples were collected weekly to determine milk FA composition. There was no interaction between NM strategy and YS supplementation on milk FA. Similarly, YS supplementation did not affect milk FA profile. However, GP had higher concentrations of C10 ($P = 0.04$), C12 ($P = 0.01$), C14 ($P = 0.02$) and medium chain FA ($P = 0.02$) vs. AP. Whereas AP had higher concentrations of the FA cis-9, C18:1 ($P < 0.01$), long chain FA ($P = 0.1$) and unsaturated FA ($P = 0.01$) and lower concentrations of saturated FA ($P = 0.01$) vs. GP. These results suggest that abruptly introducing the early lactation dairy cow to a pasture based diets positively alters the FA composition of the milk produced when compared to the milk from a dairy cow gradually introduced to pasture.

Keywords: Fatty Acid; Dairy Cow; Nutrition; Post-Partum

1. Introduction

In recent years consumers have become increasingly aware of the potential health benefits from consuming dairy products rich in polyunsaturated fatty acids (PUFA), particularly those in the n-3 (omega-3) fatty acid (FA) group and the conjugated linoleic acid (CLA) isomer *cis*-9 *trans*-11 C18:2 (Ruminic acid, RA) [1]. Therefore, when devising nutritional management strategies for the early lactation dairy cow it is important to measure the effects of said strategies on the FA profile of the milk produced. Several factors are known to influence the FA

composition of bovine milk including diet [2], rumen function [3], stage of lactation and body tissue mobilization [4]. In Ireland, early turn out to pasture post calving has been advocated as a means of maximizing milk production from grazed pasture, increasing the competitiveness of Irish dairy production systems [5]. In addition, the milk produced from pasture fed dairy cows has been reported to contain higher levels of CLA compared to cows offered total mixed ration (TMR) diets [2,6]. However, abrupt changes in the diet offered during early lactation can impair rumen function [7], depress dry matter intake (DMI) and exacerbate body tissue mobilization [8], thus altering the FA profile in milk.

Nutritional management (NM) strategies such as gradual introduction to pasture (GP) may ameliorate the negative effects of abrupt dietary changes [9]. However, the effect of this dietary strategy on milk FA profile is less well known. Supplementation with live yeast has also been suggested as a means of overcoming the risks associated with abrupt dietary change [10]. In addition, live yeast supplementation has been reported to alter rumen pH and volatile fatty acid (VFA) concentrations [11, 12]. However, there is a scarcity in literature reporting the effect of live yeast supplementation on milk FA profile.

The hypothesis of the current study is that abrupt turn out to pasture and supplementation with live yeast may improve the concentrations of CLA in the milk of early lactation dairy cows when compared to milk from dairy cows gradually introduced to pasture.

2. Materials and Methods

2.1. Animals and Management

All procedures involving the use of animals were conducted under experimental license from the Department of Health and Children, Ireland [13]. Forty multiparous dairy cows were selected from the dairy herd at UCD Lyons Research Farm, Dublin, Ireland (53°17'56"N, 6°32'18"W) and blocked by previous lactation 305 d milk yield, parity, BCS, and predicted calving date. Within each block, cows were randomly allocated to dietary treatments in a 2 × 2 factorial arrangement. Animals were either introduced to pasture gradually (GP, n = 20) or abruptly (AP, n = 20) post calving and were offered concentrates with (Y) or without (C) supplementary Yea-sacc[1026] (YS) at a rate of 2.5×10^9 cfu of *Saccharomyces cerevisiae*, strain 1026 (Yea-sacc[1026], Alltech, KY, United States of America). The experiment was conducted from d 0 to 70 postpartum (PP). From d 1 to 21 PP the GP group were housed in a free stall barn and offered 23 kg dry matter (DM)/day of a TMR containing on a DM basis; 27% maize silage, 16.5% grass silage, 3.5% wheat straw and 53% TMR concentrate with or without YS. Individual feed intake during this time was facilitated using computerized feed troughs (RIC System; Insentec B.V., Marknesse, The Netherlands). From d 21 to 28 PP, the GP group was allowed access to pasture between am and pm milking (0800 and 1530 h) and had access to the TMR between pm and am milking. During this time the GP groups were introduced to the pasture concentrates at an initial rate of 3.5 kg DM/d increasing by 0.5 kg DM/d until d 28. By d 28 the GP group were grazing full time with no supplementary TMR. Animals in the AP group were offered 20 kg DM/d of a perennial

ryegrass based pasture plus 7 kg DM/d of a pasture concentrate with or without YS throughout the experiment.

2.2. Data and Sample Collection and Analysis

Samples of the TMR, pasture and concentrates were collected weekly and stored at −20°C pending analysis. Dry matter, neutral detergent fiber (NDF), acid detergent fiber (ADF), crude protein (CP), gross energy, ether extract, starch and ash were determined as described in Whelan *et al.* [14]. The FA in feed was extracted using the one-step methylation procedure of Sukhija and Palmquist [15] as described in Palladino *et al.* [16]. Concentrations of individual FA were determined using gas chromatography (GC) (Varian CP-3800 GC; Varian Inc., Palo Alto, Canada) fitted with a 100 m capillary column with an internal diameter of 0.25 mm coated with 0.39 μm film (CP7420, Varian Inc.). Nitrogen was the carrier gas and the column flow was held at 2 ml/ min (**Table 1**). Cows were milked twice daily at 0700 and 1600 h. Milk output was recorded daily and sampled weekly using a milk metering system (Weighall, Dairymaster, Kerry, Ireland). Weekly milk samples were collected from consecutive am and pm milking and pooled according to output. Milk fat was extracted from these samples by centrifuging samples at 4°C for 20 min at 978 × g, the cream was collected and stored at −20°C pending analysis. Concentrations of individual milk FA were determined using GC (previously described) using the method of Christie [17] as described in Palladino *et al.* [16]. Data were checked for adherence to normal distribution and homogeneity of variance using histograms and formal statistical tests as part of the Univariate procedure of SAS [18]. The natural logarithmic transformations of C10:0, C12:0, C18:3 and saturated FA (SFA) were used to normalize data distribution since the preliminary analyses revealed that the distribution of values for these variables were positively skewed. Transformed data were used to calculate P-values. However, the corresponding least squares means and standard errors of the non-transformed data are presented in the results. Analysis of data was conducted using Proc Mixed of SAS [18] including terms for NM, YS supplementation, block, time and their interactions. Statistically significant differences between least squares means were tested using the PDIFF command incorporating the Tukey test for pairwise comparison of treatment means. Where interactions were not significant the term was excluded from the model. Statistical significance was assumed at a value of $P < 0.05$ and a tendency toward significance assumed at a value of $P > 0.05$ but <0.10.

3. Results and Discussion

There is considerable interest in developing dairy prod-

Short Communication: Effect of Timing of Introduction to Pasture Post Calving and Supplementation
with Saccharomyces cerevisiae on Milk Fatty Acid Profiles in Early Lactation Dairy Cows

95

Table 1. Chemical and fatty acid composition of the total mixed ration (TMR), fresh perennial ryegrass, TMR-concentrate portion, and pasture-concentrate used as a supplement for grazing cows.

Item	TMR	Fresh Grass	TMR-Conc.	Pasture-Conc.
Composition (% of DM, unless stated)				
Dry matter (%)	44.2	23.1	90.0	90.0
Crude protein	16.0	16.6	20.2	14.8
Ash	6.2	7.6	9.1	9.4
NDF[1]	48.6	38.5	26.9	26.0
ADF[1]	20.7	20.7	10.9	11.5
Starch	24.8	0.0	15.9	19.6
ME[1] (MJ/kg DM)	11.4	11.4	13.0	13.0
EE[1]	-	-	2.07	1.53
Total FA	3.71	4.56	4.69	5.45
C14:0	0.06	0.04	0.06	0.04
C16:0	1.00	0.86	1.87	1.76
cis-9 C16:1	0.02	0.13	0.03	0.05
C18:0	0.05	0.03	0.06	0.09
cis-9 C18:1	0.63	0.11	0.82	1.17
cis-9 cis-12 C18:2 n6	1.59	1.07	1.64	2.18
C18:3 n3	0.36	2.32	0.21	0.16

[1]TMR = total mixed ration, NDF = neutral detergent fiber, ADF = acid detergent fiber, ME = metabolizable energy, EE = ether extract.

ucts rich in long chain fatty acids (LCFA) and CLA because of their beneficial effects on human health [17]. In the current study, milk produced by the GP group had higher concentrations of caprinic acid (C10, $P = 0.04$), laurinc acid (C12, $P = 0.01$), meristic acid (C14, $P = 0.02$) and medium chain FA ($P = 0.02$) than that of the AP group indicating a reduction in de-novo synthesis in pasture fed cows [2] (**Table 2**). In a companion study, Al Ibrahim et al. [18] reported a lower DMI (15.8 vs. 17.0 kg/d, $P = 0.04$) and energy intake (15.1 vs. 16.3 units of energy for lactation/d, $P = 0.04$) in the AP group versus the GP group. It is possible that this reduction in DMI reduced the supply of substrates and energy required for de-novo synthesis in the AP group as evidenced by the reduction in milk fat yield (1.09 vs. 1.17 kg/d, $P = 0.08$) and consistent with the observations of Kelly et al. [2] where a similar dietary regime was imposed. The hormonal status of an animal is also known to affect FA synthesis [6]. In the companion study, Al Ibrahim et al. [18] reported higher insulin concentrations in the GP group, thus contributing to increased production of C10:0 to C14:0 FA [19]. As the concentrations of C12:0 and C14:0 have been reported to contribute to plasma low

density lipoprotein cholesterol levels, reducing the portion of these FA in bovine milk may have important human health implications [6].

Improving the concentrations of LCFA and CLA in bovine milk is also likely to benefit human health [17]. However, to improve the concentrations of LCFA and CLA in bovine milk the dairy cow must be offered diets that are also rich in LCFA as these FA cannot be synthesized de-novo [20]. The pasture offered in this experiment contained a higher portion of linolenic acid (C18:3) than the TMR diet. Additionally, pasture based diets have been reported to favor the proliferation of *Butyrivibrio fibrisolvens* in the rumen, a species responsible for the biohydrogenation of C18:3 to produce linoleic (C18:2) and oleic acid (C18:1) [21]. Analysis of the milk FA on a weekly basis shows an increase in concentrations of cis-9, C18:1 once the GP group was turned out to pasture (11.0 vs. 20.9% of milk FA for week 3 vs. week 4, $P < 0.05$). Thus, it is likely that a combination of increased dietary supply of C18:3 and rumen biohydrogenation contributed to the increased concentrations of cis-9, C18:1 ($P < 0.01$) in the milk of the AP versus the GP group.

Mobilization of body adipose tissue can contribute up

Table 2. Effect of live yeast culture (YS) supplementation and early postpartum nutritional management (NM) on milk fatty acid (FA) composition of early lactation dairy cows.

	YS[1]		SED[3]	NM[2]		SED	Significance		
	C[1]	Y[1]		AP[2]	GP[2]		YS	NM	YS × NM
FA (% of total)									
C4:0	6.8	6.6	0.42	6.5	6.9	0.62	0.68	0.51	0.98
C6:0	4.9	4.2	0.54	4.1	5.0	0.80	0.23	0.27	0.93
C8:0	2.9	2.4	0.35	2.4	2.9	0.05	0.18	0.29	0.93
C10:0[4]	4.8	4.1	0.47	3.8	5.2	0.68	0.59	0.04	0.16
C12:0[4]	3.9	3.5	0.25	3.2	4.2	0.38	0.09	0.01	0.29
C14:0	9.6	9.3	0.40	8.6	10.3	0.61	0.37	0.02	0.07
C14:1	0.5	0.5	0.05	0.5	0.5	0.07	0.94	0.92	0.99
C15:0	0.7	0.7	0.06	0.6	0.7	0.08	0.90	0.20	0.70
C16:0	23.4	24.3	0.10	23.1	24.6	0.15	0.39	0.33	0.91
C16:1	2.3	2.3	0.02	2.5	2.1	0.04	0.84	0.23	0.91
C17:0	0.6	0.6	0.05	0.6	0.5	0.07	0.60	0.12	0.43
C18:0	6.5	6.5	0.03	6.9	6.1	0.04	0.82	0.06	0.39
cis-9 C18:1	21.9	23.7	1.80	25.4	20.1	1.61	0.13	<0.01	0.68
trans-11 C18:1	1.5	1.4	0.10	1.4	1.6	0.15	0.29	0.21	0.11
cis-9 cis-12 C18:2 n6	1.5	1.5	0.15	1.5	1.5	0.22	0.79	0.89	0.29
cis-9 trans-11 C18:2	0.8	0.8	0.09	0.8	0.8	0.13	0.59	0.87	0.99
C18:3 n3[4]	0.8	0.8	0.11	0.8	0.7	0.16	0.81	0.56	0.75
SFA[4,5]	64.2	62.2	1.33	60.1	66.3	2.00	0.16	<0.01	0.43
UFA[6]	29.3	30.4	1.38	32.1	26.2	2.06	0.23	0.01	0.56
UFA:SFA	0.46	0.50	0.029	0.55	0.41	0.043	0.20	<0.01	0.48
SCFA[7]	147.3	133.1	12.93	132.5	147.9	19.30	0.20	0.40	0.71
MCFA[8]	452.9	445.8	7.03	423.3	471.4	18.97	0.60	0.02	0.21
LCFA[9]	330.7	348.3	14.97	372.1	306.9	22.36	0.26	0.01	0.49
Δ9-desaturase index	0.05	0.05	0.004	0.06	0.05	0.005	0.81	0.17	0.28

[1]YS = live yeast culture: C = control, Y = yeast supplemented (2.5 g/cow/d × 10^9 CFU of *S. cerevisiae*[1026]/g); [2]NM = nutritional management: AP = abrupt introduction to pasture at calving, GP = gradual introduction to pasture following 21 d postpartum on a TMR; [3]Standard error of the differences; [4]Level of significance correspond to transformed variable; [5]Saturated fatty acids; [6]Unsaturated fatty acids; [7]Short chain fatty acids; sum from C4:0 to C8:0; [8]Medium chain fatty acids; sum from C10:0 to C16:1; [9]Long chain fatty acids; sum from C17:0 to C22:6.

to 20% of FA in bovine milk during early lactation, particularly the LCFA [20]. In the companion study, Al Ibrahim et al. [18] reported concentrations of blood non-esterified FA (0.7 mmol/L) in the AP group that were indicative of negative energy balance, suggesting that part of the increase in LCFA in the milk from the AP group may have been a result of body tissue mobilization. This is important because managing the degree of body

tissue mobilization experienced by the dairy cow is vital in maintaining the health of the dairy cow [8].

Lastly, a lower rumen pH has also been associated with increases in LCFA, unsaturated FA and the ratio of unsaturated FA to saturated FA [22]. In a companion study Al Ibrahim et al. [9] reported an increase in rumen pH in YS supplemented cows. However, there was no effect of YS supplementation on DMI or milk yield and

Short Communication: Effect of Timing of Introduction to Pasture Post Calving and Supplementation
with Saccharomyces cerevisiae on Milk Fatty Acid Profiles in Early Lactation Dairy Cows

97

it is likely that the differences in rumen pH was not sufficient to cause a change in milk FA profile as observed in this experiment.

4. Conclusion

The findings of the present study indicate that abruptly introducing animals to pasture can alter the FA profile of bovine milk toward a product with potential human health benefits.

5. Acknowledgements

The authors would like to acknowledge the contribution of the laboratory and farm staff of UCD Lyons Research Farm for their contribution to this work. This work was co-funded by Alltech as part of an Enterprise Ireland innovation partnership.

REFERENCES

[1] R. G. Jensen, "The Composition of Bovine Milk Lipids: January 1995 to December 2000," *Journal of Dairy Science*, Vol. 85, No. 2, 2002, pp. 295-350.

[2] M. L. Kelly, *et al.*, "Effect of Intake of Pasture on Concentrations of Conjugated Linoleic Acid in Milk of Lactating Cows," *Journal of Dairy Science*, Vol. 81, No. 6, 1998, pp. 1630-1636.

[3] R. J. Dewhurst, *et al.*, "Increasing the Concentrations of Beneficial Polyunsaturated Fatty Acids in Milk Produced by Dairy Cows in High-Forage Systems," *Animal Feed Science and Technology*, Vol. 131, No. 3-4, 2006, pp. 168-206.

[4] P. C. Garnsworthy, *et al.*, "Variation of Milk Citrate with Stage of Lactation and De Novo Fatty Acid Synthesis in Dairy Cows," *Journal of Dairy Science*, Vol. 89, No. 5, 2006, pp. 1604-1612.

[5] P. A. T. Dillon, *et al.*, "Future Outlook for the Irish Dairy Industry: A Study of International Competitiveness, Influence of International Trade Reform and Requirement for Change," *International Journal of Dairy Technology*, Vol. 61, No. 1, 2008, pp. 16-29.

[6] A. L. Lock and P. C. Garnsworthy, "Seasonal Variation in Milk Conjugated Linoleic Acid and Δ9-Desaturase Activity in Dairy Cows," *Livestock Production Science*, Vol. 79, No. 1, 2003, pp. 47-59.

[7] G. A. Donovan, *et al.*, "Influence of Transition Diets on Occurrence of Subclinical Laminitis in Holstein Dairy Cows," *Journal of Dairy Science*, Vol. 87, No. 1, 2004, pp. 73-84.

[8] F. J. Mulligan and M. L. Doherty, "Production Diseases of the Transition Cow," *The Veterinary Journal*, Vol. 176,

No. 1, 2008, pp. 3-9.

[9] R. M. Al Ibrahim, *et al.*, "The Effect of Abrupt or Gradual Introduction to Pasture after Calving and Supplementation with *Saccharomyces cerevisiae* (Strain 1026) on Ruminal pH and Fermentation in Early Lactation Dairy Cows," *Animal Feed Science and Technology*, Vol. 178, No. 1-2, 2012, pp. 40-47.

[10] R. M. Al Ibrahim, *et al.*, "The Effect of Nutritional Management and Supplementation with *Saccharomyces cerevisiae* on Ruminal pH and Fermentation in Early Lactation Dairy Cows," *Animal Feed Science and Technology*, 2012 (accecpted).

[11] C. M. Guedes, *et al.*, "Effects of a *Saccharomyces cerevisiae* Yeast on Ruminal Fermentation and Fibre Degradation of Maize Silages in Cows," *Animal Feed Science and Technology*, Vol. 145, No. 1-4, 2008, pp. 27-40.

[12] M. Thrune, *et al.*, "Effects of *Saccharomyces cerevisiae* on Ruminal pH and Microbial Fermentation in Dairy Cows: Yeast Supplementation on Rumen Fermentation," *Livestock Science*, Vol. 124, No. 1-3, 2009, pp. 261-265.

[13] S.I.613, "Cruelty to Animals Act 1876 (as Amended by European Communities. Regulations 2002 and 2005)," Statutory Instruments No. 613 of 2005, Department of Health and Children, Dublin, 2005.

[14] S. J. Whelan, *et al.*, "Effect of Forage Source and a Supplementary Methionine Hydroxy Analogue on Nitrogen Balance in Lactating Dairy Cows Offered a Low Crude Protein Diet," *Journal of Dairy Science*, Vol. 94, No. 10, 2011, pp. 5080-5089.

[15] P. S. Sukhija and D. L. Palmquist, "Rapid Method for Determination of Total Fatty Acid Content and Composition of Feedstuffs and Feces," *Journal of Agricultural and Food Chemistry*, Vol. 36, No. 6, 1988, pp. 1202-1206.

[16] R. A. Palladino, *et al.*, "Fatty Acid Intake and Milk Fatty Acid Composition of Holstein Dairy Cows under Different Grazing Strategies: Herbage Mass and Daily Herbage Allowance," *Journal of Dairy Science*, Vol. 92, No. 10, 2009, pp. 5212-5223.

[17] A. Lock and D. Bauman, "Modifying Milk Fat Composition of Dairy Cows to Enhance Fatty Acids Beneficial to Human Health," *Lipids*, Vol. 39, No. 12, 2004, pp. 1197-1206.

[18] R. M. Al Ibrahim, *et al.*, "Effect of Timing of Post-Partum Introduction to Pasture and Supplementation with *Saccharomyces cerevisiae* on Milk Production, Metabolic Status, Energy Balance and Some Reproductive Parameters in Early Lactation Dairy Cows," *Journal of Animal Physiology and Animal Nutrition*, Vol. 97, 2013, pp. 105-114.

[19] D. Yin, M. J. Griffin and T. D. Etherton, "Analysis of the Signal Pathways Involved in the Regulation of Fatty Acid Synthase Gene Expression by Insulin and Somatotropin," *Journal of Animal Science*, Vol. 79, No. 5, 2001, pp. 1194-200.

[20] D. E. Bauman and J. M. Griinari, "Nutritional Regulation of Milk Fat Synthesis," *Annual Review of Nutrition*, Vol. 23, No. 1, 2003, pp. 203-227.

[21] L. Biondi, *et al.*, "Changes in Ewe Milk Fatty Acids Following Turning Out to Pasture," *Small Ruminant Research*, Vol. 75, No. 1, 2008, pp. 17-23.

[22] Y. Chilliard, *et al.*, "Diet, Rumen Biohydrogenation and Nutritional Quality of Cow and Goat Milk Fat," *European Journal of Lipid Science and Technology*, Vol. 109, No. 8, 2007, pp. 828-855.

13

Biophysical Characterization of Genistein in Its Natural Carrier Human Hemoglobin Using Spectroscopic and Computational Approaches

Biswapathik Pahari[1*], Sandipan Chakraborty[3,4*†], Bidisha Sengupta[2†], Sudip Chaudhuri[1#],
William Martin[2], Jasmine Taylor[2], Jordan Henley[2], Donald Davis[2], Pradip K. Biswas[4],
Amit K. Sharma[1], Pradeep K. Sengupta[1†]

[1]Biophysics Division, Saha Institute of Nuclear Physics, Kolkata, India; [2]Department of Chemistry, Tougaloo College, Tougaloo, USA; [3]Saroj Mohan Institute of Technology, Hooghly, West Bengal, India; [4]Department of Physics, Tougaloo College, Tougaloo, USA.

ABSTRACT

Steady state and time resolved fluorescence spectroscopy, combined with molecular dynamics simulation, have been used to explore the interactions of a therapeutically important bioflavonoid, genistein, with normal human hemoglobin (HbA). Binding constants estimated from the fluorescence studies were $K = (3.5 \pm 0.32) \times 10^4 \, M^{-1}$ for genistein. Specific interactions with HbA were confirmed from flavonoid-induced fluorescence quenching of the tryptophan in the protein HbA. The mechanism of this quenching involves both static and dynamic components as indicated by: (a) increase in the values of Stern-Volmer quenching constants with temperatures, (b) $\bar{\tau}_0 / \bar{\tau}$ is slightly > 1 (where $\bar{\tau}_0$ and $\bar{\tau}$ are the unquenched and quenched tryptophan fluorescence lifetimes (averaged) respectively). Molecular docking and dynamic simulations reveal that genistein binds between the subunits of HbA, ~18 Å away from the closest heme group of chain $\alpha 1$, emphasizing the fact that the drug does not interfere with oxygen binding site of HbA.

Keywords: Natural Drug Carrier; Fluorescence; Circular Dichroism; Molecular Dynamics; Docking

1. Introduction

Flavonoids are polyphenolic compounds which are ubiquitous in plants of higher genera [1]. Flavonoids have varied uses as therapeutic agents in health disparities, including tumor, cancer, AIDS, diabetes, neurodegenerative and cardiac disorders [1-6]. The isoflavone genistein (structure shown in **Scheme 1**) has dual action as a topoisomerase inhibitor and a tyrosine kinase inhibitor and induces cell differentiation [7,8]. Tyrosine phosphorylation plays important roles in cell proliferation and cell transformation and so tyrosine kinase-specific inhibitors might be used as anticancer agents [9,10]. Genistein, has been shown to be a specific inhibitor of the EGF stimulated tyrosine kinase activity in cultured A431 cells [7].

Genistein induces DNA strand breakage through inhibiting topoisomerase II and induces a mature phenotype in the human myeloid HL-205 and erythroid K 562 J leukemia cells [7,8]. Although consumption of the flavonoids has been associated with the prevention of several degenerative diseases [1-6], their bioavailability is often poor probably due to their low aqueous solubility and possible interaction with plasma proteins [11]. Erythrocytes (red blood cells) have been used as drug delivery systems, due to their biocompatibility, biodegradability, usability and easy loading capability [12,13]. For designing and improving an erythrocyte based drug delivery system, the binding interactions of drugs with the main component of erythrocytes, hemoglobin (HbA), have to be fully understood. Hemoglobin is the most abundant blood protein and consists of two α and two β subunits which are noncovalently associated within erythrocytes as a 64.5 kDa tetramer [14,15]. In this article we have

---

*Contributed equally to the paper.
#Permanent address: Gandhi Centenary B. T. College, Habra, India.
†Corresponding authors.

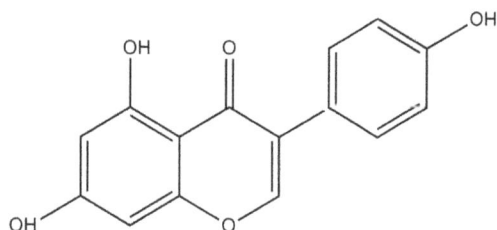

Scheme 1. Structure of the isoflavone; genistein (5,7-Dihydroxy-3-(4-hydroxyphenyl) chromen-4-one).

explored the interaction of genistein with human HbA using the intrinsic fluorescence of HbA and molecular dynamics approaches.

2. Materials and Methods

2.1. Sample Preparation

Lyophilized powder of human hemoglobin, genistein, and phosphate buffer were purchased from Sigma Chemicals, USA. Solvents used were of spectroscopic grade and obtained from Sigma-Aldrich. Purity of genistein was further confirmed by thin layer chromatography which showed only one spot under UV light. Absorption and fluorescence spectroscopic measurements were performed with genistein concentrations of 1×10^{-5} M. HbA was dissolved in pH 7.4 phosphate buffer solution (1×10^{-2} M) and the HbA stock solution (2×10^{-4} M) was kept in the dark at 277 K. The protein concentration was determined spectrophotometrically using the molar extinction coefficient of HbA at 276 nm (120,808 $M^{-1}\cdot cm^{-1}$) [16,17]. For fluorescence quenching studies, the HbA concentration was kept constant at 10^{-5} M. Varying aliquots of concentrated methanolic solution of genistein were added to obtain final concentrations ranging from 0 - 2.5×10^{-5} M. The concentrations of methanol were always kept <1% (by volume) in all samples.

2.2. Spectroscopic Measurements

Steady state absorption spectra were recorded with Shimadzu UV2550 and Cecil model 7500 spectrophotometers. Steady state fluorescence measurements were carried out with Shimadzu RF5301 (equipped with a Fisher temperature controlled accessory) and Varian Cary Eclipse spectrofluorometers. A quartz cuvette of 1 cm path length was used in all experiments. Time resolved fluorescence decay measurements were performed using Jobin-Yvon nanosecond time correlated single photon counting (TCSPC) setup. Fluorescence decay measurements were performed using a nanosecond time correlated single photon counting setup with 295 nm excitation source (nanoLED-295) having pulse FWHM ~760 ps. An emission monochromator was used to block the

scattered light and isolate the emission. Data analyses were performed using DAS6 Fluorescence Decay Analysis Software, provided with the TCSPC instrument and were fitted with a multi exponential decay function, $I(t)) = \sum a_i \exp-(t/\tau_i)$ where a_i and τ_i represent the amplitudes and decay times respectively of the individual components for multi-exponential decay profiles. The goodness of fit was estimated by using reduced χ^2 (namely x_R^2) values as well as Durbin-Watson parameters (DW). A fit is considered acceptable for a given set of observed data and chosen function, when the x_R^2 value is in the range 0.8 - 1.2 and the DW value is greater than 1.7, 1.75 and 1.8 for a single, double and triple exponential fit respectively [18]. Average lifetime is calculated using the equation, $\bar{\tau} = \dfrac{\sum\limits_i a_i \tau_i^2}{\sum\limits_i a_i \tau_i}$ where a_i and τ_i

represent the amplitude and decay time respectively of the individual components for multi-exponential decay profiles. Circular dichroism spectra were acquired with Biologic Science Instruments (France) spectropolarimeter, using a rectangular cuvette with 1 mm path length. The scan rate was 60 - 100 nm/min, and three/five consecutive spectra were averaged to produce the final spectrum. All spectral measurements were performed at ambient temperature. The highest concentration of HbA for fluorescence decay, steady state, and circular dichroism experiments were kept at 10 μM in order to avoid scattering and related artifacts [6].

2.3. Docking Study

AutoDock4 [19] was employed to gain an insight into the genistein binding with HbA. 3-D atomic coordinates of HbA were obtained from the Brookhaven Protein Data Bank (PDB ID 2D60) and prepared for docking. Hemoglobin was considered as a tetramer. All hetero atoms were deleted and non-polar hydrogens were merged. The Kollman united-atom charge model was applied to the protein. Particular attention was given to the parameterization of the porphyrin rings. Partial atomic charges for the porphyrin ring were assigned using the Gasteiger-Marsili method while the valence state of the Iron (Fe) was added manually. Atomic solvation parameters and fragmental volumes were added to the protein. Grid maps used by the empirical free-energy scoring function in AutoDock were generated. A grid box of 100 × 100 × 100 grid points in size with a grid-point spacing of 0.375 Å was considered for docking. The map was centered such that it covered the entire protein including all possible binding sites.

The 3-D structure of genistein was built using the

Biophysical Characterization of Genistein in Its Natural Carrier Human Hemoglobin
Using Spectroscopic and Computational Approaches

101

HYPERCHEM 7.5 [20] molecular builder module and optimized using the AM1 semi-empirical method to an RMS convergence of 0.001 kcal/(Å mol) with the Polak-Ribiere conjugate gradient algorithm implemented in the HYPERCHEM 7.5 package. Rotatable bonds were assigned for the ligand and partial atomic charges were calculated using the Gasteiger-Marsili method after merging non-polar hydrogens. 100 docking runs were performed and for each run, a maximum of 2,500,000 GA operations were performed on a single population of 150 individuals. The weights for crossover, mutation and elitism were default parameters of 0.8, 0.02 and 1 respectively.

2.4. Molecular Dynamics of the HbA-Genistein Complex

MD simulations on free hemoglobin and its complex with genistein were performed using GROMACS [21,22] with OPLS all atom force-field [23]. Genistein parameters were developed according to the OPLS forcefield defined atomic groups. The partial atomic charges were initially assigned according to the OPLS force field using chemical-group analogy method and then readjusted to keep the charge neutrality of the atomic groups to make the ligand charge neutral. Ligand parameter were tested by comparing the GROMACS optimized structure with 1) the parameter-independent QM optimized structure of genistein in vacuum, and 2) the QM/MM optimized structures of genistein in explicit water. CPMD [24] was used for QM calculations and QM/MM calculations were carried-out using GROMACS-CPMD [25]. For QM/MM, the genistein molecule was considered in the QM sub-system and the water was considered in MM sub-system and their interaction between the subsystems was described by the QM/MM Hamiltonian [25].

The crystal structure coordinates of HbA obtained from the protein data bank (PDB ID: 2D60) and used in the docking study, were considered for the simulation of free HbA. For HbA-genistein complex, the lowest energy docked complex obtained from the docking study was used for MD simulation. The OPLS parameters for the heme prosthetic group were taken from the previously published parameter set [26]. The coupling of heme with His-92 of HbA through Fe-N bonding was facilitated through "specbond" option implemented in GROMACS. The HbA structure was primarily subjected to molecular dynamics simulation using GROMACS to check the planarity of the porphyrin ring and proper positioning of the distal histidine residue of hemoglobin that coordinates to the Iron (Fe) of the heme moiety, an essential structural feature for the biological functioning of HbA. Then both ligand bound/unbound structures of HbA were subjected

to *in vacuo* minimization using steepest descent algorithm. Each structure was soaked in a water box containing SPC water molecules such that all the protein atoms were at a distance equal or greater than 1 nm from the box edges. Then the system was further minimized using 500 steps of steepest descent algorithm in a water box. Then each minimized system was subjected to 100 ps of position restrained dynamics at 300 K where proteins were kept fixed by adding restraining forces, but water molecules were allowed to move. Final production simulations were performed in the isothermal-isobaric (NPT) ensemble at 300 K, by coupling to an external bath with a coupling constant of 0.1 ps using Berendsen methods. Pressure was kept constant (1 bar) by using the time-constant for pressure coupling set to 1 ps. The LINCS [27] algorithm was used to constrain bond lengths, allowing the use of 2 fs time steps. Van der Waals and Coulomb interactions were truncated at 1.2 nm. Conformations generated during MD simulation were stored at every 5 ps. All the analyses were carried out using the available trajectory analysis tools of GROMACS packages.

3. Results and Discussions

3.1. Steady State and Time Resolved Fluorescence Results

Figures 1(a) and (b) present the fluorescence emission and excitation spectra and Figure 1(c) shows the absorption spectra of hemoglobin with increasing concentration of genistein. It is evident from Figure 1(a) that addition of genistein induces significant changes in the emission intensity of tryptophan in human hemoglobin with no major change in λ_{em}^{max}. The quenching of tryptophan fluorescence with increasing concentrations of the flavonoid genistein, suggest genistein is binding at a region in HbA which is close to the tryptophan residue. The excitation wavelength used for this is 280 nm where unbound genistein has little absorbance (see the blue absorption profile in Figure 1(c). Figures 1(b) and (c) show the fluorescence excitation and absorbance of HbA (10 μM) with (0 μM —, 5 μM ..., 10 μM - - -, 15 μM - . -, 20 μM - . . -) genistein in 10 mM phosphate buffer at pH 7. There is no appreciable change in the λ_{ex}^{max} in Figure 1(b). However, significant difference in the absorption profiles between unbound and conjugated HbA was observed confirming the binding of genistein with HbA. It is pertinent to mention that fluorescence excitation spectra look at the excited state of the chromophore tryptophan, whereas absorption spectroscopy studies the ground state characteristics of the whole HbA along with tryptophan and genistein.

Quenching studies of protein tryptophan fluorescence

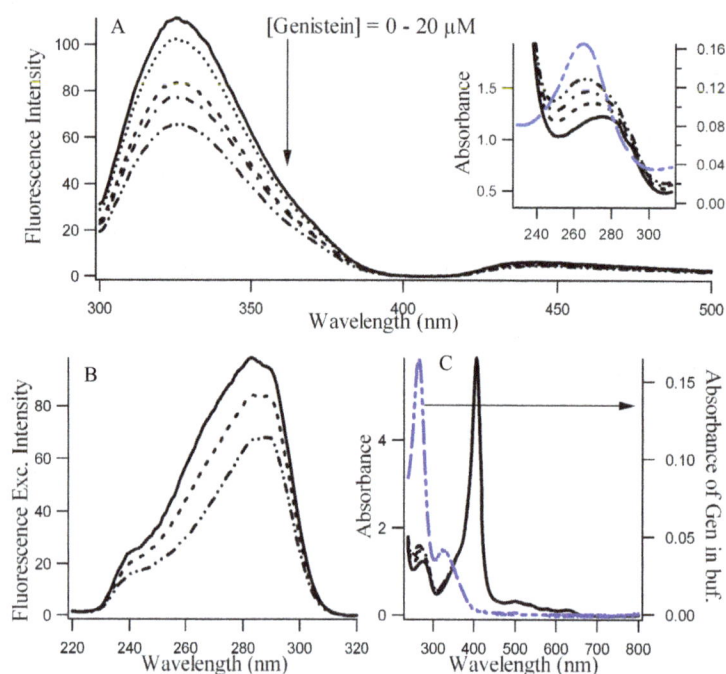

Figure 1. Fluorescence emission spectra (λ_{ex} = 280 nm), (b) Fluorescence excitation spectra (λ_{em} = 340 nm) and (c) Absorption spectra, of hemoglobin (10 μM) in presence of increasing concentrations of genistein (0 μM —, 5 μM ..., 10 μM - - -, 15 μM - . -, 20 μM - . . -). The absorption of unbound genistein in buffer (blue) is shown in Figure 1(c) using the right axis. Furthermore, in this figure the absorption of tryptophan at 280 nm as well as the Soret (~406 nm) and heme (> 500 nm) bands of HbA protein are displayed. Figure 1(a) inset highlights the absorption profiles of Figure 1(c) in the UV region to indicate the changes in the absorption spectra of tryptophan of HbA upon binding of genistein.

by ligands is a convenient means for exploring ligand-protein interactions [5,6]. The ratio F_0/F for each addition of genistein are plotted against the genistein concentrations (see **Figure 2(a)**), where F_0 and F are the fluorescence intensities of Trp in HbA in the absence and presence of genistein, respectively. It is observed that the ratio F_0/F increases linearly with the genistein concentration and a linear regression equation following Stern-Volmer relation is obtained. **Figure 2(a)** presents corresponding Stern-Volmer plots based on the equation.

$$\frac{F_0}{F} = 1 + K_{SV}\left[Flavonoid\right] \qquad (1)$$

where K_{SV} is the Stern-Volmer quenching constant for the quenching of HbA tryptophan fluorescence by flavonoid [16]. The K_{SV} values obtained are ca. $3.21 \pm 0.002 \times 10^4$ M^{-1} for genistein which indicate that HbA tryptophan fluorescence is efficiently quenched by the flavonoid. To understand the fluorescence quenching better, temperature dependent studies were performed. **Figure 2(b)** presents the variation of K_{SV} with increasing temperature where significant dependence between the two is observed within the experimental error. A linear Stern-Volmer plot (**Figure 2(a)**) can be expected to arise from

collisional and/or static quenching [5]. The increase in K_{SV} with rise in temperature clearly indicates the influence of collisional or dynamic factor in the fluorescence quenching mechanism. To clarify this aspect, time resolved studies are performed on HbA at different concentrations of genistein which is shown in **Figure 1(c)** and also discussed later.

Fluorescence lifetime measurements were used to obtain further confirmation regarding the nature of the fluorescence quenching, as is discussed above in **Figures 1** and **2**. Lifetime measurements of HbA tryptophan fluorescence both in presence and absence of the flavonoid (see **Table 1**) have been performed. We observed that in presence of genistein, a decrease in $\bar{\tau}$ occurred, which should be the case if dynamic quenching was the sole responsible factor in fluorescence quenching. However, this change in $\bar{\tau}$ is very little. Thus the fact that the average lifetimes ($\bar{\tau}$) computed from the decay parameters does not change appreciably with increase in genistein concentrations (see **Table 1** and **Figure 2(c)**), corroborate the conclusion drawn from the steady state fluorescence data that both static and dynamic mechanisms are responsible for the observed fluorescence quenching, implying that flavonoids indeed bind to hemoglobin. Since the intrinsic fluorescence of human hemoglobin originates

Biophysical Characterization of Genistein in Its Natural Carrier Human Hemoglobin
Using Spectroscopic and Computational Approaches

103

Table 1. Fluorescence decay parameters of HbA (tryptophan) in buffer and in presence of genistein.

*Sample	τ_1(ns)	τ_2(ns)	τ_3(ns)	a_1	a_2	a_3	$\bar{\tau}$ (ns)	χ^2	DW
HbA in Aq. Buffer	0.21	1.21	4.03	0.85	0.12	0.03	1.57	1.16	1.86
HbA + 10 μM Gen	0.15	1.12	3.95	0.88	0.097	0.023	1.51	1.08	1.83
HbA + 40 μM Gen	0.13	1.04	3.798	0.91	0.073	0.017	1.31	1.19	1.85

*[HbA] = 10 μM, Buffer = pH 7.43 (0.01 M) phosphate buffer. Instrument parameters: Counts = 10,000, Slit = 16 nm, λ_{ex} = 295 nm, λ_{em} = 340 nm, FWHM ~ 760 ps.

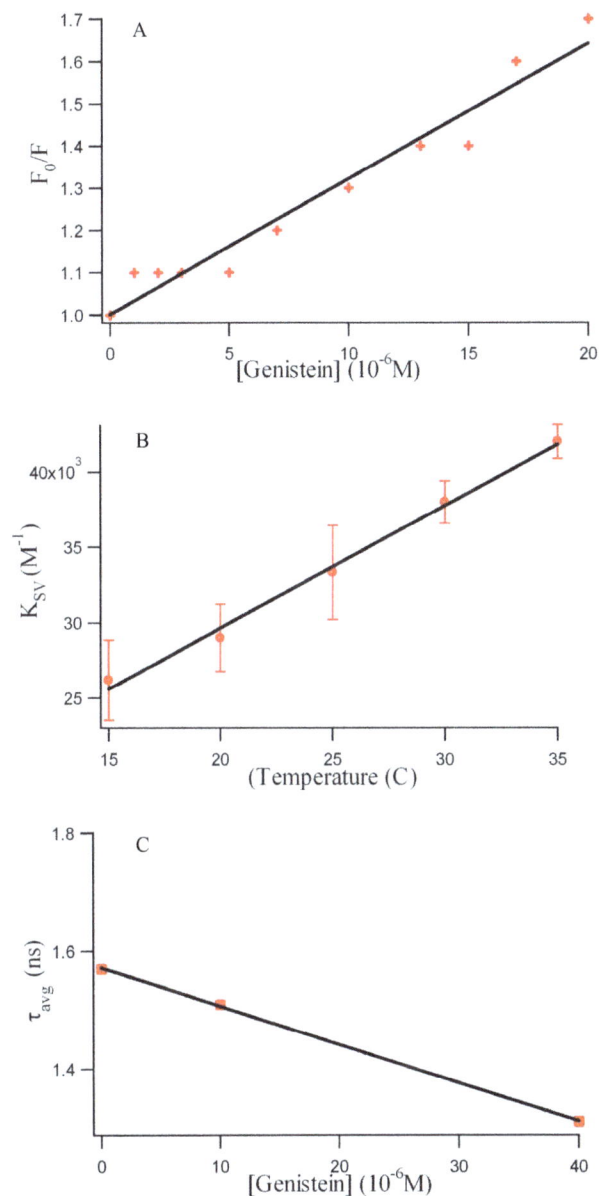

primarily from β-37 tryptophan, it seems reasonable to infer that the β-37 tryptophan residue is presumably at or near the binding site of the flavonoids. The apparent binding constant "K" and the number of binding site(s) "n" were estimated from fluorescence titration studies, using the plot of Log$(F_o - F)/F$ vs. Log$(1/([D_t] - (F_o - F))[P_t]/F_0))$ [5] (**Figure 3**) which is based on the equation:

$$\text{Log}\frac{F_0 - F}{F} = n\text{Log}K - n\text{Log}\left(\frac{1}{([D_t] - (F_0 - F))[P]/F}\right)$$

(2)

where F_0 and F are the fluorescence intensity of HbA in absence and presence of flavonoid (D) respectively, $[D_t]$ is the total flavonoid concentration and $[P_t]$ is the total protein concentration, n is the number of binding sites and K is the binding constant.

Table 2 shows the spectroscopic parameters as well as binding constants "K" and number of binding sites "n" for the binding of genistein with HbA. From **Figure 3**, we observed a K value of $(3.5 \pm 0.32) \times 10^4$ and $n \sim 1$.

The binding forces contributing to interactions of flavonoids with proteins often include a van der Waals interaction, hydrophobic force, electrostatic interactions, hydrogen bond, etc.

The free energy change (ΔG^0) of reaction provides the main line of evidence for confirming the binding force. The thermodynamic parameter is evaluated using the following equation:

$$-RT \ln K = \Delta G^0$$

(3)

where K and R are the binding constant and gas constant, respectively which is provided in **Table 2**. The spontaneity of the binding of genistein with HbA is evident from negative value of ΔG^0.

3.2. Far Ultraviolet Circular Dichroism (CD) Spectroscopic Studies

To investigate the possible effect of the flavonoid on the secondary structure of HbA, we used far-UV CD spectroscopy. The CD spectrum of HbA in aqueous buffer (in the absence of flavonoid) has two characteristic peaks of

Figure 2. (a) Stern-Volmer plots of the HbA tryptophan fluorescence quenching with increasing genistein concentration at 25°C. B: Variation of K_{sv} with temperatures. Each data point indicates the average of three experiments; error bars indicate standard deviations. C: The plot of $\bar{\tau}$ with increase in genistein concentrations.

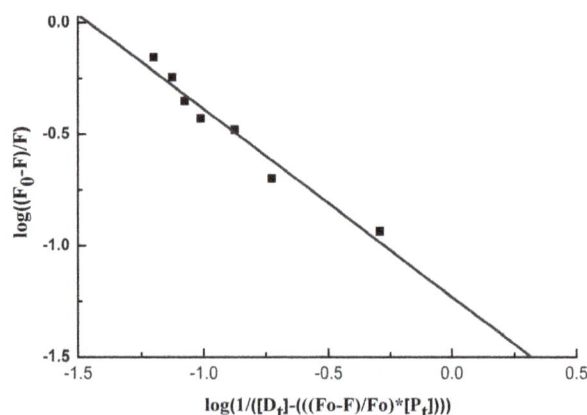

Figure 3. The plot of $\log(1/([D_t] - (F_0 - F))[P_t]/F_0)$ vs. $\log(F_0 - F)/F_0$ of the fluorescence quenching data. The concentration of HbA, $[P_t] = 10.0 \times 10^{-6}$ M.

Table 2. Binding and thermodynamic parameters for Genistein-HbA interactions at 25°C.

$^\dagger K_{SV}$, M^{-1}	$^\dagger K$, M^{-1}	$^\dagger n$	$^\dagger \Delta G^0$ (Kcal/mol)
$(3.21 \pm 0.002) \times 10^4$	$(3.5 \pm 0.32) \times 10^4$	1	−6.24

$^\dagger K_{SV}$, the Stern-Volmer quenching constant; K, the binding constant and n, the number of binding sites.

negative ellipticities at 208 nm and 222 nm indicating its predominantly α-helical secondary structure (**Figure 4**). As shown in the figure, the CD spectrum of the HbA remains essentially unchanged upon addition of different concentrations of flavonoids (hemoglobin concentration remaining fixed).

The quenching observed for HbA in presence of genistein must be due to some specific interaction that increases the local concentrations of the flavonoids around the tryptophan residue(s) in hemoglobin. The significant dependence of K_{SV} on temperature indicates that the observed fluorescence quenching is influenced by both static and dynamic factors. However the variation of $\bar{\tau}$ with increasing genistein concentrations is minor. (N.B. - For static quenching the complexed fluorophores are nonfluorescent, and the only observed fluorescence arises from the uncomplexed fraction. Therefore, the average lifetime of the uncomplexed fluorophores ($\bar{\tau}_0$) remains unchanged and consequently $\bar{\tau}_0/\bar{\tau} = 1$ ($\bar{\tau}$ is the average lifetime in the presence of quencher). By contrast, for dynamic quenching $\bar{\tau}_0/\bar{\tau} = F_0/F > 1$ (where F_0 and F are the fluorescence intensities in the absence and presence of quencher, respectively) [16]. It should be noted that although there is a dramatic increase (~60%) in K_{SV} with increase in temperature from 15°C to 35°C as is shown in **Figure 2(b)** confirming the existence of both static and dynamic factors in fluorescence quenching, the decrease in $\bar{\tau}$ with increase in genistein concentrations

is only by ~16.56% (as shown in **Figure 2(c)**). This small change in $\bar{\tau}$ reflects a local structural change in the microenvironment of tryptophan of HbA, with no appreciable global change in the protein structure, as is evident from the CD spectra in **Figure 4**.

3.3. Molecular Dynamics Studies

To explore the effect of genistein binding on the structural stability of HbA, we have analyzed the dynamic structural properties (RMSD and radius of gyration, R_g) of both the ligand free and bound HbA obtained from MD simulation and shown in **Figure 5**.

The RMSD of the backbone C_α atom of the simulated protein over the time is a reliable parameter to analyze the stability of the system. As evident from **Figure 5(a)**, during the first 3 ns of the simulation, both the systems undergo structural readjustments according to its environments, and monotonically reach the equilibrium state characterized by a stable RMSD profile. A closer look at each trajectory obtained from the MD simulation reveals that the HbA remains stable throughout the simulation for both ligand free and bound HbA. Also, genistein binding does not induce any structural perturbations in HbA. Observed RMSD for free and genistein bound HbA are highly similar, 0.59 ± 0.05 nm and 0.58 ± 0.04 nm respectively during the last 7 ns of simulation. On the other hand, R_g provides a good measure of the overall structural dimension of the protein. HbA is a globular protein with four chains. The calculated R_g value for free HbA is 2.439 ± 0.02 nm, while that of genistein bound HbA is 2.43 ± 0.011 nm. Again the high similarity values of free and ligand bound HbA signifies no such structural unfolding induced by the binding of the ligand, genistein. This agrees well the CD spectra observed in **Figure 4**.

The interaction energy components between HbA and genistein were analyzed in MD simulation during last 7 ns of the simulation and shown in **Figure 6**. Lennard-Jones (LJ) interactions between the ligand and the pro-

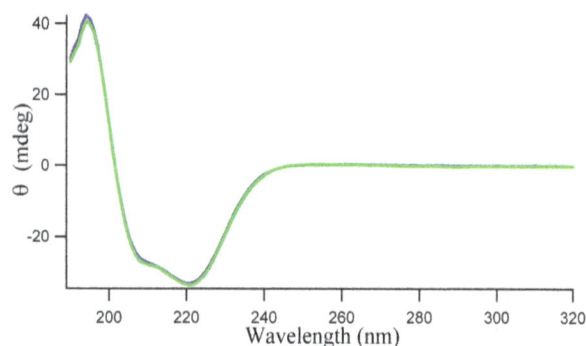

Figure 4. Circular dichroism spectra of 3.0×10^{-6} M hemoglobin (HbA) in absence (blue) and presence (green) of 10.0×10^{-6} M genistein in 0.01 M phosphate buffer, pH 7.4.

Biophysical Characterization of Genistein in Its Natural Carrier Human Hemoglobin
Using Spectroscopic and Computational Approaches

105

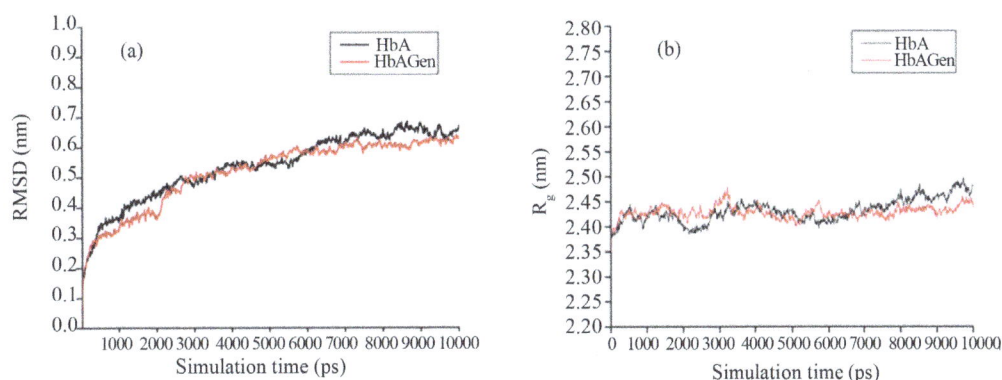

Figure 5. Variation of (a) RMSD (nm) and (b) Radius of gyration (Rg, nm) with simulation time for HbA (black) and HbA with bound genistein (red) obtained from molecular dynamics simulation.

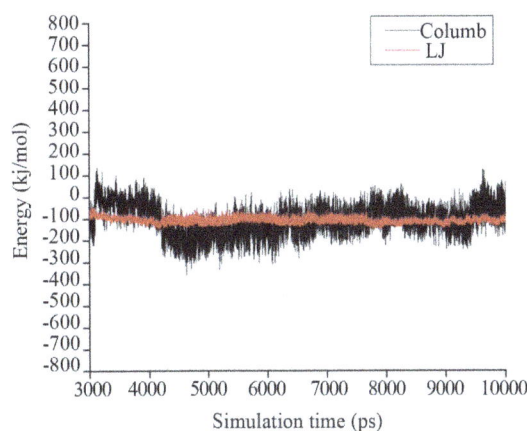

Figure 6. Analysis of different energetic component between HbA and genistein during molecular dynamics simulation. Black and red line represents Columb and van der Waals (Leanard-Jones, LJ) interactions between protein and ligand.

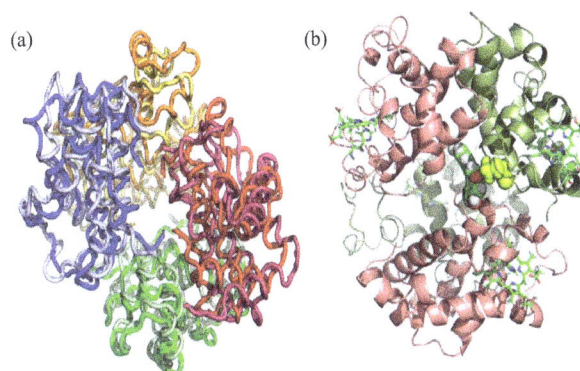

Figure 7. (a) Backbone trace of the MD average structure of HbA (light color) and HbA-genistein complex (dark color). All the four chains have been colored differently. (b) Ribbon representation of HbA-genistein complex. Two α chains and two β chains are colored pink and green accordingly. Four heme groups are represented in sick representation, while genistein and Trp-37 are represented in CPK representation. Trp-37 is represented in green CPK mode.

tein are more stable compared to the Coulombic interactions. The average LJ interaction energy obtained from MD simulation is found to be −107.9 ± 13.7 KJ/mole while Columbic interactions are found to be −104.7 ± 67.1 KJ/mole averaged over the last 7 ns of the simulation. The high standard deviation of average Coulombic interactions indicates less specificity of the Coulomb interactions in the binding of genistein with HbA while the stable LJ interactions pattern signifies its contribution to the binding process. **Figure 7(a)** reveals the backbone trace of the MD average structure of HbA (light color) and HbA-genistein complex (dark color). As evident from the figure, the tetrameric structure of HbA constituted by two α chains and two β chains is highly preserved during simulation and the structure of HbA in presence and absence of genistein is highly similar with calculated RMSD of 0.4 nm over the entire protein backbone. Also the helical nature of each chain is unperturbed during the simulation for both the systems.

Figure 7(b) reveals the details of the genistein binding with HbA in atomistic details. The lowest energy docked conformation and the MD average structure starting from the lowest energy docked complex reveals very similar binding mode where genistein readily enters the central cavity formed by four subunits of HbA, namely α1, α2, β1, β2 (shown in pink and deep green respectively). It is to be mentioned that this binding site is similar to the binding site of another bio-flavonoid daidzein within HbA [8]. Particularly, genistein strongly binds at the interface of α2 and β2 domains, very close to the Trp-37, which is evident from the figure (**Figure 7(b)**). There is a strong van der Waals overlap between genistein and Trp-37. This further confirms that the nearby Trp-37 serves as good optical probe for monitoring genistein binding to HbA. Genistein docked ~5 Å apart from Trp-37 in β2 chain and the average distance between them remains stable during the dynamics, as evident from **Figure 8(a)**. This observation is supported by the fact that the fluorescence of β-37 Trp is exploited to sense the ligand binding to that pocket, as is shown in **Figures 1-3**.

Another important aspect, we have characterized from our MD simulation is the distances between genistein and Fe atom of the porphyrin moiety. It is to be mentioned that the heme group plays a determining role in the biological functioning of HbA. We have calculated the minimum distances between genistein and the Fe atom of the four heme groups during MD simulation and the results are shown in **Figure 8(b)**. Genistein binds between the subunits of HbA, which is sufficiently away from the heme. The closest distance of genistein to any of the four heme group is ~18 Å (heme group of α1 and β2), empha-

sizing the fact that the ligand does not interfere with oxygen binding and biological functioning of HbA.

In the tetrameric form of normal adult hemoglobin, the binding of oxygen is a cooperative process. Binding of oxygen to heme group in one chain induces conformational changes which are transmitted to the remaining three monomers in the tetramer, where it induces a similar conformational change in the other heme sites such that binding of oxygen to these sites becomes easier. Thus the inter-domain contacts are highly important to modulate the biological functioning of HbA. We have characterized the mean inter-chain contacts of both free HbA and genistein bound HbA obtained from MD simulation and shown in **Figure 9**. As evident from the figure,

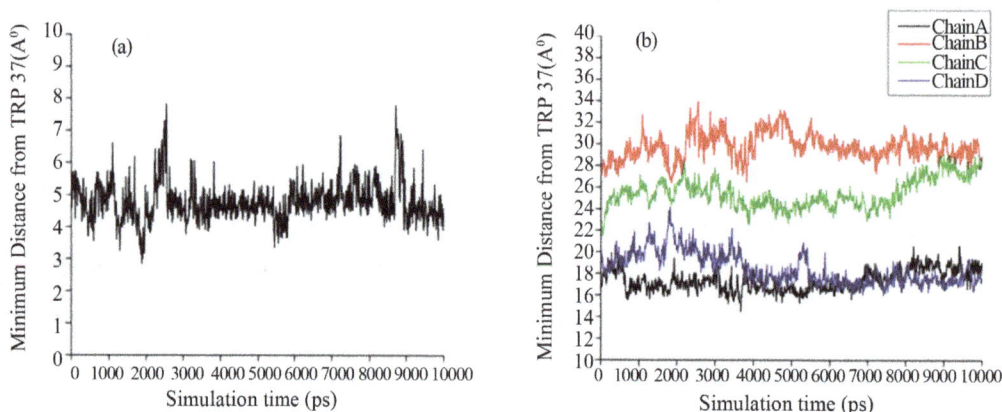

Figure 8. (a) Minimum distance between Trp-37 and genistein during MD simulation. (b) Distances between four Fe atoms in four chains and genistein during MD simulation.

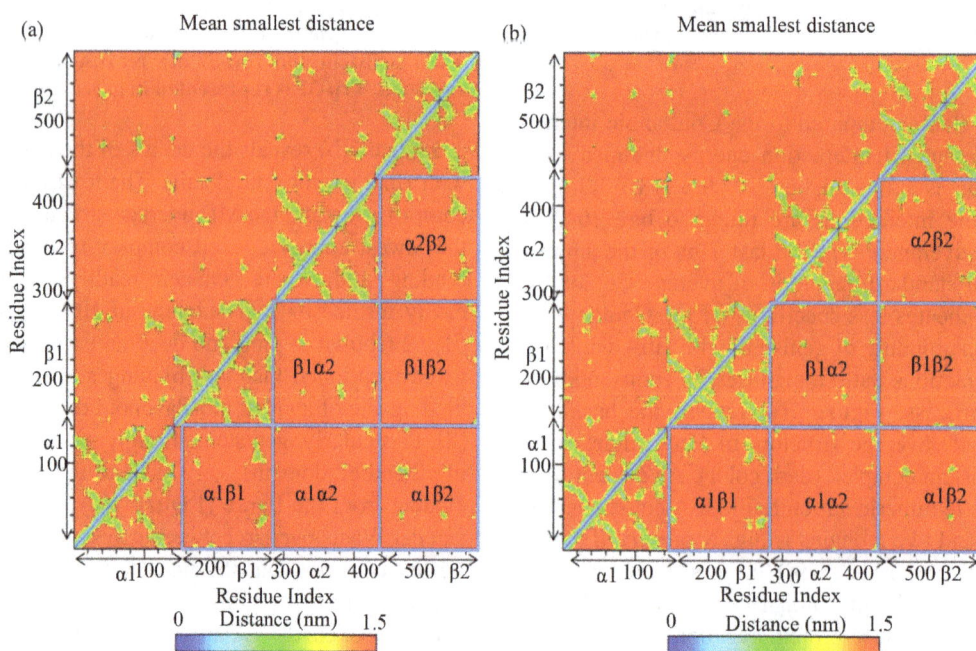

Figure 9. Mean contact distances between residues observed during MD simulation. (a) HbA only; (b) HbA-genistein complex.

Biophysical Characterization of Genistein in Its Natural Carrier Human Hemoglobin
Using Spectroscopic and Computational Approaches

107

genistein binding does not influence the interchain contacts significantly. Only slight variations are observed in the interdomain contacts of α1 and β1 chain and also in between α1 and β2. But, the overall essential contacts between any two chains remain unperturbed upon genistein binding to HbA.

4. Concluding Remarks

Since the pharmacological actions of flavonoids *in vivo* are closely related to their binding with cellular targets including proteins, the investigation on binding of flavonoids with proteins is very significant. HbA has been found to play an important role in the distribution and bioavailability of flavonoids [5,6]. Besides, flavonoids are emerging as potentially useful drugs for cardiac and neurodegenerative disorders which involves oxidative stress [1-6,28-30]. With this scenario in mind we have explored the interactions of the isoflavonoid genistein with HbA. We have fruitfully exploited the intrinsic protein tryptophan fluorescence for this study. Molecular modeling studies have been further employed to provide detailed insights at atomistic level regarding the flavonoid recognition by HbA. The rationale of this study is that the small flavonoid molecules can easily penetrate the erythrocyte membrane and reaches inside the cells where hemoglobin is present in high concentrations.

In summary, our present study using steady state and time resolved fluorescence and circular dichroism spectroscopic and molecular modeling approaches, shows how genistein interacts with HbA. The spectroscopic data suggest that the flavonoid bind with hemoglobin and quench its fluorescence. Molecular docking studies reveal that the position of Trp-37 (β chain) residue, the emitting fluorophore in HbA is present just ~5 Å away from the binding cavity. This approach can easily be applied to other flavonoid derivatives, which would open the door to new avenues for the "screening and design" of the most suitable flavonoid derivatives from among numerous structural variants of this new generation of rapidly emerging therapeutic drugs of immense importance in modern medicine.

5. Acknowledgements

BSG is thankful for the travel award from AAAS WIRC @ MSIs/NSF which provided the travel and part of the research supply support for this international collaboration, as is presented in the present study. BSG also likes to acknowledge the financial and research supports from NIH/NCMHHD/RIMI grant # 1P20MD002725; HBCU-UP Grant-NSF ID: 0811638; JHS program with the contract # N01-HC-95172; HHMI grant # 52007562 at Tougaloo College. PKB and S. Chakraborty thank the financial support from MS-INBRE (#USM-GR04015-05-9; NIH/NCRR #P20RR016476). PKS acknowledges CSIR, India for the award of an Emeritus Scientist grant (no. 21(0864) 11/ EMR-II) and Prof. Milan K. Sanyal, Director, Saha Institute of Nuclear Physics, for providing the opportunity to continue his scientific endeavors at this institute and for extending necessary facilities. We thank Prof. Abhijit Chakrabarti of Structural Genomics Division, and Prof. Samita Basu of Chemical Sciences Division of SINP for letting us use the Circular Dichroism spectrometer and TCSPC fluorescence lifetime facility respectively.

REFERENCES

[1] St. Rusznyák and A. Szent-Györgyi, "Vitamin P: Flavonols as Vitamins," *Nature*, Vol. 138, No. 3479, 1936, p. 27.

[2] S. Chaudhuri, A. Banerjee, K. Basu, B. Sengupta and P. K. Sengupta, "Interaction of Flavonoids with Red Blood Cell Membrane Lipids and Proteins: Antioxidant and Antihemolytic Effects," *International Journal of Biological Macromolecule*, Vol. 41, No. 1, 2007, pp. 42-48.

[3] B. Sengupta, T. Uematsu, P. Jacobsson and J. Swenson, "Exploring the Antioxidant Property of Bioflavonoid Quercetin in Preventing DNA Glycation: A Calorimetric and Spectroscopic Study," *Biochemical and Biophysical Research Communications*, Vol. 339, No. 1, 2006, pp. 355-361.

[4] B. Sengupta and J. Swenson, "Properties of Normal and Glycated Human Hemoglobin in Presence and Absence of Antioxidant," *Biochemical and Biophysical Research Communications*, Vol. 334, No. 3, 2005, pp. 954-959.

[5] S. Chakraborty, S. Chaudhuri, B. Pahari, J. Taylor, P. K. Sengupta and B. Sengupta, "A Critical Study on the Interactions of Hesperitin with Human Hemoglobin: Fluorescence Spectroscopic and Molecular Modeling Approach," *Journal of Luminescence*, Vol. 132, No. 6, 2012, pp. 1522-1528.

[6] G. Rusak, H. O. Gutzeit and J. L. Müller, "Structurally Related Flavonoids with Antioxidative Properties Differentially Affect Cell Cycle Progression and Apoptosis of Human Acuteleukemia Cells," *Nutrition Research*, Vol. 25, No. 2, 2005, pp. 143-155.

[7] A. Constantinou, K. Kiguchi and E. Huberman, "Induction of Differentiation and DNA Strand Breakage in Human HL-60 and K-562 Leukemia Cells by Genistein," *Cancer Research*, Vol. 50, No. 9, 1990, pp. 2618-2624.

[8] D. A. Frank and A. C. Sartorelli, "Alterations in Tyrosine Phosphorylation during the Granulocytic Maturation of HL-60 Leukemia Cells," *Cancer Research*, Vol. 48, No. 1, 1988, pp. 52-58.

[9] T. Akiyama, J. Ishida, S. Nakagawa, H. Ogawara, S.-I. Wa-

tanabe, N. Itoh, M. Shibuya and Y. Fukami, "Genistein, a Specific Inhibitor of Tyrosine-Specific Protein Kinases," *Journal of Biological Chemistry*, Vol. 262, No. 12. 1987, pp. 5592-5595.

[10] K. Kiguchi, A. I. Constantinou and E. Huberman, "Genistein-Induced Cell Differentiation and Protein-Linked DNA Strand Breakage in Human Melanoma Cells," *Cancer Communications*, Vol. 2, No. 8, 1990, pp. 271-277.

[11] A. Bolli, M. Marino, G. Rimbach, G. Fanali, M. Fasano and P. Ascenzi, "Flavonoid Binding to Human Serum Albumin," *Biochemical and Biophysical Research Communications*, Vol. 398, No. 3, 2010, pp. 444-449.

[12] F. K. Alanzi, G. E.-D. I. Harisa, A. Maqboul, M. A. Hamid, S. H. Neau and I. A. Alsarra, "Biochemically Altered Human Erythrocytes as a Carrier for Targeted Delivery of Primaquine: An *in Vitro* Study," *Archives of Pharmacal Research*, Vol. 34, No. 4, 2011, pp. 563-571.

[13] M. Hamidi, H. Tajerzadeh, A. R. Dehpour, M. R. Rouini and S. Ejtemaee-Mehr, "*In Vitro* Characterization of Intact Human Erythrocytes Loaded by Enalaprilat," *Drug Delivery*, Vol. 8, No. 4, 2001, pp. 223-230.

[14] M. L. Doyle, J. M. Holt and G. K. Ackers, "Effects of NaCl on the Linkages between O_2 Binding and Subunit Assembly in Human Hemoglobin: Titration of the Quaternary Enhancement Effect," *Biophysical Chemistry*, Vol. 64, No. 1-3, 1997, pp. 271-287.

[15] X. Yang, J. Chou, G. Sun, H. Yang and T. Lu, "Synchronous Fluorescence Spectra of Hemoglobin: A Study of Aggregation States in Aqueous Solutions," *Microchemical Journal*, Vol. 60, No. 3, 1998, pp. 210-216.

[16] J. R. Lakowicz, "Principles of Fluorescence Spectroscopy," 3rd Edition, Springer-Verlag, New York, 2006.

[17] W. B. Gratzer, "Medical Research Council Labs," Holly Hill, London.

[18] D. Sahoo, P. Bhattacharya and S. Chakravorti, "Quest for Mode of Binding of 2-(4-(dimethylamino) Styryl)-1-Methylpyridinium Iodide with Calf Thymus DNA," *Journal of Physical Chemistry B*, Vol. 114, No. 5, 2010, pp. 2044-2050.

[19] G. M. Morris, D. S. Goodsell, R. S. Halliday, R. Huey, W. E. Hart, R. K. Belew and A. J. Olson, "Automated Docking Using a Lamarkian Genetic Algorithm and an Empirical Binding Free Energy Function," *Journal of Computational Chemistry*, Vol. 19, No. 14, 1998, pp. 1639-1662.

[20] Hyperchem, Hypercube, Inc., 2002.

[21] D. V. D. Spoel, B. Hess, G. Groenhof, A. E. Mark and H. J. Berendsen, "GROMACS: Fast, Flexible, and Free," *Journal of Computational Chemistry*, Vol. 26, No. 16, 2005, pp. 1701-1718.

[22] E. Lindahl Erik, B. Hess and D. V. D. Spoel, "GROMACS 3.0: A Package for Molecular Simulation and Trajectory Analysis," *Journal of Molecular Modelling*, Vol. 7, No. 8, 2001, pp. 306-317.

[23] W. L. Jorgensen and J. Tirado-Rives, "The OPLS Force Field for Proteins. Energy Minimizations for Crystals of Cyclic Peptides and Crambin," *Journal of the American Chemical Society*, Vol. 110, No. 6, 1988, pp. 1657-1666.

[24] R. Car and M. Parrinello, "Unified Approach for Molecular Dynamics and Density-Functional Theory," *Physical Review Letters*, Vol. 55, No. 22, 1985, pp. 2471-2474.

[25] P. K. Biswas and V. Gogonea, "A Regularized and Renormalized Electrostatic Coupling Hamiltonian for Hybrid Quantum-Mechanical—Molecular-Mechanical Calculations," *Journal of Chemical Physics*, Vol. 123, No. 16, 2005, pp. 164114-164122.

[26] V. Gogonea, J. M. Shy and P. K. Biswas, "Electronic Structure, Ionization Potential, and Electron Affinity of the Enzyme Cofactor (6R)-5,6,7,8-Tetrahydrobiopterin in the Gas Phase, Solution, and Protein Environments," *Journal of Physical Chemistry B*, Vol. 110, No. 45, 2006, pp 22861-22871.

[27] B. Hess, H. Bekker, H. J. C. Berendsen and J. G. E. M. Fraaije, "LINCS: A Linear Constraint Solver for Molecular Simulations," *Journal of Computational Chemistry*, Vol. 18, No. 12, 1997, pp. 1463-1472.

[28] B. Pahari, B. Sengupta, S. Chakraborty, B. Thomas, D. McGowan and P. K. Sengupta, "Contrasting Binding of Fisetin and Daidzein in γ-Cyclodextrin Nanocavity," *Journal of Photochemistry and Photobiology B: Biology*, Vol. 118, No. 1, 2013, pp. 33-41.

[29] B. Pahari, S. Chakraborty, S. Chaudhuri, B. Sengupta and P. K. Sengupta, "Binding and Antioxidant Properties of Therapeutically Important Plant Flavonoids in Biomembranes: Insights from Spectroscopic and Quantum Chemical Studies," *Chemistry and Physics of Lipids*, Vol. 165, No. 4, 2012, pp. 488-496.

[30] B. Sengupta, B. Pahari, L. Blackmon and P. K. Sengupta, "Prospect of Bioflavonoid Fisetin as a Quadruplex DNA Ligand: A Biophysical Approach," *PloS One*, Vol. 8, No. 6, 2013, pp. 1-11.

14

Physico-Chemical Properties of Milk Whey Protein Agglomerates for Use in Oral Nutritional Therapy

Luciano Bruno de Carvalho-Silva[1,2*], **Fernanda Zaratini Vissotto**[3], **Jaime Amaya-Farfan**[2]

[1]Faculty of Food Engineering, University of Campinas, Campinas, Brazil; [2]School of Food Engineering, University of Campinas, Campinas, Brazil; [3]Cereal and Chocolate Technology Center, Institute of Food Technology, Campinas, Brazil.

ABSTRACT

Agglomerates based on milk whey proteins and modified starch (MS) were developed for patients with dysphagia. Calcium caseinate (CaCas), whey protein isolate (WPI), concentrate (WPC) and hydrolysate (WPH) were used. The sources were agglomerated with the MS and an increase in the porosity and viscosity of the agglomerates were observed. In all the systems evaluated, the WPI agglomerate at a concentration of 112 g/L showed a viscosity between 2122 and 5110 cP, and the agglomerates of WPC and WPH between 1115 - 2880 cP and 2600 - 6651 cP, respectively. CaCas exhibited high values in water and milk of 3200 cP and 6651 cP, respectively, and low values of 640 cP in juice. In sensory tests, the 70% WPI: 30% MS juice obtained a score 6.97, an improvement in relation to the other agglomerates, but not differing (p = 0.681) from the commercial thickener, 6.91 (p = 0.380). Based on these results, the 70% WPI: 30% MS was suggested for use in the nutritional therapy of patients with dysphagia.

Keywords: Dysphagia; Viscosity; Supplementation; Milk Whey Proteins; Malnutrition

1. Introduction

Many of the nutritional complications that occur with dysphagia sufferers are due to the low viscosity of the diet. Fluid liquids are difficult to swallow by patients with reduced deglutition control. This happens because such liquids can be swallowed very quickly and do not maintain any defined shape in the oral cavity, thus allowing part of the liquid food to penetrate the airways, which are still open. As a result, accidental, bronchoaspiration will occur, and ensuing pneumonia can worsen the nutritional state of the patient leading to dehydration, as the disease progresses. Proper thickening of the food can minimise such intercurrences [1,2].

In this context and aiming to optimise the nutritional care of dysphagic patients sufferers, the American guide "National Dysphagia Diet: Standardization for Optimal Care" established standards and limits for the variation in viscosity, with the objective of making the diet adequate for patients with compromised swallowing capacity. Such standards and limits do not necessarily represent the expected variation, but serve as a basis for discussion

and for an analysis of the prescribed diet. These values are classified in centipoisess (cP) as "thin" (1 - 50 cP—water), "nectar" (51 - 350 cP—mango juice), "honey" (351 - 1750 cP) and "pudding" (>1750 cP) [3].

Ingredients such as milk whey proteins, when chemically or physically modified, can alter the viscosity of food systems. However, on their own, they are unable to reach the viscosity standardised by the American Dietetics Association (ADA) [3] for fluids, but agglomerated with carbohydrate sources such as modified starch, they could increase the viscosity of food systems to values close to those cited by the ADA.

Based on the high malnutrition indices and difficulty in nutritional handling of patients with dysphagia, the main objective of this study was to develop and sensory test a protein supplement based on milk whey proteins and modified starch, with adequate thickening capacity, for nutritional therapy.

2. Materials & Methods

2.1. Raw Material

Whey protein isolate (WPI), concentrate (WPC) and hy-

*Corresponding author.

drolysate (WPH) from Glanbia Foods Inc. (Monroe, Wisconsin, USA) were used, donated by Integralmédica Teaching and Research Institute, São Paulo, Brazil; calcium caseinate (CaCas) and modified starch (Thick & Easy—Hormel Health Labs).

2.2. Proximate Composition of the Protein Sources

The moisture, total solids, ash and protein contents were determined [4]. Total lipids were determined [5], and total per-cent carbohydrates were estimated by difference, subtracting the sum of the values obtained in the other determinations from 100.

2.3. Determination of the Degree of Hydrolysis

The degree of hydrolysis (DH) was determined based on the Adler-Niessen [6] method, which consists of the spectrophotometric measurement of the chromophore formed in the reaction between trinitrobenzenesulphonic acid (TNBS) and amino groups, under alkaline conditions. After 1 hour of incubation, the reaction was interrupted by lowering the pH with 0.1 M HCl. The sample was dispersed in sodium dodecyl sulphate (SDS) and the reaction occurred in 0.2125 M phosphate buffer, pH 8.2 L-leucine (0 to 2.0 mM) was used as the standard and the readings made at 340 nm.

2.4. Solubility of the Protein Sources

The protein solubility (% PS) was determined according to the method of Morr *et al.* (1985). The effects of pH (2.5 to 7.5) were studied for all the protein sources: WPI, WPC, CaCas and WPH.

2.5. Determination of Total Amino Acids

The total amino acids were determined by reversed phase liquid chromatography [7] after a 24 h acid hydrolysis step, in 20% HCl plus phenol, followed by derivatisation with phenylisothiocyanate.

2.6. Agglomerate Production

Different concentrations of modified starch, calcium caseinate and milk whey protein isolate, concentrate and hydrolysate were used.

Preliminary physicochemical tests were carried out using modified starch as the standard. Concentrations varying from 10% to 50% protein plus modified starch were tested and evaluated for their viscosity profile. The formulation showing a viscosity closest to that of the commercial thickener was submitted to the process of agglomeration. The formulation selected for this study was 70% protein: 30% modified starch.

2.7. Agglomeration Process

Agglomeration of the formulations was carried out using a pilot plant scale model RCR instantiser (capacity of 5 kg) from ICF Industrie S.p.a. (Maranello-MO, Italy).

The following parameters were used for agglomeration: round 1.5 μm mesh grill, with 50% opening and a rotary dryer at 95°C with a vapour pressure of 2 bar. After agglomeration and cooling, the products were packed into 500 g packages and stored at a temperature of approximately 10°C.

2.8. Granulometric Distribution

The particle size distribution of the protein sources and agglomerated products was determined using the vibratory Granutest equipment vibrating for 5 minutes with the 150, 250, 350 and 500 μm sieves. The amounts retained on each sieve were weighed and expressed as percentages.

2.9. Water Activity of the Agglomerated Products

This was determined [4] at zero time (for samples at room temperature). A portable pawkit portable water activity, model 950 NE from Decagon® was used for the measurements. The equipment was calibrated using saturated solutions of magnesium and lithium chlorides (water activities of 0.1 - 0.4) and potassium sulphate (water activity of 0.975). The measurements were made in duplicate at room temperature (25.5°C).

2.10. Water Absorption Capacity

The water absorption capacity of the protein sources and agglomerated products was measured in triplicate [8,9]. This technique consists of using a horizontal capillary connected to a Buchner funnel by flexible tubing. 100 g-samples were placed on a Whatman n° 1 filter paper, and the water absorbed at equilibrium expressed as the mLs of water absorbed/g of protein.

2.11. Bulk Density

This was determined in triplicate, based on the definition: the mass of particles occupying a defined volume. The analysis consists of standardising the product distribution, based on the volume it occupies [10], placing 3 g of sample in a 25 mL graduated cylinder with the help of a funnel, thus standardising the distribution and minimising the effects of agglomerate compacting.

2.12. Particle Density

This was determined in triplicate using the pycnometer [11], with toluene as the inert liquid ($p = 0.866$ g·cm^{-3}).

2.13. Porosity

The porosity (%P) of the protein sources and agglomerated products was determined according to Peleg [10].

2.14. Viscosity

The apparent viscosities (η) were determined in a Brookfield DV—III rheometer with a shear rate of 30 rpm (similar to that obtained in the processes of chewing and swallowing a food), using 30 s reading time and spindles 16, 18 and 31, the results being expressed in centipoises (cP). Protein concentrations of 18 and 28 g were used for women and men respectively, representing 30% of the individual's needs. The protein sources or agglomerates were added to the following food systems: water, whole UHT milk and commercial orange juice (pH 4.0 and 1.5% total solids). The viscosity was determined in triplicate at 25°C.

2.15. Analysis of Preference of the Foods Containing Added Agglomerates

Two samples were used: control (juice/milk + commercial thickener) and experimental (juice/milk + agglomerated products), 18 g and 28 g for women and men, respectively. The samples were served to the panellist (patients) individually in a monadic way and balanced order, using 50 mL disposable plastic cups coded with 3 digit algorithms. The participation of the 30 patients in the test was voluntary after agreeing to take part and freely signing a consent form. A 9-cm structured hedonic scale was used [12]. The protocol was approved by the Ethics Committee (n° 428/2006) of the University of Campinas School of Medicine.

2.16. Statistical Analyses

All the results were analysed by multivariate analysis of variance and the differences amongst the means by Tukey's test. The software used was the Statistica-Basic Statistics and Tables and SPSS for windows 15.1.

3. Results & Discussion

3.1. Proximate Composition of the Protein Sources

With the exception of WPC, all the materials used presented a protein concentration above 80% (**Table 1**). There was no significant difference in protein content between WPI and WPH. The sample with the lowest protein concentration was WPC. With respect to the lipid content, WPC showed the highest values and there was no significant difference (p = 0.835) between WPI and WPH or between CaCas and WPH. For the ash content, there was a significant difference between WPI and WPC (p = 0.7110); WPI and CaCas (p = 0.673); and between WPC and CaCas (p = 0.742). With respect to moisture content, there was no significant difference (p = 0.0613) between the samples CaCas and WPH, but WPI and WPC were statistically different from the other samples. WPC showed the highest carbohydrate content and WPI the lowest. WPI is obtained by removing the carbohydrates, lipids and salts from WPC. For this reason, the WPI showed a higher protein concentration and lower carbohydrate and ash contents.

3.2. Determination of the Degree of Hydrolysis

The DH found was 10.11 mM·g^{-1}, classified as a protein source of medium degree of hydrolysis (7 to 15 mM·g^{-1}).

3.3. Solubility of the Protein Sources

Milk whey protein concentrates and isolates show good solubility throughout a wide range of pH values, temperatures, protein concentrations, water activities and ionic strengths [13]. Thus this property was shown to be stable in the various pH ranges applied. With CaCas a decrease in solubility was found at pH values close to the isoelectric point (pH 4.6), as shown in **Table 2**.

In foods, protein solubility is affected by the pH, ionic strength, temperature, solvent polarity, isolation method, processing conditions, interactions with other compo-

Table 1. Proximate composition of the protein sources: isolate (WPI), concentrate (WPC) and hydrolysate (WPH) milk whey protein and calcium caseinate (CasCa).

Sources	Protein (%)[1,2,3]	Fat (%)[1,2,3]	Ash (%)[1,2,3]	Water (%)[1,2,3]	Carbohydrate (%)[2,4]
WPI	92.94 ± 0.01[a]	0.57 ± 0.02[b]	1.58 ± 0.69[ab]	4.69 ± 0.03[c]	0.22
WPC	77.30 ± 0.43[c]	0.70 ± 0.03[a]	2.15 ± 0.33[a]	6.36 ± 0.01[a]	13.49
CasCa	89.31 ± 0.11[b]	0.45 ± 0.02[c]	1.74 ± 0.31[a]	5.25 ± 0.02[b]	3.25
WPH	92.38 ± 0.06[a]	0.51 ± 0.03[bc]	1.39 ± 0.88[b]	5.16 ± 0.08[b]	0.56

[1]Values correspond to means (± SD) of three determinations; [2]Values expressed in dry basis; [3]Values not sharing similar letter in the same column are different (p < 0.05) in Tukey test; [4]Calculated by difference = 100 − (protein + total far + ash + water).

nents and mechanical treatments [14]. These factors affect the solubility of the proteins, mainly causing alterations in the hydrophilic and hydrophobic interactions of the surface groups of the protein with the solvent [15].

Whey proteins remain soluble around their isoelectric point (pI), that is, in the pH range between 4 and 5 or between 4 and 6 [16,17]. In the pH range between 4 and 6, no decrease in the values for solubility was found for the sources WPC and WPI.

3.4. Total Amino Acid Contents of the Protein Sources

The protein sources used present amino acid compositions that attend all the recommendations for essential amino acids based on the Institute of Medicine standard [18] for all stages of life (**Table 3**).

Caseins have a high (35% - 45%) apolar amino acid content (Val, Leu, Ile, Phe, Tyr, Pro) and a low sulphur amino acid content, which limits their biological value [19]. Nevertheless, as can be seen in **Table 3**, the essential amino acid content of all the samples studied was in agreement with the IOM [18] reference standard.

Due to the profile presented by the milk whey proteins, they can be recommended for the formulation of various special products such as infant formulas [20], and for muscle metabolism and physical performance, particularly because of the high content of branched chain essential amino acids, such as leucine and isoleucine [21]. These peculiarities are extremely important for ALS patients, due to their hypermetabolism and progressive loss of lean mass with the clinical evolution of the disease.

3.5. Agglomeration Process Yield

The agglomeration process yield was 94.36, 94.53, 93.86 and 92.32% for WPI, WPC, CaCas and WPH, respectively.

Table 2. Solubility of the protein sources: isolate (WPI), concentrate (WPC) and hydrolysate (WPH) milk whey protein and calcium caseinate (CasCa) at 25°C.

pH	WPI[1,2]	WPC[1,2]	CasCa[1,2]	WPH[1,2]
2.5	71.01 ± 1.12[Da]	71.56 ± 1.45[Ca]	86.47 ± 0.94[Aa]	71.41 ± 1.03[Ab]
3.5	82.68 ± 0.45[Aa]	80.93 ± 0.53[Aa]	73.06 ± 0.31[Cb]	58.29 ± 0.31[Bc]
4.5	77.94 ± 0.34[Ca]	77.54 ± 0.11[Ba]	1.82 ± 0.21[Fb]	58.11 ± 0.14[Bb]
5.5	80.63 ± 0.12[Ba]	80.71 ± 0.28[Aa]	38.51 ± 0.47[Eb]	35.03 ± 0.29[Db]
6.5	80.56 ± 0.65[Ba]	81.18 ± 0.56[Aa]	56.45 ± 0.71[Db]	45.39 ± 0.61[Cc]
7.5	80.27 ± 0.16[Ba]	79.88 ± 0.72[Aa]	80.84 ± 0.03[Ba]	71.06 ± 0.85[Ab]

[1]Values correspond to means (+ SD) of three determinations; [2]Values sharing similar capital letter in the same column and small letter in the same line are not different (p > 0.05) in Tukey test.

Table 3. Total amino acids (g per 100 g of protein) of the protein sources: isolate (WPI), concentrate (WPC) and hydrolysate (WPH) milk whey protein and calcium caseinate (CasCa), compared to reference IOM (2002).

Amino acids (g/100g of protein)	IOM[2]-2002 Pre-school[1]	IOM[2]-2002 Adults[1]	WPI	WPC	CasCa	WPH
Threonine	2.7	2.4	6.06	5.76	4.31	6.56
Methionine + Cysteine	2.5	2.3	5.05	2.84	4.68	4.59
Valine	3.2	2.9	5.20	4.38	5.36	4.94
Leucine	5.5	5.2	14.24	8.92	9.24	10.66
Isoleucine	2.5	2.3	5.57	4.43	4.24	6.24
Phenylalanine + Tyrosine	4.7	4.1	8.69	5.47	9.57	6.34
Lysine	5.1	4.7	10.06	6.35	6.74	8.87
Histidine	1.8	1.7	1.76	1.47	3.06	1.37
Trytophan	0.8	6.0	*	*	*	*

[1]Values based on EAR (estimated average requirement): EAR amino acids/EAR protein; Children (1 to 3 years) EAR protein = 0.88 g/kg/day; adults (>18 years). EAR protein = 0.66 g/kg/day; [2]IOM: Institute of Medicine. *Amino acid not determinated.

3.6. Granulometric Distribution

Before starting the agglomeration process, the milk whey and casein protein sources had more than 90% of their particles retained on sieves with mesh below 150 μm, a size known to be characteristic of spray dried products (**Table 4**).

The agglomerates of WPI, WPC and WPH presented mostly particles smaller than 250 μm, although with a substantial increase in diameter of particles above 500 μm, generally of 40%. The majority of the CaCas particles were larger than 500 μm. WPH showed the smallest particles, justified by the smaller granulometry found in the granulometric distribution of the protein sources presented in **Table 4**.

3.7. Water Activity

The water activity is defined as the ratio between the water vapour pressure in equilibrium with a food and the vapour saturation pressure at the same temperature [22]. It is a measurement used in the quality control of foods, including powdered foods. In order to retard alterations in this type of product due to undesirable changes such as the exponential growth of microorganisms, enzymatic reactions or enzymatic browning, the value for a_w should be below 0.6. **Table 5** shows the values for water activity of the agglomerated products. The lowest values for a_w were found in the WPH agglomerate and the values for a_w increased for all the agglomerates during the 30 days of storage.

For all the agglomerates the values found for a_w at the beginning of the study (T_0) for all the formulations developed, were within the values reported in the literature for powdered (dehydrated) foods, which, by their very nature, present low a_w values, generally below 0.30. However at the end of this study (T_{30}), the a_w had increased to approximately 0.30, suggesting that the packaging used did not offer an adequate barrier to water vapour, and that a packaging material showing greater protection should therefore be used to store the agglomerated products.

3.8. Water Absorption Capacity of the Protein Sources and the Agglomerates

The water absorption capacity of the protein sources varied from 1.82 to 6.11 mL water absorbed/g protein. **Table 6** presents the WAC values of the protein sources.

Of all the sources, CaCas showed the highest WAC values. No statistical differences were observed between the WAC values of WPH and WPI (p = 0.0712) or WPC (p = 0.0604). Nevertheless the samples WPI and WPC were statistically different (p = 0.0021), WPI showing the higher values.

These findings are in agreement with the literature, where the difference between the WAC of protein sources has been attributed to protein denaturation. Protein sources containing more denatured protein and showing decreased solubility, exhibited higher WAC values [23, 24] This was clearly shown in the case of the CaCas samples, which presented the lowest values for solubility (independent of the isoelectric point) and highest values for WAC. Similar behaviour was shown by the WPC samples which, when compared to WPI, showed higher WAC and lower solubility at pH values of about 6.5, considering 5% probability.

The WAC values were shown to increase for all the products after the agglomeration process, as can be seen in **Table 6**. The agglomerate WPH did not differ statistically from the agglomerates WPI (p = 0.0860) and WPC (p = 0.0968) with respect to WAC. However, the agglomerates WPI and WPC were statistically different (p = 0.0490). The CaCas based agglomerate showed the highest values for WAC of all the products evaluated.

In part, the greater WAC observed for CaCas could be due to its granulometry, as compared to the other agglomerates. Increases in granule size improve WAC. To the contrary, products with very small granulometry (<125 μm) show reduced WAC, favouring the formation of clusters on the surface or deposition of residues. Agglomeration results in products with a porous structure that absorb liquids quicker, dissolving in an instantaneous way [25,26].

Table 4. Granulometric distribution of the protein sources and agglomerate: isolate (WPI), concentrate (WPC) and hydrolysate (WPH) milk whey protein and calcium caseinate (CasCa) using screen of <150, 250, 350 and 500 μm.

Components	<150 μm[1,2]		250 μm[1,2]		350 μm[1,2]		500 μm[1,2]	
	Source	Agglomerate	Source	Agglomerate	Source	Agglomerate	Source	Agglomerate
WPI	99.65 ± 0.3[a]	42.57 ± 1.1[b]	0.35 ± 0.9[b]	4.02 ± 1.4[a]		14.80 ± 1.2[a]		38.48 ± 0.7[b]
WPC	98.48 ± 0.4[a]	43.50 ± 0.7[b]	1.52 ± 0.8[a]	3.13 ± 0.4[a]		13.74 ± 1.8[a]		39.63 ± 0.4[b]
CasCa	99.88 ± 0.5[a]	42.45 ± 0.8[b]	0.12 ± 0.9[b]	4.07 ± 0.5[a]		9.86 ± 1.3[b]		43.62 ± 0.5[a]
WPH	99.37 ± 0.6[a]	46.84 ± 0.9[a]	0.25 ± 0.9[b]	3.86 ± 0.2[a]		9.82 ± 0.9[b]		39.48 ± 0.3[b]

[1]Values correspond to means (±SD) of three determinations; [2]Values not sharing similar letter in the same column are different (p < 0.05) in Tukey test.

Table 5. Water activity of the agglomerates: isolate (WPI), concentrate (WPC) and hydrolysate (WPH) milk whey protein and calcium caseinate (CasCa) at initial time (T_0) and final time, after thirty days (T_{30}), at 25°C.

Agglomerates	$a_w (T_0)$[1,2]	$a_w (T_{30})$[1,2]
	Temp. 25°	
WPI	0.24 ± 0.03^{Ab}	0.30 ± 0.01^{Aa}
WPC	0.25 ± 0.94^{Ab}	0.31 ± 0.10^{Aa}
CasCa	0.25 ± 0.03^{Ab}	0.31 ± 0.01^{Aa}
WPH	0.18 ± 0.06^{Bb}	0.23 ± 0.08^{Ba}

[1]Values correspond to means (± SD) of three determinations; [2]Values sharing similar capital letter in the same column and small letter in the same line are not different (p > 0.05) in Tukey test.

Table 6. Water absorption capacity (WAC) of the protein sources and agglomerate: isolate (WPI), concentrate (WPC) and hydrolysate (WPH) milk whey protein and calcium caseinate (CasCa) at 25°C during 30 minutes.

Components	WAC (mL absorbed water/g protein)[1,2]	
	Source	Agglomerates
WPI	1.82 ± 0.18^{Db}	4.02 ± 0.14^{Da}
WPC	3.77 ± 0.28^{Bb}	6.30 ± 0.23^{Ba}
CasCa	6.11 ± 0.10^{Ab}	8.17 ± 0.13^{Aa}
WPH	2.63 ± 0.09^{Cb}	5.15 ± 0.17^{Ca}

[1]Values correspond to means (± SD) of three determinations; [2]Values sharing similar capital letter in the same column and small letter in the same line are not different (p > 0.05) in Tukey test.

The physical-chemical alterations occurring during agglomeration can alter the WAC. One of the peculiarities of the agglomeration process is to moisten the fine particles of the powder with vapour, such that the particles enter into contact or collide with each other, forming porous agglomerates subsequently dried in hot air [25]. Alterations in the protein conformation resulting from this process can affect the thermodynamics of water binding by altering the availability of polar sites or hydration sites. The transition of the compact globular conformation of the protein molecule to a random conformation results in an increase of the available surface area and the exposure of peptides and amino acid side chains, that were natively hidden, thereby increasing their interaction with water [27].

3.9. Apparent Density, Particle Density and Porosity

The protein sources WPI and WPC were not statistically different from each other (p = 0.9315). The CaCas and WPH were significantly different (p = 0.0324) from each other and from the other protein sources (**Table 7**).

The apparent density depends on the intensity of the attractive forces between the particles, the particle size and the number of points of contact. Powdered foods have apparent densities between 0.3 and 0.8 g·cm^{-3}. The ρ_{ap} of powdered milk whey is 0.52 g·cm^{-3}, whilst for WPI with protein contents of 85%, this value was 0.38 g·cm^{-3} [28]. It could be seen that after agglomeration a significant difference was observed between the protein sources (raw material) and the agglomerates, the differences found between the samples of CaCas and WPH remaining (p = 0.033).

The particle density (ρ_{part}) reflects the existence of internal pores in the powder granules, and is defined as the mean weight of the particles per unit volume, excluding the volume occupied by interstitial air [29]. It is known as the measurement of true density, and is important in situations where one must obtain the relationship between the weight of the particles and the forces between them [10].

As in the case of the values obtained in the analyses for ρ_{ap} for the protein sources and agglomerated products, the WPH showed higher values for ρ_{part} for both the protein sources and the agglomerates (**Table 7**).

The particle density of the majority of powdered foods

Table 7. Apparent density (ρ_{ap}), particle density (ρ_{part}) and porosity P (%) of the protein sources and agglomerate: isolate (WPI), concentrate (WPC) and hydrolysate (WPH) milk whey protein and calcium caseinate (CasCa).

Components	ρ_{ap} g·cm^{-3} [1,2,3]		ρ_{part} g·cm^{-3} [1,2,4]		P (%)[1,2,5]	
	Source	Agglomerate	Source	Agglomerate	Source	Agglomerate
WPI	0.394 ± 0.042^{Ba}	0.344 ± 0.005^{Bb}	1.088 ± 0.006^{Bb}	1.334 ± 0.020^{Ba}	64.12^{Bb}	74.13^{Ba}
WPC	0.391 ± 0.018^{Ba}	0.341 ± 0.003^{Bb}	1.087 ± 0.003^{Bb}	1.332 ± 0.073^{Ba}	64.02^{Bb}	74.16^{Ba}
CasCa	0.366 ± 0.009^{Ca}	0.319 ± 0.036^{Cb}	1.085 ± 0.005^{Bb}	1.329 ± 0.037^{Ba}	66.27^{Ab}	75.27^{Aa}
WPH	0.413 ± 0.004^{Aa}	0.363 ± 0.012^{Ab}	1.093 ± 0.006^{Ab}	1.342 ± 0.001^{Aa}	62.21^{Cb}	74.89^{Ca}

[1]Values correspond to means (±SD) of three determinations; [2]Similar letter in the same column are not different (p > 0.05) in Tukey test; [3]Similar letter in the same line referent to ρ_{ap} are not different (p > 0.05) in Tukey test; [4]Similar letter in the same line referent to ρ_{part} are not different (p > 0.05) in Tukey test; [5]Similar letter in the same line referent to % P are not different (p > 0.05) in Tukey test.

is between 1.4 and 1.5 $g \cdot cm^{-3}$, depending on the moisture content [10]. However, for milk whey products, the value for ρ_{part} was 1.0 $g \cdot cm^{-3}$ [28], corroborating with the present study. The results of the present study corroborate with the papers cited, the variation being from 1.08 to 1.14 $g \cdot cm^{-3}$.

Porosity is a function of particle size, size distribution and form. The use of porosity allows for and facilitates the treatment and comparisons between powdered foods that could have different particle densities [10]. As in the case of apparent density, the results for % P of the protein sources and agglomerates showed significant differences between the % P of WPH and of CaCas (p = 0.092) and those of the other sources used. Amongst the agglomerates, the greatest % P was found for the CaCas samples (**Table 7**).

The values found in the present study agree with those found by Peleg [10], who showed that powdered foods with ρ_{part} of about 1.4 $g \cdot cm^{-3}$ showed internal, external or both porosities between 40 and 80%. The % P can be a parameter showing the efficiency of the agglomeration process. The interstitial space shown, in an irregular array, by large particles, favours wettability (ability of the powder to bind water on the surface), whereas small, symmetrical particles show reduced interstices that hamper water penetration [26,29].

3.10. Apparent Viscosity of the Protein Sources and Agglomerated Products

The viscosity of a food is one of the most important variables in swallowing. Thin liquids make swallowing difficult for patients with reduced oral control, since they are swallowed quickly and fail to maintain any form inside the oral cavity. Part of the liquid food may slip prematurely to the pharynx and thus penetrate the still open airways, that is, before swallowing actually occurs. To avoid this effect, the ideal viscosity for swallowing to occur safely should be determined [2].

Considering the protein sources of different food systems as a base, and using protein concentrations of 18 and 28 g, it can be seen that the sources WPI, WPC and WPH showed similar behaviour in water and in milk. CaCas showed higher values in whole milk and lower values in orange juice (**Table 8**).

Considering the food system in water as the base, no differences of viscosity were observed between WPC and WPH (p = 0.961), the same being observed in milk (p = 0.955) and juice (p = 0.738) at a concentration of 18 g of protein. Similar behaviour was observed at a final concentration of 28 g of protein. As in water, the values found for CaCas in milk were greater than the others. In orange juice, WPI showed the highest values and CaCas the lowest (**Table 8**).

According to the standards established by ADA [3], both for the use of 18 g (recommended for female adult) and 28 g (recommended for a male adult), the protein sources WPC and WPH were classified as thin liquids (1 - 50 centipoise—cP) in all the food systems. CaCas was classified as nectar (51 - 350 cP) in water and milk, and in orange juice this same source was classified as a thin liquid (1 - 50 cP—water). The low values for the viscosity found with CaCas in the orange juice are related to the system pH of about 4.5, close to the isoelectric point of the casein, which favours precipitation of these sources. WPI was classified as a nectar in all the food systems.

Table 8. Apparent viscosity (η'), in centipoise (cP), of the protein sources: isolate (WPI), concentrate (WPC) and hydrolysate (WPH) milk whey protein and calcium caseinate (CasCa) at the concentration of 18 at 28 g in 250 mL of foods (water, whole milk and orange juice) at 25°C.

Protein source		η' (cP)[1,2]		
		Water	Whole milk	Orange juice
[ ] 18 g of protein	WPI	90.91 ± 0.18^{Db}	90.46 ± 0.37^{Db}	103.06 ± 0.67^{Ba}
	WPC	5.38 ± 0.28^{Fb}	6.01 ± 0.53^{Fb}	11.30 ± 0.35^{Fa}
	CasCa	114.4 ± 1.04^{Cb}	192.33 ± 0.34^{Ba}	20.29 ± 0.41^{Dc}
	WPH	5.30 ± 0.87^{Fb}	5.45 ± 0.65^{Fb}	10.90 ± 0.07^{Fa}
[ ] 28 g of protein	WPI	140.70 ± 0.61^{Bb}	140.46 ± 0.79^{Cb}	160.7 ± 0.52^{Aa}
	WPC	10.11 ± 0.20^{Ea}	10.35 ± 0.45^{Eb}	18.60 ± 0.12^{Ea}
	CasCa	181.5 ± 1.32^{Ab}	302.36 ± 0.47^{Aa}	32.15 ± 2.15^{Cc}
	WPH	10.14 ± 0.02^{Ea}	10.15 ± 0.03^{Eb}	18.25 ± 0.98^{Ea}

[1]Values correspond to means (± SD) of three determinations; [2]Values sharing similar capital letter in the same column and small letter in the same line are not different (p > 0.05) in Tukey test.

After the agglomeration, which was the main purpose of the present study, it was verified that the apparent viscosity increased for all the agglomerates (**Table 9**). In relation to the behaviour of the agglomerates in the different systems, it can be seen that those of WPI showed the highest values of viscosity in the orange juice. Those from CaCas showed the highest values in milk and lowest in the orange juice.

Comparing the various agglomerates in a single system, with a final concentration of 18 g protein in water, the highest viscosity values were found for CaCas and the lowest for WPI and WPH. Similar behaviour was shown with a final concentration of 28 g of protein. In milk the highest values were again found for CaCas at the two protein concentrations, followed by WPI. The agglomerates WPC and WPH presented no significant differences at the protein concentrations of 18 g (p = 0.52) and 28 g (p = 0.22). In orange juice the WPI agglomerates showed the highest values at both concentrations and the CaCas agglomerates the lowest values.

At the concentration of 18 g protein, the WPI agglomerates showed viscosity values varying from 1361.33 to 3283.31 cP. In water these agglomerates were classified as honey (351 - 1750 cP), and in milk and orange juice presented the consistency of pudding (>1750 cP), showing significant differences in all the systems. Lower values were found for the WPC and WPH agglomerates, being classified as honey in water and milk and pudding in orange juice. The CaCas agglomerates showed the highest values in the water and milk systems (pudding), but in orange juice the viscosity was only 410.33 cP (honey).

When used at the concentration of 28 g (**Table 9**), the WPI agglomerates were classified as pudding in all the

systems. In water and milk, WPC showed the same behaviour, being classified as honey, but in orange juice the viscosity was 2980.73 cP (pudding). The CaCas agglomerates showed higher values than the other agglomerates in water and milk and lower values than the others in the orange juice. WPH agglomerates showed a viscosity of 1005.00 cP in water and 1017 cP in milk, being classified as honey. In orange juice the viscosity of these agglomerates was classified as pudding.

3.11. Analysis of Preference of the Foods Containing Added Agglomerates

Figure 1 shows the results of the preference tests carried out with ALS patients. It can be seen that the means for preference of the milk and orange juice systems with added WPI agglomerate differed statistically (p = 0.007), being between 5.61 and 6.97 (between "liked slightly" and "liked moderately"). No significant difference (p = 0.804) was observed between the use of this agglomerate and the commercial thickener.

The means found for the agglomerates based on WPC, CaCas and WPH in orange juice were 3.61 (±1.26), 1.37 (±0.59) and 1.23 (±0.42). In milk the means were 3.25 (±1.26), 1.63 (±0.77) and 1.37 (±0.49). For the WPC agglomerates the scores given on the hedonic scale were between "disliked moderately" and "disliked slightly" and for the CaCas and WPH agglomerates between "disliked intensely" and "disliked a lot".

It is important to point out that in the systems with added 70% WPI: 30% MS and with added commercial thickener, means above the cut-off point of 5 [30] were obtained, indicating preference for these products.

Table 9. Apparent viscosity (η'), centipoise (cP), of the agglomerates: isolate (WPI), concentrate (WPC) and hydrolysate (WPH) milk whey protein and calcium caseinate (CasCa) at the concentration of 18 at 28 g in 250 mL of foods (water, whole milk and orange juice) at 25°C.

Agglomerates		η' (cP)[1,2]		
		Water	Whole milk	Orange juice
[ ] 18 g of protein	WPI	1361.33 ± 1.10^{Dc}	2701.38 ± 1.50^{Db}	3283.31 ± 2.30^{Ba}
	WPC	864.33 ± 1.15^{Gb}	850.57 ± 0.50^{Fc}	2850.00 ± 1.02^{Da}
	CasCa	2051.67 ± 2.89^{Cb}	4273.33 ± 2.39^{Ba}	410.33 ± 0.58^{Hc}
	WPH	862.66 ± 1.53^{Gb}	854.00 ± 0.21^{Fc}	1851.00 ± 1.00^{Fa}
[ ] 28 g of protein	WPI	2122.66 ± 2.31^{Bc}	4001.32 ± 1.14^{Cb}	5110.66 ± 1.15^{Aa}
	WPC	1115.00 ± 0.34^{Eb}	1018.18 ± 0.32^{Eb}	2980.73 ± 0.62^{Ca}
	CasCa	3200.60 ± 0.58^{Ab}	6651.66 ± 1.53^{Aa}	640.16 ± 0.29^{Gc}
	WPH	988.00 ± 0.32^{Fc}	1017.00 ± 0.48^{Eb}	2600.00 ± 0.01^{Ea}

[1]Values correspond to means (± SD) of three determinations; [2]Values sharing similar capital letter in the same column and small letter in the same line are not different (p > 0.05) in Tukey test.

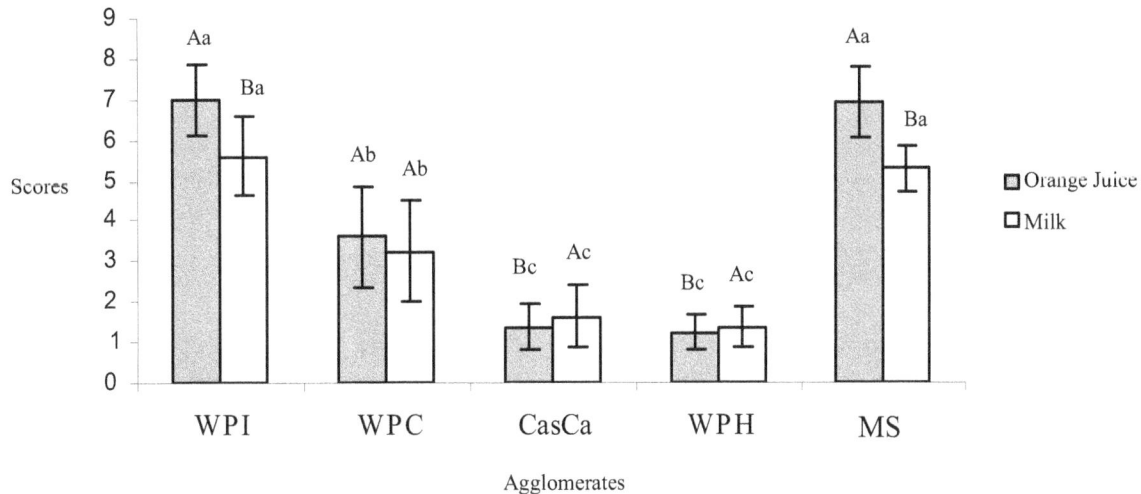

Figure 1. Analysis of preference of the foods containing added agglomerates isolate (WPI), concentrate (WPC) and hydrolysate (WPH) milk whey protein, calcium caseinate (CasCa) and modified starch (MS) at the concentration of 18 at 28 g in 250 mL at 25°C. *Similar capital letter in the same column that represent the same agglomarate are not different (p > 0.05) in Tukey test. Similar small letter in the same column that represent the same food are not different (p > 0.05) in Tukey test.

4. Conclusions

With respect to the proximate composition of the protein sources, the protein concentration of the WPI was higher than those of the other sources. Of all the protein sources, WPI and WPC exhibited the highest values for solubility, independent of pH. The CaCas showed higher solubility at the extreme pH values, but solubility at the isoelectric point was close to 0. WPH showed intermediate solubility with the minimum values at pH values between 3.5 and 6.5. All the sources satisfied the recommendations in terms of essential amino acids, according to the IOM recommendation of 2002.

The elaborated formulations showed good yields, varying from 92.53% to 94.53%, and an increase in granule size of the products after agglomeration, with more than 38% of the particles presenting diameters greater than 500 μm. Despite the low initial water activity values, the agglomerates showed increased values after 30 days of storage, suggesting that the packaging used did not offer an adequate barrier to water vapour. The greatest values for water absorption capacity were noted for the protein sources and agglomerated products containing CaCas, and an expressive increase in this property was observed for the milk whey protein based agglomerates (WPI, WPC and WPH) after the agglomeration process.

There was no significant difference (p > 0.05) in apparent density between the protein sources WPI and WPC, and the values for CaCas were lower than those of WPH. The values for apparent density of the protein sources decreased after agglomeration with modified starch. The mean density of the protein source particles was about 1.09 $g \cdot cm^{-3}$, increasing to 1.3 $g \cdot cm^{-3}$ after agglomeration, close to the expected value for powdered foods. Considering the values for the apparent and particle densities, the values for porosity were obtained, showing an increase after agglomeration and resulting in highly porous products, as desired for this type of product.

Of the protein sources, the WPI and CaCas samples at concentrations of 18 and 28 g, presented higher values for viscosity. The former showed similar behaviour in all the systems, but in orange juice the CaCas showed low values for viscosity because of the system pH (close to the isoelectric point of the caseins). After the agglomeration process, the products were able to modify the system viscosity, approaching that of the ADA standard (2002). The WPI agglomerate stood out because of the values obtained and its versatility in all the systems evaluated. In the sensory preference tests, the WPI agglomerate obtained the highest scores from the patients, not differing from those obtained with the commercial thickener. The WPC, CaCas and WPH samples received below average scores for acceptance.

Considering the set of physical-chemical and nutritional properties of the 70% WPI: 30% MS agglomerate as a whole, for use as a food thickener for feeding patients suffering from ALS, it was concluded that it would be an advantageous substitute for the current commercial thickeners in managing of the disease. The cost factor should be added to the benefits of a nutritional therapy.

5. Acknowledgements

The authors are grateful to the Integralmedica Teaching and Research Institute, São Paulo, Brazil, for donating the protein sources; to CNPq, the Brazilian National Research Council, for the doctoral scholarship to LBCS.

REFERENCES

[1] L. B. Carvalho-Silva and C. M. Ikeda, "Nutrition Care in Dysphagia: An Alternative for the Maximization of the Nutritional State," *Brazilian Journal of Clinical Nutrition*, Vol. 24, No. 3, 2009, pp. 203-210.

[2] L. B. Carvalho-Silva, "Amyotrophic Lateral Sclerosis," In: M. H. Maurer, Ed., *Rijeka*, Vol. 1, 2012, pp. 595-612.

[3] ADA National Dysphagia Diet, "The Manual of Clinical Dietetics," 5th Edition, The American Dietetics Association, Chicago, 1996, p. 56.

[4] Association of Official Analytical Chemists, "Official Methods of Analysis," In: W. Horwtz, Eds., Washington DC, 1990, p. 98.

[5] E. G. Bligh and W. J. Dyer, "A Rapid Method of Total Lipid Extraction and Purification," *Canadian Journal of Biochemistry and Physiology*, Vol. 37, No. 8, 1959, pp. 911-917.

[6] J. Adler-Nissen, "Determination of the Degree of Hydrolysis of Food Protein Hydrolysates by Trinitrobenzenesulfonic Acid," *Journal of Agriculture and Food Chemistry*, Vol. 27, No. 6, 1979, pp. 1256-1262.

[7] S. R. Hagen, B. Frost and J. Augustin, "Pre-Column Phenylisothiocyanate derivatization and Liquid Chromatography of Amino Acids in Food," *Journal of AOAC*, Vol. 72, No. 6, 1989, pp. 912-916.

[8] H. Baumann, "Appatur Nach Baumann zur Besting der Flússig-Keitsaunahme von Pulvrigen Substanzen," *Fette*, Vol. 68, No. 9, 1966, p. 741.

[9] H. Torgensen and R. T. Toledo, "Physical Properties of Protein Preparations Relates to Their Functional Characteristics in Comminuted Meat Systems," *Journal of Food Science*, Vol. 42, No. 6, 1977, pp. 1615-1618.

[10] M. Peleg, "Physical Properties of Foods," In: M. Peleg and E. B. Bagley, Eds., AVI Publishing Co. Inc., Westport, 1983, pp. 293-321.

[11] Y. Pomeranz and C. E. Meloan, "Food Analysis: Theory and Practice," 3rd Edition, AVI Publishing Co. Inc., Westport, 1994.

[12] M. Meilgaard, G. V. Civille and B. T. Carr, "Sensory Evaluation Techniques," 3rd Edition, CRC Press, New York, 1989.

[13] L. M. Huffman, "Processing Whey Protein for Use as a Food Ingredient," *Food Technology*, Vol. 50, No. 2, 1996, pp. 49-52.

[14] F. Vojdani, "Methods of Testing Protein Functionality," In: G. M. Hall, Ed., Blackie Academic & Professional, London, 1996, pp. 11-60.

[15] S. Damodaran, "Food Proteins," In: J. E. Linsella, Ed., Champaign, 1989, pp. 21-51.

[16] C. V. Morr, B. German, J. E. Kinsella, J. P. Regenstein, V. Buren, A. Kilara, B. A. Lewis and M. E. Mangino, "Collaborative Study to Develop a Standardized Food Protein Solubility Procedure," *Journal of Food Science*, Vol. 50,

No. 6, 1985, pp. 1715-1718.

[17] J. Giese, "Proteins as Ingredients: Types, Functions, Applications," *Food Technology*, Vol. 48, No. 10, 1994, pp. 50-60.

[18] IOM (Institute of Medicine), "National Academy of Sciences on Dietary Reference Intakes (DRI's). Dietary Reference Intakes for Energy, Carbohydrates, Fiber, Fat, Protein and Amino Acids (Macronutrients), National Academy Press, Washington, 2002.

[19] C. G. De Kruif and C. Holt, "Advanced Dairy Chemistry," In: P. F. Fox and P. L. H. CaSweeney, Eds., 3rd Edition, Kluwer Academic/Plenum Publishers, New York, Vol. 1, 2003, pp. 233-276.

[20] L. Hambraeus, "Nutritional Aspects of Milk Proteins," In: P. F. Fox, Ed., *Development of Dairy Chemistry*, Applies Science, London, 1982, pp. 289-313.

[21] R. D. Steele and A. E. Harper, "Present Knowledge in Nutrition," In: M. L. Brown, Ed., 6th Edition, *Nutritional Foundation*, Washington DC, 1990, pp. 67-79.

[22] A. J. Fontana, "Water Activity: Why Is It Important for Food Safety," *Proceedings of the 1st NSF International Conference of Food Safety*, Albuquerque, November 1998.

[23] E. L. Arrese, D. A. Sorgentini, J. R. Wagner and M. C. Añon, "Eletrophorctic, Solubility, and Functional Properties of Commercial Soy Protein Isolates," *Journal of Agricultural and Food Chemistry*, Vol. 39, No. 6, 1991, pp. 1029-1032.

[24] G. Remondetto, M. C. Añon and R. J. González, "Hydratation Properties of Soybean Protein Isolates," *Brasilian Archives of Biology and Technology*, Vol. 44, No. 4, 2001, pp. 425-431.

[25] J. M. Aguilera, J. M. Valle and M. Karel, "Caking Phenomena in Amorphous Food Powders," *Trends in Food Science & Technology*, Vol. 6, No. 5, 1995, pp. 149-155.

[26] K. Masters, "Spray Drying. Chemical and Process Engineering Series," Leonard Hill Books, London, 1972.

[27] J. E. Kinsella, "Milk Proteins: Physicochemical and Functional Properties," *Critical Reviews in Food Science and Nutrition*, Vol. 21, No. 3, 1994, pp. 197-287.

[28] G. V. Barbosa-Cánovas and P. Juliano, "Physical and Chemical Properties of Food Powders," In: C. Onwulata, Eds., *Encapsulated and Powdered Foods*, Taylor & Francis, New York, 2005, pp. 39-71.

[29] A. B. R. Maia and M. Golgher, "Parâmetros para a Avaliação da Qualidade de Reconstituição do leite em pó Desidratado em Secador de Aspersão ('Spray Dryer')," *Boletim SBCTA*, Vol. 17, 1983, pp. 235-254.

[30] H. Stone and J. L. Sidel, "Sensory Evaluation Practices," Academic Press, New York, 1993, pp. 69-96.

15

Traditional Food Culture (Local Cuisines, Japanese Sake) That Has Been Nurtured by the Rich Nature of the Region: The Case of the Coastal Area in Chiba Prefecture, Japan

Korehisa Kaneko[1*], Keiko Oshida[2], Hajime Matsushima[3]

[1]Hokuso Creature Association, Tokyo, Japan; [2]Department of Town Planning and Design, College of Science and Technology, Nihon University, Chiba, Japan; [3]Research Faculty of Agriculture, Hokkaido University, Sapporo, Japan.

ABSTRACT

In Chiba Prefecture, Japan, during the Edo period (1603-1867), the development of waterway traffic by ships and the management of ports, highways and post towns around the ports progressed with the prosperity of the Edo (present-day Tokyo), which became heavily populated and the center of politics. We estimated that the demand of Japanese sake, which is luxury grocery item, was high. The freshwater layer that is abundant in mineral water to a depth of approximately 10 m is formed in coastal sand dunes. The fresh water layer is hard water, in which the concentrations of minerals such as calcium and magnesium are high. When the fresh water layer is used as the preparation water, the working rice malt and yeast in the sake brewing process become active. Japanese sake trends to be dry with a full-bodied taste. In addition, the main ingredients of local cuisines are fish and shellfish; many local cuisines are seasoned using soy sauce, miso and salt, and these local cuisines pair well with the type of Japanese sake described above. The local cuisines have been nurtured in harmony with the region's rich nature and heritage. In the future, we need to conserve the rich natural environment of the tidal flat, coast, seaweed beds, and marine, which have been producing the main local cuisine in Chiba Prefecture, and the water source area (a successive environment on the plateau from the coast, which was previously called the coastal dune area) of the preparation water for making Japanese sake. We also need to proactively develop local production for local consumption activities. Thus, we hypothesize that if the Japanese food life is secured and the traditional food culture is continued, the region will become revitalized by the development of the exchanges in the region.

Keywords: Local Cuisines; Japanese Sake; Traditional Food Culture; Coastal Environment; Blessing of Nature

1. Introduction

Chiba Prefecture is located in the central portion of the Japanese archipelago and is surrounded by the sea on three sides by Tokyo Bay and the Pacific Ocean. On the Tokyo Bay side, the tidal flats and shallows were spread until the period of rapid economic growth (1950) after World War II, and the region on the Pacific Ocean side is blessed with nature, where the Kurosio (warm current) and Oyashio (cold current) are confluent and the seafood is rich. In particular, since Ieyasu Tokugawa opened sho-gunate at Edo (present-day Tokyo) in 1603, Edo flourished as a central city of politics, economics and culture in Japan and became the major food consumption area with the rapid increase in population. In Chiba Prefecture, which is located close to Edo, large-scale development of rice paddies was conducted to increase rice production. Moreover, fishing and fishing villages have developed significantly with the rapid increase in migrant fishing populations from the southwestern region in Japan, which developed the fishery regions at the time, and fishing technology progressed. Consequently, a new settlement was created, and the population of Chiba Prefec-

*Corresponding author.

ture increased [1]. Through ties with the capital of Edo, land and water transportation routes in Chiba Prefecture developed, and post towns, ports, and the highway developed. [2] reported that during the Edo period, Japanese sake was actively made, and in the coastal and rural regions in Chiba Prefecture, high quantities of Japanese sake were consumed. In particular, there are many towns where the function of the port was developed [3]; the coastal sand dune zone near the port formed a layer of fresh water approximately 10 m underground originating from rain filtered through the sand stratum and the land area. The fresh water layer is hard water, in which the concentrations of minerals such as calcium and magnesium are high; hard water activates rice-malt and yeast, making the fresh water suitable for sake brewing [4,5]. We suggested that the demand of Japanese sake, which is luxury grocery item, was high. This origin of local cuisines is unknown; however, it is known that fishermen had to eat on the ship. As inferred from the history and culture, the main local cuisine in Chiba Prefecture contains many fish dishes [6]. The local cuisine was used as a relish and has evolved as a traditional food culture and in the regional industry. The relationship between nature and traditional food culture is very important to support the foundation population survival. However, few examples of this relationship have been examined in detail, which is an issue to be considered.

In this study, we report the relationship between nature and traditional food culture (including local cuisines and Japanese sake).

2. Materials and Methods

Study Site

Chiba Prefecture is located in the central portion of the Japanese archipelago and is surrounded by the sea on three sides by Tokyo Bay and the Pacific Ocean (**Figure 1**). On the Pacific Ocean side, where the Kurosio (warm current) and Oyashio (cold current) are confluent, there are rich fishing grounds that the environment has varied. Fisheries are one of the largest industries in the country that focuses on coastal fishing and offshore fishing. Based on fishing methods, such as unloaded fish and topographical features of the coast, the area was divided into 4 areas as follows: Choshi·Kjuukuri (Choshi City to Ichinomiya City: coastline extension 83.1 km), Sotobo (Isumi City to Minami-Boso City: coastline extension 132.7 km), Uchibo (Tateyama City to Futthu City: coastline extension 85.0 km), and Inner Bay of Tokyo Bay (Kisarazu City to Urayashu City: coastline extension 197.0 km) [3]. For sea fishery catches in 2012, Chiba was ranked 7th in Japan with 162,634 tons [7].

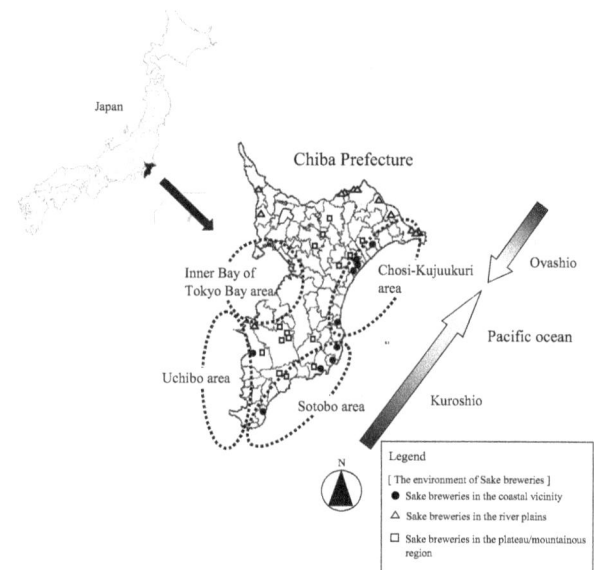

Figure 1. Study sites.

3. Results

3.1. The Main Local Cuisines in Chiba Prefecture

The main local cuisines in Chiba Prefecture are mainly fish dishes such as a mactra chinensis dish, a sardine dish, gomazuke, kusari-zushi (sushi seasoned with vinegared fish), sliced raw fish, sanga broil, namerou (cuisine chopped with miso and spice fresh fish), finely chopped horse mackerel, a bonito dish, a conger eel dish, turbo cooked in its own shell, and a saury dish, among others [8].

The Choshi·Kujuukuri region is the producing area of local cuisine, such as the sardine dish and the saury dish; the Kujuukuri, Sotobo, and Uchibo areas produce the finely chopped horse mackerel, bonito dish, turbo cooked in its own shell, nuta of turban shell, and magocha. The Sotobo and Uchibo areas produce namerou and sanga broil, and the Inner Bay of Tokyo Bay produces the conger eel dish, manila clam dish, and the mactra chinensis dish (**Table 1**).

As a taste of local cuisine, the mactra chinensis dish and the sardine dish (sliced raw sardine, finely chopped horse sardine) are mainly slightly plain in taste; the sanga broil, finely chopped horse mackerel, and bonito dish are either slightly or heavily seasoned using soy sauce, miso, vinegar, or other flavorings. The conger eel dish, Namero, and turbo cooked in its own shell are heavily seasoned to taste using salt, soy sauce, and secret sauce.

3.2. The Catch of Major Fish and Shellfish That Are Ingredients of Local Cuisine

We investigated the catch of the main fish and shellfish

that are ingredients of local cuisine in each area [9-12] with respect to the composition rate of the catch in each area. Sardines and saury accounted for approximately 85% of the whole catch in the Choshi·Kujukuri area. Tunas and pompanos were plentiful in the Choshi·Kujukuri and Sotobo areas; the total catch in these areas was

more than 80%. Bonitos and turban shell were high at 50% to 80% of the total catch in the Sotobo area. Conger and little necks were the highest in the Inner Bay of Tokyo Bay and Uchibo; they were more than 80% of the total catch (**Table 2**, **Figure 2**).

Regarding changes in the average catch of fish and

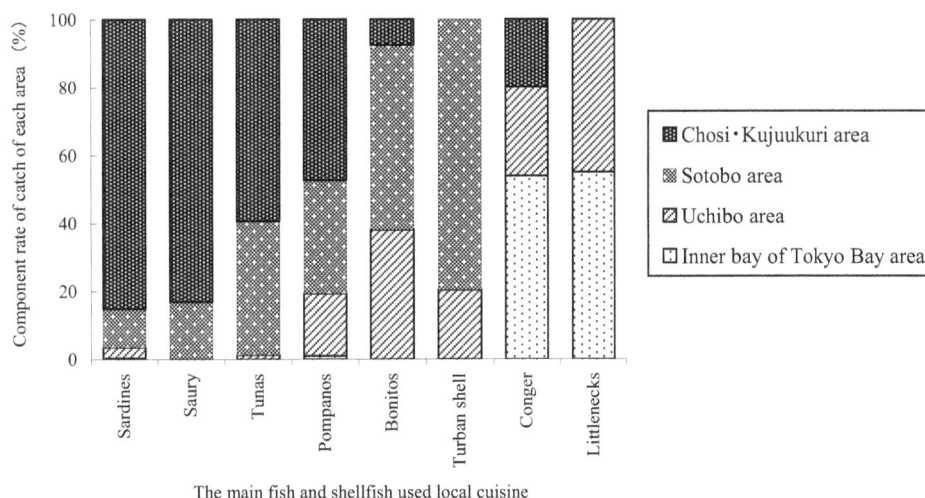

The main fish and shellfish used local cuisine

Figure 2. The composition rate of catch in each area of Chiba Prefecture. ※ Data quoted from Ministry of Agriculture, Forestry and Fisheries (2012), and Agriculture, Forestry and Fisheries statistics annual report (2010-2011).

Table 1. The main local cuisine in Chiba Prefecture.

Area division	Dish name	Food ingredients	Dish summary
Choshi·Kujuukuri	Sardine dish	Sadine, Anchovy, etc.	Sardine dish is counted more than 100 kinds as sliced row fish, chopped fish, sesame pickle etc.
	Saury dish	Saury	Grilled with salt, sliced row fish, etc.
	Finely chopped horse mackerel	Horse mackerel	Cooking picked to vinegar Namerou.
	Bonito dish	Bonitos	Sliced row fish, hammering finish, bonito rice, etc.
Kujuukuri·Sotobo·Uchibo	Turbo cooked in its own shell	Turban shell	This dish roast fire directly turban shell, season with soy source, and eat out after backed. The origin is unknown, but had appeared in the (1603-1867) Edo period.
	Nuta of turban shell	Turban shell	Cooking mixed with mustard vinegared miso and vinegared miso the turban shell.
			Nuta means that reminds a marsh and field because miso source is semi-solid.
Sotobo·Uchibo	Magocha	Tunas·Bonitos	Chazuke using tunas and bonitos pickled soy souce.
	Namerou	Horse mackerel, sardine, saury, mactra chinensis, etc.	Cooking mixed with miso, ginger, chopped leeks and green perilla after chopped fish found in coastal waters of Sadine and hoese mackerel, etc. Raw fish dish that can be eaten without the soy sauce.
	Sanga boil	Horse mackerel, sardine, saury, mactra chinensis, etc.	Cooking method was molded into a hamburger shape knead well Namero, and baked it.
	Conger eel dish	Conger	Conger eel rice, Teppou boiled of conger eel etc.
Inner Bay of Tokyo Bay	Manila clam dish	Manila clam	Manila clam rice, Manila clam spits etc.
	Mactra chinensis dish	Mactra chinensis	Sliced row fish, Tempura, Vinegared food.

※ Data quoted the hundred best local cuisines (Chiba Prefecture): http://www.rdpc.or.jp/kyoudoryouri100/ryouri/12.html

Table 2. The catch of each municipalities of the main fish and shellfish using local cuisine in Chiba Prefecture (2009).

Area division	Municipalities	Fish and sellfish							
		Sardines	Saury	Tunas	Pompanos	Bonitos	Turban shell	Conger	little necks
Chosi-Kujuukuri	Chosi City	**21,176**	**15,098**	**358**	284	93	-	22	-
	Asahi City	**37,004**	-	-	483	-	-	0	-
	Sousa City	**22,184**	-	-	306	-	-	0	-
	Yokoshibahikari Town	0	-	-	0	-	-	0	-
	Sanmu City	0	-	-	0	-	-	1	-
	Kujuukuri Town	0	-	-	406	-	-	-	-
	Oamishirasato Town	0	-	-	0	-	-	-	-
	Shirako Town	6018	-	-	172	6	-	-	-
	Chosei Village	2	-	-	4	-	-	-	-
	Ichinomiya Town	1	-	-	22	-	-	-	-
Sotobo	Isumi City	2909	-	-	76	0	2	0	-
	Onjuku Town	0	-	9	0	**221**	1	-	-
	Katsuura City	0	265	118	6	**264**	74	-	-
	Kamogawa City	6997	193	91	412	128	**86**	-	-
	Minamiboso City	1596	2590	20	**688**	108	87	0	-
Uchibo	Tateyama City	1100	4	4	308	196	26	2	-
	Kyonan Town	1770	26	4	231	**306**	27	1	-
	Futtsu City	381	1	-	110	0	11	27	**165**
	Kisarazu City	0	-	-	6	-	-	62	155
	Sodegaura City	0	-	-	0	-	-	-	X
Inner Bay of Tokyo Bay	Narashino City	0	-	-	0	-	-	-	4
	Funabashi City	357	-	-	30	-	-	0	27
	Ichikawa City	0	-	-	0	-	-	-	16
	Urayasu City	0	-	-	0	-	-	0	X

shellfish from 1960 to 2009, tunas, saury, littlenecks, and conger were markedly decreased. Sardines and turban shell were at their highest of 1980 to 1989; these fish have decreased since 1989. Bonitos were increasing and decreasing by the period; however, during the period of 2000 to 2009, the fish decreased compared with that of 1990 to 1999. Pompanos was the highest in the period of 1990 to 1999; however, they decreased in the period of 2000 to 2009 (**Figure 3**).

3.3. Sake Breweries in Chiba Prefecture

The oldest sake brewery in Chiba Prefecture was established in the early Edo period, and many sake breweries established in the Meiji period. Notably, the opening of the brewery was free during the early Meiji period, and the number of sake breweries was 800 in 1879 and was more than 1000 in 1883. However, with the ban of home brewed sake in 1896, the number of sake breweries decreased to approximately 200 by the end of the Meiji period. Since then, the number of sake breweries has decreased by half because of a large tax increase on liquor for the expenses of war, particularly the maintenance of enterprises during the Second World War [13]. With the impact of the reduced demand for Japanese sake after the Second World War, the number of sake breweries in Chiba Prefecture is currently 39 (December 2012). The 39 breweries included home brewing, which has not been

Traditional Food Culture (Local Cuisines, Japanese Sake) That Has Been Nurtured by the Rich
Nature of the Region: The Case of the Coastal Area in Chiba Prefecture, Japan

123

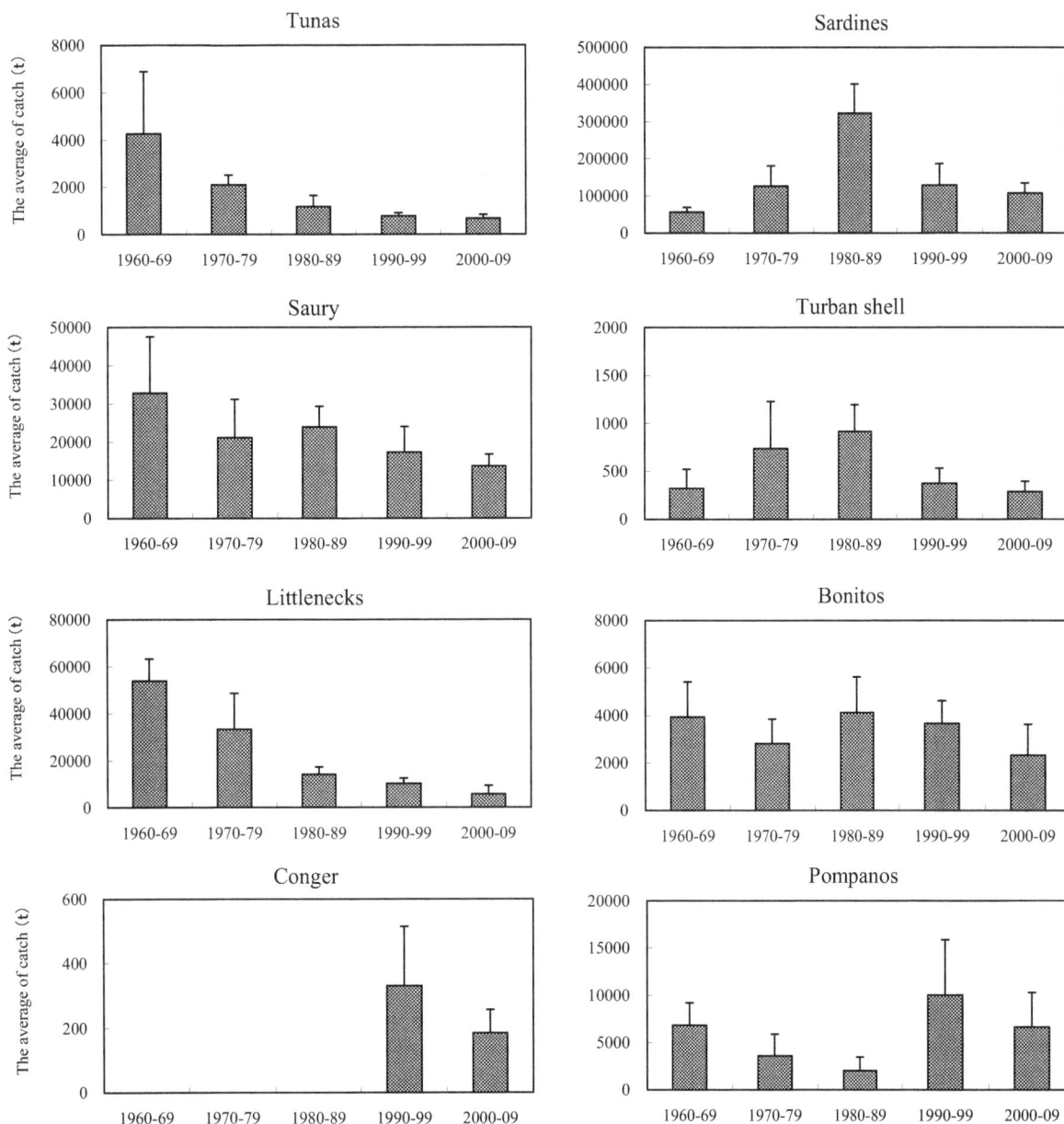

Figure 3. The change in the average catch of fish and shellfish that is used in the main local cuisine in Chiba (1960-2009). ※Data quoted from Chiba Prefecture, Agriculture, Forestry and Fisheries Division, fisheries station (2012). ※Data quoted from Ministry of Agriculture, Forestry and Fisheries, Kanto Regional Agricultural Administration Office, Agricultural administration office Chiba Department of Statistics (2007-2009).

conducted for several decades. There is an environmental distinction of sake breweries in Chiba Prefecture; there are 16 sake breweries in the plateau/mountainous region, 13 in the river plains, and 10 in the coastal vicinity [4,5].

3.4. Affinity with Japanese Sake and Local Cuisines

Based on an article from [14], we researched the relationship between tastes of the main local cuisines and the Japanese sake that is made in their production areas. The

slightly plain taste of the mactra chinensis dish and the sardine dish matched the Japanese sake of the taste that is slightly crispy and dry; the fatty meat and slightly heavy-tasting cooking of sanga broil, kusari-zushi, finely chopped horse mackerel, and the bonito dish matched the Japanese sake of a slightly deep taste that is dry and mellow. The heavy-tasting cooking of the namerou, the conger eel dish, turbo cooked in its own shell, and the saury dish matched the Japanese sake of deep taste that is slightly dry and deep (**Figure 4**). Moreover, the preparation water of the sake brewery that is located in the production areas of the above-mentioned local cuisines had an average hardness of 15 dh, which is considered to be very hard water by the World Health Organization rankings (**Figure 5**).

We investigated the relationship between the sake meter value and acidity with the Japanese liquor Junmaishu (sake made without added alcohol or sugar) of each sake brewery in the coastal area of Chiba Prefecture (In the liquor tax law of the United States, because sake without added alcohol is brewed as Japanese sake, it has selected the above types). The sake meter value was 2.86 ± 0.73, and the acidity was 1.61 ± 0.12, and these values were the taste of deep and dry (**Figure 6**).

- Sake meter value: The value is a measure to represent the dryness and sweetness of Japanese sake. When the sugar content is higher, the specific gravity becomes heavy. When measured using a sake meter, the value of water is 0. When the sake meter value is negative, it indicates a sweet taste and that the specific gravity of the sake is heavier than that of water.

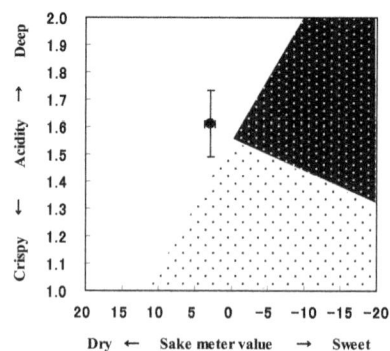

Deep and dry: Acidity and sugar content are many, the flavor is richness and sharpness.
Deep and sweet: Acidity and sugar content are many, the flavor is richness and mellowness.
Crispy and dry: Acidity and sugar are less, the flavor is sharpness and refreshing.
Crispy and sweet: Acidity and sugar content are less, the flavor is mellowness and refreshing.

Figure 6. The relationship between the sake meter value and the acidity of the sake brewery (Junmai course) in the coastal area of Chiba Prefecture. *In the Liquor Tax Law of the United States, sake that is not Junmai course is not considered a brew of Japanese sake. *Junmai course: Sake made without added alcohol or sugar

When the sake meter value is positive, it indicates a dry taste and that the specific gravity of the sake is lighter than that of water.

- Acidity: The value representing the amount of acid contained in the Japanese sake. Japanese sake of high acidity tastes dry and deep, and that of low acidity tastes sweet and crispy.

4. Discussion

In the local cuisine of Chiba Prefecture, many dishes use fish and shellfish ingredients; these dishes have been eaten by fishermen on ships and in the port. We hypothesized that the development of fishery and the marine products industry since the Edo periods has greatly influenced the local food culture. The Choshi·Kujuukuri areas produces the sardine dish and the saury dish, the Kujuukuri, Sotobo, and Uchibo areas produce the finely chopped horse mackerel, bonito dish, turbo cooked in its own shell, nuta of turban shell, and magocha. The Sotobo and Uchibo areas produce namerou and sanga broil, and the Inner Bay of Tokyo Bay produces the conger eel dish, manila clam dish, and the mactra chinensis dish (**Table 1**). The catch of the fish and shellfish that are the ingredients of the local cuisines was high in relation to the producing area (**Table 2**). We found that the blessings of nature in each area affect the food culture. Currently, the number of sake breweries near the coast of Chiba Prefecture is 10 (February 2012), while the number of sake breweries in 1925 was three times that [13]. We estimated that the demand for Japanese sake was high in the city, which was influenced by the development of the

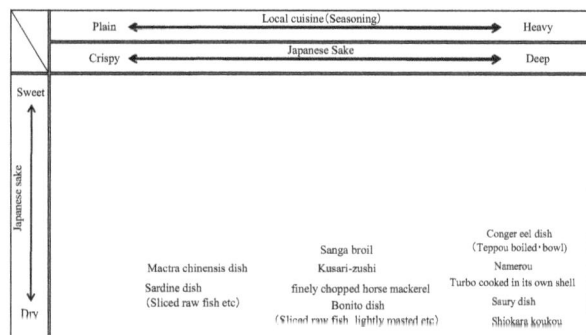

Figure 4. Taste of Japanese sake compared to the seasoning of the main local cuisine in Chiba Prefecture. *Data quoted Food and Beverage Specialist Organization (NPO) (2009).

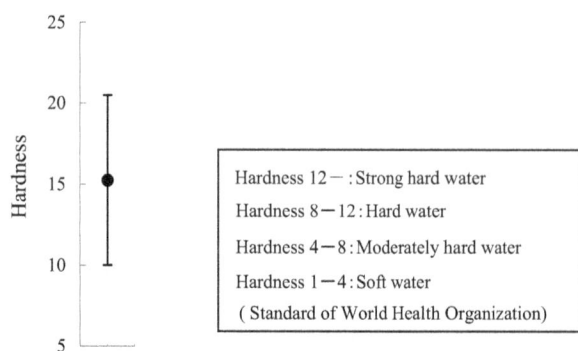

Figure 5 The average hardness of the preparation water of the sake brewery that is located in the coastal zone in Chiba Prefecture. * ● indicate the average value. The vertical bars indicate the standard deviation.

port function, land and water transportation line.

The similarity in taste between local cuisine and Japanese sake that was brewed in the vicinity of the producing area was analyzed. The slightly bland taste of the mactra chinensis dish and sardine dish match the Japanese sake taste, which is slightly crispy and dry. When the seasoning of the dish becomes heavier, the trend was to pair the food with a Japanese sake of a full-bodied and dry taste, which is deep and delicious (**Figure 4**). Moreover, the preparation water of the sake brewery that is located in the producing area of the above local cuisines was found to be very hard water based on the World Health Organization rankings (**Figure 5**). The hardness of the preparation water in Chiba Prefecture was high in the coastal areas of the alluvium region and was low in the inland side of the diluvium region [4,5]. The very hard water becomes the nutrient source for yeast and invigorates the fermentation. Japanese sake that is brewed using hard water easily becomes full-bodied, deep and dry. Japanese sake that is brewed using soft water easily has a mellow taste and is slightly sweet [13]. We evaluated the sake based on the sake meter value and the acidity of sake (Japanese liquor junmaishu) made without added alcohol or sugar that was brewed in the sake brewery located in the coastal dunes. The Japanese sake from that sake brewery had a deep and dry taste (**Figure 6**). [14] reported that a full-bodied and dry Japanese sake that is brewed with hard water tends to match with a heavily seasoned local cuisine. Moreover, [15] reported that amino acid and inosinic acid containing fish and shellfish are suitable with succinic acid, which is the component that provides the sake's deep taste. In Japanese sake that is brewed in the coastal area and contains high succinic acid, we consider that Japanese sakes in these regions have been brewed to match the delicious taste of fish and shellfish.

Consequently, we consider that local cuisine using fish and shellfish ingredients and the Japanese sake that is brewed with water from the fresh water layer under the coastal dune are of a high value as a traditional food culture, which has developed the local production and consumption, and has composed in coexistence with people and a rich natural environment for many years. However, regarding changes in the average catch from 1960 to 2009, all fish and shellfish during 2000-2009 mostly decreased compared with the peak catch period. In particular, tunas were 6% of the peak catch period; saury, turban shell and littlenecks were approximately 20% of the peak catch period, and sardines and congers were 34% - 39% of the peak catch period (**Figure 3**).

In the future, if excessive harvesting of fish and shellfish, water pollution and environmental destruction progress and these habitats deteriorate, we predict that the catch will be further decreased, which means that the local cuisine could potentially disappear. Moreover, since the Edo period, large-scale development of farmland has been conducted mainly in the plain areas. In recent years, the conversion to residential land from farmland in the plain areas, the construction of industrial parks and land-scale multi-unit apartments on the plateau has increased [3]. We suggest that these actions exacerbated the water quality of the fresh water layer in the basement of the coastal dunes and would adversely affect the taste of Japanese sake.

On this subject, it is necessary to conserve the rich natural environment in the tidal flats, coast, sea grass beds, and sea that produce the local cuisines of Chiba Prefecture, the water source (a successive environment on the plateau from coast to plateau that was previously called the coastal sand dune area) of the preparation water for making Japanese sake that matches the local cuisines, and the activities of aggressive local production for local consumption. Thus, we believe that if the Japanese food culture is safe and secure, the traditional food culture will continue, and it would be able to promote regional revitalization through the development of the exchanges in the region.

5. Acknowledgements

We wish to sincerely express our gratitude to the people in the sake breweries in Chiba Prefecture, Japan, who provided information.

REFERENCES

[1]   Chiba Prefecture, Planning Department, Citizen Section, "Chiba Prefectural History (Meiji Edition Offprint)," Chiba Prefecture, Chiba, 1962.

[2]   N. Takahashi, "Sake of Boso in Edo Period—To Sales and Consumption from Manufacturing," Material for Prefectural History Lecture in Chiba, Prefectural Archives, 2012.

[3]   Chiba Historical Materials Study Foundation, "Chiba Prefectural History, Distinction, Geographical Book 1, General Remarks, Prefectural History Series 36," Chiba Prefecture, Gyosei, 1996.

[4]   K. Kaneko, K. Oshida and H. Mathushima, "Ecosystem Services on Coastal Sand Beach, the Significance as Provisioning, Regulating, Cultural Services," *Japanese Journal of Landscape Ecology*, Vol. 17, No. 1, 2012, pp. 19-24.

[5]   K. Kaneko, K. Oshida and H. Mathushima, "Ecosystem Services of Coastal Sand Dunes Saw from the Aspect of Sake Breweries in Chiba Prefecture, Japan: A Comparison of Coastal and Inland Areas," *Open Journal of Ecology*, Vol. 3, No. 1, 2013, pp. 48-52.

[6] U. Naruse, "47 Administrative Divisions of Japan—Traditional Food Encyclopaedia," In: U. Naruse, Ed., *Part II Administrative Divisions—Traditional Food and Characteristics*, Maruzen Co., Ltd., Tokyo, 2009, pp. 102-107.

[7] Chiba Prefecture, Agriculture, Forestry and Fisheries Division, Fisheries Station, "Chiba Fisheries Handbook (Web Edition)," Chiba Prefecture, Chiba, 2012.

[8] Rural Development Planning Commission, "The Hundred Best Local Cuisines, Chiba Prefecture," 2007. http://www.rdpc.or.jp/kyoudoryouri100/ryouri/12.html

[9] Ministry of Agriculture, Forestry and Fisheries, "Agriculture, Forestry and Fisheries Statistics Annual Report (2012-2013)," Department of Fisheries, Kanto Regional Agricultural Administration Office Department of Statistics, Tokyo, 2012, pp. 216-259.

[10] Ministry of Agriculture, Forestry and Fisheries, "Agriculture, Forestry and Fisheries statistics. 2006 Sea Fishery Fishery Statictics (Chiba Prefecture—Rough Number)," Kanto Regional Agricultural Administration Office, Agricultural Administration Office Chiba Department of Statistics, Chiba, 2007, pp. 1-6.

[11] Ministry of Agriculture, Forestry and Fisheries, "Agriculture, Forestry and Fisheries Statistics. 2007 Sea Fishery Fishery Statictics (Chiba Prefecture—Rough Number)," Kanto Regional Agricultural Administration Office, Agricultural administration office Chiba Department of Statistics, Chiba, 2008, pp. 1-9.

[12] Ministry of Agriculture, Forestry and Fisheries, "Agriculture, Forestry and Fisheries Statistics, 2008 Sea Fishery Fishery Statictics (Chiba Prefecture—Rough Number)," Kanto Regional Agricultural Administration Office, Agricultural administration office Chiba Department of Statistics, Chiba, 2009, pp. 1-6.

[13] K. Suzuki, "Sake History of Chiba—Sake Brewing—A Heart and Climate and the History," Chiba Brewing Association, Chiba, 1997.

[14] Food and Beverage Specialist Organization, "Motoi of Japanese Sake—To You Who Aims at the Professional, Japanese Sake Sommelier of Japanese Sake, Lecture Text," Sake Service Institute, Tokyo, 2009.

[15] U. Naruse, "Science of Sake and Relish," In: U. Naruse, Ed., *Delicious Sake and Relish*, 4th Section, First Chapter, Relish Agree with Japanese Sake, Science I Shinsho, Tokyo, 2009, pp. 67-85.

16

Walnut Trim down Lipid Profile and BMI in Obese Male in Different Ethnic Groups of Quetta Population, Pakistan

Rehana Mushtaq[1], Rubina Mushtaq[1], Sobia Khwaja[1], Zahida Tasawar Khan[2]

[1]Federal Urdu University of Arts, Science & Technology, Gulshan Iqbal Campus, Karachi, Pakistan; [2]Institute of Pure and Applied Biological Sciences, Bahauddin Zakriya University, Multan, Pakistan.

ABSTRACT

A total of 64 male obese subjects were randomized to observe the effect of 40 g of walnut in daily breakfast on lipid profile and Body Mass Index (BMI), total Cholesterol (CHO), Triglyceride (TG) High Density Lipoprotein (HDL) cholesterol and Low Density Lipoprotein (LDL) cholesterol in obese male subjects of various ethnics *i.e.* Baloch (B), Pathan (P), Hazara (H) and Punjabi (PU) residing in Quetta region of Balochistan for this purpose four weeks controlled study was designed. A batch of 32 obese male subjects 8 from each ethnic group as a control and another batch of 32 obese males 8 from each ethnic group as treated were selected. Twelve hour fasting blood samples a day after stoppage of walnut were taken from obese control and obese treated subjects. Daily walnut consumption in obese male subjects evidently demonstrated reduced BMI in all ethnic groups. Walnut supplementation in obese exhibited significant reduction in cholesterol level in Baloch ($P < 0.001$) and Punjabi ($P < 0.01$) males. There was profound and statistically significant [B ($P < 0.05$), P ($P < 0.05$), H ($P < 0.01$) and PU ($P < 0.05$)] elevation in HDL-C in all male ethnic groups. In male sub-population LDL-C was significantly [P ($P < 0.01$), H ($P < 0.001$) and PU ($P < 0.05$)] reduced in these groups. In these subjects walnut supplementation showed pronounced reduction [B ($P < 0.001$), P ($P < 0.05$), H ($P < 0.001$) and PU ($P < 0.001$)] in triglyceride levels. The constructive influence of walnut on lipid profile suggests that walnut rich diet may have advantageous effects beyond changes in plasma lipid level.

Keywords: Walnut; Cholesterol; Triglyceride; High Density Lipoprotein; Low Density Lipoprotein

1. Introduction

Obesity and Overweight are a global health problem [1, 2]. It is related with a number of acute diseases and lipoprotein disorder which is fears of possible weight gain. Obesity outcomes from a continuous left over of energy intake contrast to energy disbursement, which escort to storage of unnecessary amounts of triglycerides in adipose tissue [3]. In the US, the recent incidence of plumpness among adults is about 33% [4] similarly in the United Kingdom obesity rate has increased to about 23% [5] and these incidence statistics are reflected all over the rest world [6-8]. Obesity is now documented as an epidemic [9] and in spite of present intensive efforts to decrease obesity via diet, exercise, education, surgery and medicinetherapies; an effective long term solution to this pandemic is yet to be provided. Obesity, overweight, family history of diabetes mellitus, and hypertension are main factors of type 2 diabetes in Pakistan [10]. Rate of obesity increase in every age, sex, race, and smoking status, and data shows that divisions of persons in the highest weight category (*i.e.*, BMI > 40 kg/m) have increased more than those in lower BMI category (BMI < 35 kg/m) [11].

It has been reported that blood pressure in youth, serum lipid levels, and body mass index (BMI) strongly associated with middle age [12].

Obese people tend to have relatively high triglyceride (TG), low HDL-C and increased LDL-C. Hyperglycemia prevails despite high levels of insulin referred as hyperinsulinemia. These all constitute major factors in the pathogenesis of Cardiac Heart Disease. The recent investigations indicate that these parameters are essential in study of characteristics of obesity. Branchi has studied serum lipids in patients of metabolic syndrome while subjecting them meals of different composition [13].

Plasma cholesterol levels are fairly reduced when low cholesterol diets are used [14]. The result of dietary cholesterol on plasma cholesterol levels may be predisposed by the types of fatty acid consumed which may be saturated or unsaturated [15].

Long term nut consumption is associated with lower body weight and lower risk of obesity also high light by Sabate [16]. Further these findings are support and recommended by Bes-Rastrollo [17] that nut consumption is an important component of a cardio protective diet and those participants who fulfill risk factors for obesity and they ate nuts two or more times per day had a significantly lower risk of also allay fears of possible weight gain.

Nuts have unique composition *i.e.* vitamins, minerals, mono- and polyunsaturated fatty acids, fiber, arotenoids, phenols (particularly flavonoids), phytosterols, squalene and tocopherols which slowing the pathogenesis of chronic disease, antioxidant and anti-inflammatory activity as well as the capacity to promote detoxification, reduce cell proliferation, and/or lower serum low-density lipoprotein (LDL) cholesterol [18]. Nuts are main element of healthy diets such as the Mediterranean diet. Outcome of several epidemiological studies and mores suggest that there may be an association between frequent nut consumption and a reduced rate of CHD [19].

There was substantiation of decreased total cholesterol and LDL cholesterol in diets of at-risk subjects supplemented with two to three servings of walnuts per day, with no net gain in body weight [20]. Lipid profile of patients with type 2 diabetes improved after including 30 g walnuts/day [21]. The previous findings show [22,23] that the average American diet, both the LA and the ALA diets including walnuts lowered total cholesterol 11%, LDLs 11% - 12% and triglycerides 18%.

The population of city of Quetta and some other towns in Balochistan well represent different ethnic groups of [Pathan (P), Baloch (B), Hazara (H) and Punjabi (PU)]. Thus a study on the basis of ethnicity was possible in such populations.

The region of Quetta city which is situated in North West of Pakistan is inhibited by various ethnic groups since many decade and present at high altitude of 1600 meter.

Thus, with the above concept of interactive mechanism the present study was planned to investigate the effect of dietary lipids supplementation on lipid profile and BMI in obese male. In order to investigate this point, walnuts were incorporated into the diets fed to male obese subjects that provided.

2. Materials and Methods

The study was designed as intervention-controlled clinical pattern trials with participant volunteers were selected from the local community, primarily through pasting posters in hospitals, universities, colleges. Newspaper advertisements, telephonic messages, emails and by coun-

seling in different communities study was carried out in selected ethnic groups [Pathan (P), Baloch (B), Hazara (H) and Punjabi (PU)] in Quetta, Balochistan participant and also in various medical camps.

Volunteers were screened and those were barred from the trial that had known nut allergies, consumed nuts frequently, smoked cigarettes and had history of atherosclerotic or hypertension or metabolic disorder. Discussions with the selected volunteers were carried out to explain the protocol of the study. Thirty two male obese subjects (experimental) 8 from each ethnic group, participated in the study, and thirty two control male obese subjects were selected. Assortment of the subjects were according to the WHO, 1998 criteria where BMI = 30 - 34.9 is considered as obese I (at a moderate risk of co-morbidities), BMI = 35 - 39.9 is obese II (at a severe risk of co-morbidities), and BMI ≥ 40 is obese III (at a very severe risk of co morbidities).

The control volunteers were asked to give 12 hrs fasting blood sample, at the start of the study and only experimentally selected subjects were ask to take daily 40 g walnut in their breakfast along with normal eating habits for four weeks. At the end of the fourth week they were sampled for blood after a 12 h fast and general data was collected. BMI in the general observations and blood samples before and after the walnut consumption were subjected for estimation of total cholesterol, triglycerides, HDL cholesterol and LDL cholesterol with commercial kits.

Statistical analysis was undertaken with statistical program of Sigma Stat 3.5. Student t test was used for comparison between normal and obese subject groups and P < 0.05 was considered as statistically significant.

3. Results

3.1. Ages

The age of the control obese subjects ranged between 37.9 ± 2.7, 43.6 ± 5.2, 44.5 ± 4.3 and 41.1 ± 3.7 years and in walnut supplemented batch of obese male subjects it was ranging at 41.9 ± 3.0, 45.1 ± 4.9, 44.1 ± 3.5 and 43.8 ± 3.4 years in Baloch, Pathan, Hazara and Punjabi sub-populations respectively. The subjects were in close range of age and well-matched for the comparison.

3.2. Body Mass Index (BMI)

The body mass index of obese Baloch was 40.9 ± 1.4 kg/m² and 40.6 ± 0.5 kg/m² in control and walnut supplemented obese subjects respectively. The average value of BMI were 40.6 ± 1.3 kg/m² in control Pathan volunteers and 39.6 ± 0.7 kg/m² in walnut taking volunteers. In Hazara the mean values of BMI were 40.9 ± 1.3 kg/m² in controls while it was 38.5 ± 0.6 kg/m² in walnut supplemented groups. In Punjabi control volunteers the BMI was 41.3 ± 1.2 kg/m² and in walnut consumers it was

39.6 ± 1.0 kg/m^2 (**Table 1**).

Body mass index in walnut supplemented obese subjects was found to be lower in all ethnic sub population in comparison to controls obese, however, non significant finding were experienced in all ethnic group. It is quite obvious from these results that walnut intakes had no effect on BMI.

3.3. Total Cholesterol

In Baloch control obese volunteers of this batch the cholesterol concentration was 268.6 ± 4.8 mg/dl. Walnut supplemented subjects (239.8 ± 3.7 mg/dl) manifested remarkable and statistically significant ($P < 0.001$) reduction of 10.8% in walnut taking subjects. The cholesterol concentration in control Pathan subjects of this ethnicity level observed to be 261.9 ± 4.0 mg/dl, comparably in walnut supplemented volunteers it was 256.6 ± 2.5 mg/dl. The minor reduction of only 2.2% was noticed. An average total cholesterol fraction of 251.3 ± 4.4 mg/dl and 245.1 ± 3.6 mg/dl was estimated in controls of obese Hazara volunteers and walnut taking Hazara subjects respectively. Walnut supplementation did not bring any significant change. Obese Punjabi volunteers exhibited a concentration of 259.9 ± 3.6 mg/dl and 243.4 ± 3.0 mg/dl total cholesterol in controls and walnut taking respectively. The fraction was statistically significant ($P < 0.004$) lower with 6.4% in walnut supplemented subjects (**Figure 1**).

Although the cholesterol concentration estimated was beyond the normal range in walnut treated subjects, but it demonstrated beneficial effects in Baloch and Punjabi volunteers compare to controls.

3.4. Low Density Lipoprotein Cholesterol

The overall LDL cholesterol response is presented in **Figure 2** in male subjects experiment.

In Baloch obese control male subjects average value of LDL cholesterol concentration was 191.8 ± 2.1 mg/dl and in walnut supplemented it was 188.9 ± 1.3 mg/dl. The difference was non significant with minor reduction of only 1.6%. A concentration of 191.6 mg/dl and 175.8

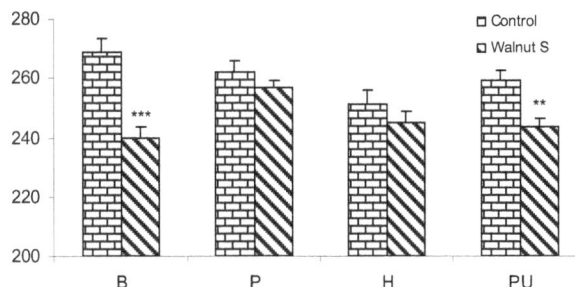

Figure 1. Serum total cholesterol mg/dl in obese M (males), of controls and walnut supplementation (S) in B (Baloch), P (Pathan), H (Hazara) and PU (Punjabi) ethnic groups. $^{**}P < 0.01$, $^{***}P < 0.001$.

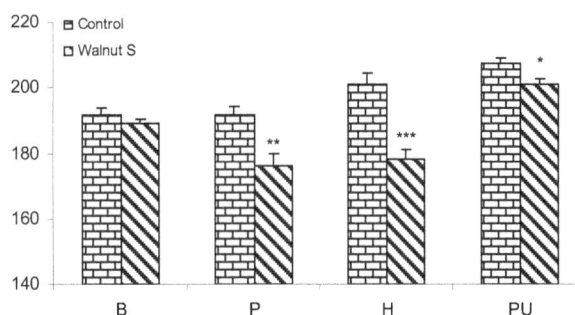

Figure 2. Serum LDL cholesterol mg/dl in obese M (males), of controls and walnut supplementation (S) in B (Baloch), P (Pathan), H (Hazara) and PU (Punjabi) ethnic groups. $^{*}P < 0.05$, $^{**}P < 0.01$, $^{***}P < 0.001$.

± 4.0 mg/dl LDL cholesterol was estimated in the controls and in walnut supplemented Pathan volunteers respectively. The noticeable and highly significant ($P < 0.01$) reduction was determined in walnut taking obese volunteers with value of 8.3%. Obese Hazara control volunteers showed an average value of 201.1 ± 3.1 mg/dl LDL cholesterol and it was 178.1 ± 3.1 mg/dl in walnut consuming Hazara subjects. Approximately 21.5% lower LDL cholesterol concentration was estimated in walnut supplemented group which was highly significant ($P < 0.001$). In Punjabi volunteers of this group the concentration of LDL cholesterol was 207.3 ± 1.9 in the controls and was 201.1 ± 1.5 mg/dl in walnut consuming subjects. The walnut supplementation showed marked and statistically significant ($P < 0.025$) reduction of 3% (**Figure 2**).

A significant response of reduction in LDL cholesterol concentration had been noticed in walnut supplemented subjects in all ethnic sub population except in Baloch where minor decline noticed.

3.5. High Density Lipoprotein

The **Figure 3** presents the overview of the HDL cholesterol response in the male subjects of the trials.

In Baloch control obese male subjects the concentra-

Table 1. Average ages and BMI kg/m^2 in obese male population of all ethnic groups in controls and with walnut supplementation. Con: control obese; Walnut S: walnut supplementation.

Ethic Groups	Age (Con)	Age (Walnut S)	BMI kg/m^2 (Con)	BMI kg/m^2 (Walnut S)
Baloch	37.9 ± 2.7	41.9 ± 3.0	40.9 ± 1.4	40.6 ± 0.5
Pathan	43.6 ± 5.2	45.1 ± 4.9	40.6 ± 1.3	39.6 ± 0.7
Hazara	44.5 ± 4.3	$44.1 \pm .3.5$	40.9 ± 1.3	38.5 ± 0.6
Punjabi	41.1 ± 3.7	43.8 ± 3.4	41.3 ± 1.2	39.6 ± 1.0

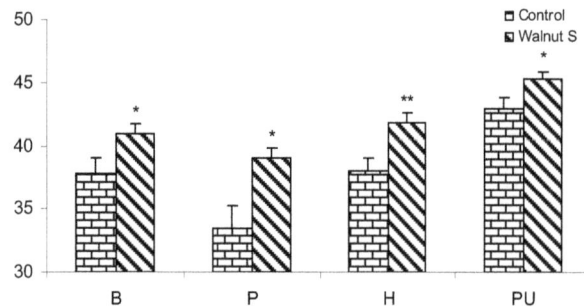

Figure 3. Serum HDL cholesterol mg/dl in obese males, of controls and walnut supplementation (S) in B (Baloch), P (Pathan), H (Hazara) and PU (Punjabi) ethnic groups. *P < 0.05, **P < 0.01.

tion of HDL cholesterol is 37.8 ± 1.2 mg/dl comparably in walnut taking subjects of this ethnicity the level was observed to be 41 ± 0.7 mg/dl. Walnut supplemented subjects exhibited considerable and significant 8.4% (P < 0.037) increase in HDL. A concentration of 33.5 ± 1.8 mg/dl HDL cholesterol was estimated in obese Pathan controls and 39.1 ± 0.7 mg/dl in walnut consumers. Comparison of two group showed noticeable and significant increase of 16.7% (P < 0.011) in HDL cholesterol. Obese Hazara volunteers of this group exhibited the concentration of cholesterol at 38.0 ± 1.0 mg/dl in the controls and 41.8 ± 0.8 mg/dl in walnut taking subjects. The walnut consumers group manifested statistically significant elevation of 10% (P < 0.008). An average value of the fraction at 43.1 ± 0.8 mg/dl in the controls and 45.3 ± 0.6 mg/dl in walnut supplemented Punjabi subjects was observed. The fraction was increased 5% in walnut supplemented subjects and was statistically significant (P < 0,045) (**Figure 3**). In the obese males of the entire ethnic sub populations the HDL cholesterol concentration estimated in controls were in unhealthy range, however after walnut supplementation it exhibited a good response in all ethnic groups, although in Punjabi subjects it was already in healthy range, even than a significant raise was observed.

3.6. Triglycerides

In Baloch obese male volunteers the concentration of triglycerides was 222.8 ± 3.2 mg/dl while it was 201.8 ± 1.9 mg/dl in walnut supplemented subjects. The fraction is found to be appreciably 9.5% lowered in walnut supplemented subjects than controls and was significant statistically (P < 0.001). Triglyceride concentration of 227.5 ± 2.9 mg/dl and 218.1 ± 2.5 mg/dl was estimated in the obese controls and obese walnut consumers Pathan volunteers. Walnut supplementation caused marked lowering of 11.3% in Tg fraction was significant statistically (P < 0.028). In Hazara volunteers of this group the concentration of triglycerides was 220.1 ± 2.2 mg/dl in con-

trols and 196.8 ± 2.5 mg/dl in walnut supplemented Hazara subjects. The fraction was found to be markedly lower 10.6% in the walnut taking subjects and was significant statistically (P < 0.001). An average fraction was at 219.8 ± 2.6 mg/dl in the control and 195.5 ± 2.8 mg/dl in walnuts supplemented Punjabi subjects. Walnut supplementation caused noticeable effects on Tg concentration with decreased levels and statistically significant (P < 0.001) (**Figure 4**).

Walnut manifested its beneficial effects on triglyceride concentration on obese males of all ethnic sub population the supplementation exhibited significant lowering characteristics comparable to controls.

4. Discussion

Walnut is an inexpensive and safe treatment, which significantly reduces lipid profile and raises HDL-c. These effects perhaps are due to the reduction of SREBP-1c expression and the rise of PPARα expression in diabetic rat [24].

Overweight and obesity lead to serious health consequences, with the risk increasing progressively with body mass index (BMI) and consequently risk of heart disease, type 2 diabetes and some cancers like endometrial, breast and colon cancer [25]. Swiftly increasing obesity incidence rates require weight management to be precedence for the anticipation and cure of chronic diseases. Nutrition supplementation and enhanced activity have been the notable approaches in this regard and numerous studies have been appearing on these aspects. In the present study walnut supplementations effects have been investigated in obese male volunteers of the ethnic sub-populations.

The results of present study clearly have demonstrated that daily walnut consumption in obese male subjects reduced BMI in all ethnic groups. In all other studied sub-population non significant lowering was noticed in BMI. These findings agree with the previous result [26] and associated with a slightly lower risk of obesity.

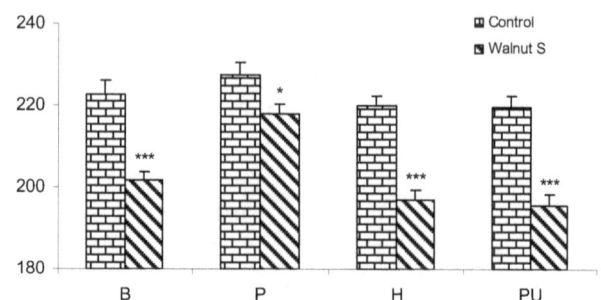

Figure 4. Serum triglyceride mg/dl in obese M (males), of controls and walnut supplementation (S) in B (Baloch), P (Pathan), H (Hazara) and PU (Punjabi) ethnic groups. *P < 0.05, ***P < 0.001.

A randomized controlled experiment of a moderate fat, low-energy diet compared with a low fat, low-energy diet for weight loss in overweight adults. Participants following a Mediterranean-style moderate fat weight loss diet, including peanuts and tree nuts like walnuts were able to improve weight loss and keep weight off for a longer period than people following the usually suggested low fat diet. The previous studies suggest that the diet sensation of the partakers may have been ascribed to the satiety of the tree nuts (*i.e.* walnuts) a key factor in weight loss [27].

Two reviews summarize the findings of 15 human intervention trials that evaluated the effects of nut consumption on body weight changes and concluded that self-selected diets that included nuts in free-living populations did not have a tendency to increase body weight [28,29].

Higher phospholipid proportions of oleic, alpha-linolenic (ALA), and docosahexaenoic acids showed inverse associations to carotid artery thickness in subjects with primary dyslipidemia. The researchers concluded that high intakes of specific unsaturated fatty acids might explain, in part, the low incident of IHD in the Spanish population [30].

The results of present study demonstrated that in obese subjects, walnut in take exhibits significant reduction in cholesterol level in Baloch ($P < 0.001$) and Punjabi ($P < 0.01$) males. Walnut consumption exhibited profound and statistically significant [B ($P < 0.001$), P ($P< 0.05$), H ($P < 0.01$) and PU ($P < 0.05$)] reduction of triglyceride in all male ethnic groups. There was profound and statistically significant [B ($P < 0.05$), P ($P < 0.05$), H ($P < 0.01$) and PU ($P < 0.05$)] elevation in HDL-C in all male ethnic groups.

In male sub-population LDL-C was significantly reduced [P ($P < 0.01$), H ($P < 0.001$) and PU ($P < 0.05$)] in the majority groups. Although it is known that, for every 1% decrease in plasma LDL cholesterol, there is 2% reduction in cardiovascular disease (CVD) risk [31]. There arc suggestions that walnut-rich diets may have beneficial effects beyond changes in plasma lipid levels. The investigators also concluded that high-walnut-enriched diets considerably reduced total and LDL cholesterol for the period of the short-term experiments. Larger and longer-term tests are needed to address the effects of walnut consumption on cardiovascular risk and body weight [32].

Walnuts are rich in linoleic acid and ALA, which are identified to lower cholesterol when they replace SFA or MUFA in the diet probably by increasing the receptor-mediated uptake of LDL cholesterol [33]. Previous investigation demonstrated that walnut diet significantly reduced total cholesterol and LDL cholesterol [23].

Consumption of walnut has favorable effects on human serum lipid profiles, with a decrease in total and LDL cholesterol as well as triglycerides [34-37] and an increase in HDL cholesterol and apolipoprotein A1 [6].

The findings obtained in the present study further support to the existing recommendations that walnut rich diet may have beneficial effects beyond changes in plasma lipid level.

5. Conclusion

Obesity is positively and appreciably associated with BMI and lipid profile, our findings have important public health implications for the management and prevention of obesity related chronic obesity and lipid disorders. Further research aiming at elucidating the temporal relationship between obesity and those diseases is required.

REFERENCES

[1] World Health Organisation, "Global Strategy on Diet, Physical Activity and Health," 2010. http://www.who.int/dietphysicalactivity/en

[2] T. Lobstein, L. Baur and R. Uauy, "Obesity in Children and Young People: A Crisis in Public Health," *Obesity Reviews*, Vol. 5, Suppl. 1, 2004, pp. 4-85.

[3] R. Sturm, "Increases in Morbid Obesity in the USA: 2000-2005," *Public Health*, 2007, Vol. 121, No. 7, pp. 492-496.

[4] C. L. Ogden, S. Z. Yanovski, M. D. Carroll and K. M. Flegal, "The Epidemiology of Obesity," *Gastroenterology*, 2007, Vol. 132, No. 6, pp. 2087-2102.

[5] P. G. Kopelman, I. D. Caterson, M. J. Stock and W. H. Dietz, "Clinical Obesity in Adults and Children: In Adults and Children," Blackwell Publishing, Hoboken, 2005, p. 493.

[6] F. Lavedrine, D. Zmirou, A. Ravel, F. Balducci and J. Alary, "Blood Cholesterol and Walnut Consumption: A Cross-Sectional Survey in France," *Preventive Medicine*, Vol. 28, No. 4, 1999, pp. 333-339.

[7] A. Berghofer, T. Pischon, T. Reinhold, *et al.*, "Obesity Prevalence from a European Perspective: A Systematic Review," *BMC Public Health*, Vol. 8, 2008, p. 200.

[8] A. D. Sniderman, R. Bhopal, D. Prabhakaran, *et al.*, "Why Might South Asians Be So Susceptible to Central Obesity and Its Atherogenic Consequences? The Adipose Tissue Overflow Hypothesis," *International Journal of Epidemiology*, Vol. 36, No. 1, 2007, pp. 220-225.

[9] World Health Organization, "Obesity: Preventing and Managing the Global Epidemic: Report of a WHO Consultation on Obesity," World Health Organization, Geneva, 1998.

[10] J. Zafar, F. Bhatti, N. Akhtar, U. Rasheed, R. Bashir, S. Humayun, A. Waheed, F. Younus, M. Nazar and Umiai-

mato, "Prevalence and Risk Factors for Diabetes Mellitus in Selected Urban Population of a City in Punjab," *Journal Pakistan Medical Association*, Vol. 61, No. 1, 2011, pp. 40-47.

[11] S. M. Wright and L. J. Aronne, "Causes of Obesity," *Abdom Imaging*, Vol. 37, No. 5, 2012, pp. 730-732.

[12] J. Juhola, C. G. Magnussen, J. S. Viikari, M. Kähönen, N. Hutri-Kähönen, A. Jula, T. Lehtimäki, H. K. Åkerblom, M. Pietikäinen, T. Laitinen, E. Jokinen, L. Taittonen, O. T. Raitakari and M. Juonala, "Tracking of Serum Lipid Levels, Blood Pressure, and Body Mass Index from Childhood to Adulthood: The Cardiovascular Risk in Young Finns Study," *Journal of Pediatrics*, Vol. 159, No. 4, 2011, pp. 584-590.

[13] A. Branchi, A. Torri, C. Berra, E. Colombo and D. Sommariva, "Changes in Serum Lipids and Blood Glucose in non Diabetic Patients with Metabolic Syndrome after Mixed Meals of Different Composition," *Journal of Nutrition and Metabolism*, 2012, Article ID: 215052.

[14] P. P. Toth, "High Density Lipoprotein, Cardiovascular Risk," *Circulation*, Vol. 109, 2004, pp. 1809-1812.

[15] R. Mcpherson and G. A. Spiller, "Effect of Dietary Fatty Acids and Cholesterol on Cardiovascular Disease Risk Factors in Men," In: G. A. Spiller, Ed., *Handbook of Lipids in Human Nutrition*, CRC Press, 1996, p. 41.

[16] J. Sabate and Y. Ang, "Nuts and Health Outcomes: New Epidemiologic Evidence," *American Journal of Clinical Nutrition*, Vol. 89, No. 5, 2009, pp. 1643S-1648S.

[17] M. Bes-Rastrollo, J. Sabate, E. Gomez-Gracia, A. Alonso, J. A. Martinez and M. A. Martinez-Gonzalez, "Nut Consumption and Weight Gain in a Mediterranean Cohort: The SUN Study," *Obesity (Silver Spring)*, Vol. 15, No. 1, 2007, pp. 107-116.

[18] C.-Y. O. Chen and J. B. Blumberg, "Phytochemical Composition of Nuts," *Asia Pacific Journal of Clinical Nutrition*, Vol. 17, No. 1, 2008, pp. 329-332.

[19] M. Kornsteiner, K. H. Wanger and I. Elmadfa, "Tocopherols and Total Phenolics in 10 Different Nut Types," *Food Chemistry*, Vol. 98, No. 2, 2006, pp. 381-387.

[20] E. B. Feldman, "The Scientific Evidence for a Beneficial Health Relationship between Walnuts and Coronary Heart Disease," *Journal of Nutrition*, Vol. 132, No. 5, 2002, pp. 1062S-1101S.

[21] L. C. Tapsell, L. J. Gillen, C. S. Patch, *et al.*, "Including Walnuts in a Low-Fat/Modified-Fat Diet Substituting Walnuts for Monounsaturated Fat Improves HDL Cholesterol-To-Total Cholesterol Ratios Improves the Serum Lipid Profile of in Patients with Type 2 Diabetes," *Diabetes Care*, Vol. 27, No. 12, 2000, pp. 2777-2783.

[22] G. Zhao, T. D. Etherton, K. R. Martin, S. G. West, P. J. Gillies and P. M. Kris-Etherton, "Dietary Alpha-Lino-

lenic Acid Reduces Inflammatory and Lipid Cardiovascular Risk Factors in Hypercholesterolemic Men and Women," *Journal of Nutrition*, Vol. 134, No. 11, 2004, pp. 2991-2997.

[23] E. Ros, I. Núñez, A. Pérez-Heras, M. Serra, R. Gilabert, E. Casals and R. Deulofeu, "A Walnut Diet Improves Endothelial Function in Hypercholesterolemic Subjects: A Randomized Crossover Trial," *Circulation*, Vol. 109, No. 13, 2004, pp. 1609-1614.

[24] A. O. Ebrahim, N. S. Arash and R. Ali, "Effects of Walnut on Lipid Profile as Well as the Expression of Sterol-Regulatory Element Binding Protein-1c (SREBP-1c) and Peroxisome Proliferator Activated Receptors α (PPARα) in Diabetic Rat," *Food and Nutrition Sciences*, Vol. 3, No. 2, 2012, 255-259.

[25] World Health Organization, "Obesity: Preventing and Managing the Global Epidemic: Report of a WHO Consultation," *WHO Technical Report Series*, World Health Organization, Geneva, 2000.

[26] M. Bes-Rastrollo, N. M. Wedick, M. A. Martinez-Gonzalez, T. Y. Li, S. Laura and F. B. Hu, "Prospective Study of Nut Consumption, Long-Term Weight Change and Obesity Risk in Women," *American Journal of Clinical Nutrition*, Vol. 89, No. 6, 2009, pp. 1913-1919.

[27] K. McManus, L. Antinoro and F. Sacks, "A Randomized Controlled Trial of a Moderate-Fat, Low Energy Diet Compared with a Low Fat, Low-Energy Diet for Weight Loss in Overweight Adults," *International Journal of Obesity*, Vol. 25, No. 10, 2001, pp. 1503-1511.

[28] S. Rajaram and J. Sabate, "Nuts, Body Weight and Insulin Resistance," *British Journal of Nutrition*, Vol. 96, Suppl. S2, 2006, pp. S79-S86.

[29] P. Garcia-Lorda, I. M. Rangil and J. Salas-Salvado, "Nut Consumption, Body Weight and Insulin Resistance," *European Journal of Clinical Nutrition*, Vol. 57, Suppl. 1, 2003, pp. S8-S11.

[30] A. Sala-Vila, M. Cofán, A. Pérez-Heras, I. Núñez, R. Gilabert, M. Junyent, R. Mateo-Gallego, A. Cenarro, F. Civeira and E. Ros, "Fatty Acids in Serum Phospholipids and Carotid Intima-Media Thickness in Spanish Subjects with Primary Dyslipidemia," *American Journal of Clinical Nutrition*, Vol. 92, No. 1, 2010, pp. 186-193.

[31] G. N. Levine, J. F. Keaney, Jr and J. A. Vita, "Cholesterolreduction in Cardiovascular Disease. Clinical Benefits and Possible Mechanisms," *New England Journal of Medicine*, Vol. 332, No. 8, 1995, pp. 512-521.

[32] D. K. Banel and F. B. Hu, "Effects of Walnut Consumption on Blood Lipids and Other Cardiovascular Risk Factors: A Meta-Analysis and Systematic Review," *American Journal of Clinical Nutrition*, Vol. 90, No. 1, 2009, pp. 56-63.

[33] T. A. Mori, V. Burke, I. B. Puddey, G. F. Watts, D. N. O'Neal, J. D. Best and L. J. Beilin, "Purified Eicosapentaenoic and Docosapentaenoic Acids Have Differential Effects on Serum Lipids and Lipoproteins, LDL-C Particle Size, Glucose and Insulin in Mildly Hyperlipidemic Men," *American Journal of Clinical Nutrition*, Vol. 71, 2000, pp. 1085-1094.

[34] D. Zambón, J. Sabaté, S. Muñoz, *et al.*, "Substituting Walnuts for Monounsaturated Fat Improves the Serum Lipid Profile of Hypercholesterolemic Men and Women. A Randomized Crossover Trial," *Annals of Internal Medicine*, Vol. 132, No. 7, 2000, pp. 538-546.

[35] A. Chisholm, J. Mann, M. Skeaff, C. Frampton, W. Suth-erland, A. Duncan and S. Tiszavari, "A Diet Rich in Walnuts Favourably Influences Plasma Fatty Acid Profile in Moderately Hyperlipidaemic Subjects," *European Journal of Clinical Nutrition*, Vol. 52, No. 1, 1998, pp. 12-16.

[36] J. Sabate, G. E. Fraser, K. Burke, S. F. Knutsen, H. Bennett and K. D. Lindsted, "Effect of Walnuts on Serum Lipid Levels and Blood Pressure in Normal Men," *New England Journal of Medicine*, Vol. 238, 1993, pp. 603-607.

[37] M. Abbey, M. Noakes, G. B. Belling and P. J. Nestel, "Partial Replacement of Saturated Fatty Acids with almonds or Walnuts Lowers Total Plasma Cholesterol and Low-Density-Lipoprotein Cholesterol," *American Journal of Clinical Nutrition*, Vol. 59, 1994, pp. 995-999.

17

Virgin Olive Oil Acceptability in Emerging Olive Oil-Producing Countries

Adriana Gámbaro, Ana Claudia Ellis, Laura Raggio

Sensory Evaluation Laboratory, Food Science and Technology Department, School of Chemistry, Universidad de la República, Montevideo, Uruguay.

ABSTRACT

A sample of 99 habitual consumers sensory-evaluated 2 extra virgin and 2 ordinary olive oils in terms of overall liking and willingness to purchase based on 9-point structured scales and responded to a check-all-that-apply question comprising a list of 18 positive and negative attributes. In the second session, the same consumers evaluated the same oils also based on their respective commercial specifications and sensory profiles previously prepared by a panel of 9 trained tasters. Two consumer clusters with contrasting behavior were identified. Whereas 52% of respondents gave high overall liking scores to the extra virgin oils and scores below commercially acceptable limits to those of ordinary virgin quality, 48% gave low overall liking scores to the extra virgin oils and high overall liking scores to those oils that were defective. Consumers of neither cluster were influenced by the information made available in Session 2. Although a slight majority of consumers described the oils consistently with the sensory profiles available from the tasting panel, an alarmingly large number of respondents described the two extra virgin oils in terms of defective, bad-tasting, strange-tasting, poor quality and rancid, and those oils that were defective in terms of good quality, tasty, sweet, aromatic, mild-flavored, delicious and fresh. These results highlight the need for the implementation of relevant consumer sensitization programs in emerging olive-producing countries like Uruguay, where virgin olive oils of varied quality are locally available.

Keywords: Olive Oil; Consumer Study; Olive Oil Quality

1. Introduction

A typical component of the Mediterranean diet, olive oil has only recently reached significant consumption levels in countries outside Europe, like Uruguay. In recent years, Uruguay's olive oil production has increased markedly and national brands have emerged on the local market, consistently with a two-fold increase (from 0.2 L to 0.4 L) in average per-capita consumption between 2006 and 2012 [1].

Uruguayan consumers perceive olive oil differently from the rest of locally available edible vegetal oils, describing it as an expensive high-quality gourmet oil associated with beneficial health effects and evoking positive feelings [2]. In other countries where olive growing is a new activity, as is the case with the USA, the health benefits and the flavor associated with olive oil have been reported to be primary drivers of consumption [3].

The quality of olive oil can be defined from a commercial, nutritional or sensory standpoint [4]. The nutritional value of olive oil is associated with its high oleic acid content and the presence of minor components, such as phenolic compounds, while its flavor is strongly influenced by the presence of volatiles [5]. The sensory profile of an olive oil will vary according to olive variety, soil characteristics, climate, tree health, fruit maturity at the time of harvest, olive collection process, olive storage conditions, oil extraction process, olive oil storage method prior to packaging, packaging means and preservation method and/or additives [6,7].

The sensory quality of virgin olive oil may be quantified by evaluating the sensations defined by smell, aroma and taste, in addition to pungent and astringent mouth sensations. Healthy olives introduce positive attributes (fruity, bitter and pungent) in an oil, whereas the processes occurring after harvest tend to lessen the intensity of such attributes and induce the appearance of defects, *i.e.*, negative attributes that are to the detriment of product quality [8].

According to the International Olive Council (IOC), commercial grading of olive oil is based on physico-chemical and sensory analyses [9]. According to the results of sensory analysis, olive oils are classified on a 10-point scale as extra virgin (the median of defects is 0 and the median of the fruity attribute is greater than 0), virgin (the median of defects is greater than 0 but no greater than 3.5 and the median of the fruity attribute is greater than 0), ordinary virgin (the median of the defects is greater than 3.5 but no greater than 6.0, or the median of the defects is no greater than 3.5 and the median of the fruity attribute is 0) and lampante virgin (the median of the defects is greater than 6.0). Olive oils classified as lampante virgin cannot be sold and must be refined, losing their virgin quality [10].

Markedly defective olive oils (between 3 and 6 on the IOC scale), classified as "ordinary virgin" are available on non-Mediterranean markets where consumers lack the necessary knowledge to perceive olive oils defects and have become accustomed to consuming them. On the other hand, individual beliefs and attitudes towards foods depend on a number of factors, including cultural traditions, education and culinary habits, and may be modified by means of information [11].

Based on a sample of habitual olive oil consumers, this paper addresses the acceptability of and willingness to virgin olive oil purchase according to commercial grade. The influence of consumer acquaintance with the sensory profile and commercial grade of a virgin olive oil on its overall liking and willingness to purchase ratings was also studied.

2. Material and Methods

2.1. Sample Characterization

Four locally available virgin olive oils of different quality were used for this research: 2 labeled as extra virgin oils (A and B), and 2 labeled as ordinary virgin oils (C and D).

The commercial quality of the above oils was confirmed and a panel of 9 tasters recruited and trained according to IOC standards [12], analyzed the oils and produced a descriptive profile. The oils were evaluated in duplicate over two consecutive work sessions. In order to minimize the possibility of systematic error, oil samples were presented on a random basis.

15 mL of oil were poured into blue-colored tasting glasses [13] in order to exclude the visual factor. Oil samples coded with 3-digit random numbers were presented at 28°C ± 2°C and rated on 10-cm unstructured scales according to each of the following positive and negative attributes: fruity (green/ripe notes), bitter, pungent, green (grassy/leafy), fig leaf notes, tomato notes (plant-, leaf-, fruit-flavored), apple notes, banana notes,

almond/nut notes, sweet, and astringent, among other positive attributes; and fusty/muddy sediment, musty/humid/earthy, winey/vinegary/acid/sour, frostbitten olives (wet wood-tasting), and rancid, among other negative attributes. Evaluations were conducted in a tasting room with 5 individual cabins furnished with temperature control (22°C to 24°C) and ventilation [14].

2.2. Consumer Test

The study was conducted in the city of Montevideo, Uruguay. A total of 99 habitual consumers of olive oil (defined as consuming olive oil every day or several times a week) were randomly recruited at shopping areas, universities, restaurants and other public places. The sample included 49 female (49.5%) and 50 male (50.5%) participants, ranging in age from 18 to 62. Taking into account the areas where the participants were recruited, the sample was assumed to represent the general Uruguayan middle income groups.

Participants were initially (Session 1) presented with the 4 samples of olive oil in monadic sequential fashion. The samples were served at room temperature in white plastic containers coded with three random digits. Sliced white bread was used as tasting vehicle and mineral water, natural yogurt and green apples as palate cleansers. The 4 oil samples were presented in a balanced order.

Participants were asked to taste each olive oil sample and rate them in terms of overall liking and willingness to purchase using 9-point structured hedonic scales ranging from extreme dislike to extreme liking and from definite unwillingness to definite willingness, respectively. No information regarding the commercial grade of the tasted oils was made available at this stage.

The test was completed with a check-all-that-apply (CATA) question consisting of a list of 18 terms or attributes selected on the basis of previous work [2] and the results of the evaluation by the trained tasting panel, as follows: tasty, aromatic, bitter, pungent, fruity, grassy/leafy, strange-tasting, fresh, bad-tasting, strong-flavored, mild-flavored, sweet, delicious, rancid, defective, good quality, poor quality and cooked-olive smell.

At the end of Session 1, consumers responded to a brief survey of sociodemographic data (gender, age, marital status, number of persons and number of children in the household, and education level) and olive oil consuming habits.

After a minimum 30-min break, consumers were again (Session 2) presented with the 4 olive oil samples in monadic sequential fashion accompanied by a written description of the sensory profile of each oil. The description was based on the information provided by the panel of sensory tasters and that available from the product label. Samples bore the same codes as in Session 1 with a view to assessing the influence of the availability of sensory descriptions on the overall liking and willingness to purchase scores for each oil.

Consumers were finally asked to rate each oil sample in terms of overall liking and willingness to purchase using the same scales as used in Session 1.

The tests were carried out in a sensory laboratory designed in compliance with ISO 8589 [15].

2.3. Data Analysis

Data comprising the sensory profiles of the studied olive oils was subjected to an analysis of variance (ANOVA) considering the different oil types, the tasters, and the interaction between the two as fixed sources of variation. Where differences were significant, honestly significant differences were determined based on the Tukey test ($p < 0.05$).

2.4. Cluster Analysis Based on Overall Linking

Hierarchical cluster analysis of overall liking data from Session 1 enabled the identification of consumer groups according to overall attitude towards the tested products. The formation of clusters was based on Ward's aggregation criterion and the calculation of Euclidean distances between data points. The chi-square test was used to determine differences in gender, age and education frequency distributions between clusters.

Finally, an ANOVA was conducted on the overall liking and willingness to purchase data, considering description availability, oil sample and cluster, as well as combinations of two of the three, as factors. Mean ratings and honestly significant differences were determined based on the Tukey test ($p \leq 0.05$).

2.5. Analysis of CATA Question Data

For each term in the CATA question, frequency of mention was determined by counting the number of consumers that used that term to describe each oil sample. Cochran's Q test was carried out for each of the 18 terms, considering oil sample and consumer as sources of variation to evaluate if the CATA question was able to detect differences in consumers' perception of the tested oils. Cochran's Q test is a nonparametric test used for the analysis of two-way randomized block designs to determine whether k treatments have identical effects when the response variable is binary.

Multiple factor analysis (MFA) is a factor analysis method dealing with data sets composed of both quantitative variables and frequency tables [16]. Multiple factor analysis (MFA) was based on the frequencies of mention from the results of the CATA question and the ratings provided by the tasting panel as active variables, and overall liking as supplementary variable, in order to identify relationships between the terms and the oil sam-

ples and to generate a sensory map of the oil samples [17]. All statistical analyses were performed using XL-Stat 2011 software (Addinsoft, NY) and R language (R Development Core Team, 2007).

3. Result

3.1. Olive Oil Sensory Profiles

Whereas the IOC relies on the medians (Me) of positive attributes and defects, mean ratings associated with the studied oils and the results of the analysis of variance are shown in **Table 1**. Significant differences ($p < 0.05$) among the four olive oils were identified by consumers in terms of fusty/muddy sediment, musty/humid/earthy, winey/vinegary and rancid, other negative attributes, and fruity, bitter, pungent, green (grassy/leafy notes), tomato, banana and almond/nut notes, other fruity attributes, sweet and astringent.

3.2. Consumer Study

The information provided by the sensory panel and that contained on the product labels was used to generate the

Table 1. Olive oil sensory profiles.

Attribute	Oil type			
	A	B	C	D
Fusty/muddy sediment	0.0[a]	0.0[a]	4.6[c]	3.5[b]
Musty/humid/earthy	0.0[a]	0.0[a]	0.3[b]	0.6[b]
Winey/vinegary	0.0[a]	0.0[a]	1.0[b]	1.2[b]
Frostbitten olives	0.0[a]	0.0[a]	0.0[a]	0.0[a]
Rancid	0.0[a]	0.0[a]	1.6[b]	1.9[b]
Other negative attributes	0.0[a]	0.0[a]	0.6[b]	0.3[a,b]
Fruity	4.7[c]	4.5[c]	2.0[a]	3.1[b]
Bitter	3.4[c]	5.1[d]	0.9[a]	2.5[b]
Pungent	4.1[c]	3.8[c]	1.1[a]	2.4[b]
Green (grassy/leafy notes)	3.4[c]	3.3[c]	0.3[a]	1.0[b]
Fig leaf notes	0.0[a]	0.0[a]	0.0[a]	0.0[a]
Tomato notes	1.2[b]	2.4[c]	0.4[a]	0.6[a]
Apple notes	0.0[a]	0.0[a]	0.5[a]	0.5[a]
Banana notes	1.6[b]	1.2[b]	0.5[a]	0.5[a]
Almond/nut notes	2.3[b]	2.0[b]	0.8[a]	2.3[b]
Other fruity attributes	0.2[a]	0.7[b]	0.2[a]	0.2[a]
Sweet	2.6[a]	2.6[a]	3.0[a,b]	3.2[b]
Astringent	1.3[b]	1.4[b]	0.6[a]	0.7[a,b]

A = Colinas, B = Picual, C = Sibarita and D = Buena Vista. Values in a row with different superscripts are significantly different according to the Tukey test ($p \leq 0.05$).

following descriptions, which were presented to the consumers at the time of oil tasting in Session 2 of the consumer study.

Sample A: Bi-varietal blend of Arbequina and Coratina olives, of extra virgin quality. It has a green fruitiness of medium intensity, with green cut grass and leafy notes. Green banana and nut flavors are also present. When tasted, it is slightly sweet, with medium bitterness and persistent pungency. This profile is consistent with that of a well-balanced oil of good quality.

Sample B: Picual extra virgin olive oil. It has medium intensity fruitiness, between green and ripe, with tomato, banana and dried fruit (almond and nut) notes. When tasted it is slightly sweet, with medium to intense bitterness and medium pungency. This profile is consistent with that of a good quality oil.

Sample C: An oil with only slight fruitiness and almost imperceptible bitterness and pungency. It has several defects, such as fusty/muddy sediment (cooked-olive smell, glue, nail polish, cheese rind, garbage can), winey (vinegar) and rancid attributes.

Sample D: An oil with slight ripe fruitiness and slight bitterness and pungency. It has several defects, such as musty/humid/earthy, fusty/muddy sediment (cooked-olive smell, glue, nail polish, cheese rind, garbage can), winey (vinegar) and rancid attributes.

3.3. Cluster Analysis

Hierarchical cluster analysis of overall liking data from Session 1 led to the identification of Clusters 1 and 2, composed of 51 and 48 consumers (51.5% and 48.5% of the consumer sample), respectively. Mean overall liking scores according to cluster are shown in **Table 2**.

Cluster 1 consumers assigned highly significantly different ($p < 0.001$) scores to extra virgin olive oils (oils A and B) and virgin olive oils of ordinary quality (oils C and D), as shown in **Table 2**. Considering a minimum rating of 6 on a 9-point hedonic scale as the lowest acceptable score in appraising a product's commercialization potential [18], Cluster 1 consumers awarded high

overall liking scores to the extra virgin olive oils and scores below the commercially acceptable limit to those of ordinary virgin quality. These consumers were capable of identifying the quality of the tasted oils and readily differentiating defective, ordinary virgin olive oils from non-defective, extra virgin ones.

Although highly significant differences ($p < 0.001$) were also found in the scores assigned to the oils by Cluster 2 consumers, the group's behavior was in contrast to that of Cluster 1, since the overall liking of extra virgin oils was rated low and that of defective oils was rated high by Cluster 2 consumers.

Except for the evaluation made of oil D by Cluster 2 individuals, consumers in neither cluster were influenced by the sensory description of the oil made available in Session 2. Overall, the sensory profiles of the different oils made available in Session 2 were used by consumers to confirm their initial evaluations made in Session 1 without the profiles. This behavior was expected only for Cluster 1 consumers, as they had been able to discern the quality of the different oils.

Informed descriptions of the good or defective quality of the oil samples, except for oil D, did not affect overall liking ratings for Cluster 2 consumers. Consumer acquaintance the defects of oil D significantly lowered the overall liking scores for this oil among Cluster 2 individuals. Nonetheless, providing Cluster 2 consumers with a written description of the defects of oil D did not influence the behavior of these individuals to an extent such that it led to a reduction in the overall liking of this oil to a value lower than that assigned by these consumers to the extra virgin oils (oils A and B). Members of Cluster 2, amounting to *ca.* 50% of the consumer sample, did not change their initial assessment of the different oils based on newly available knowledge of their extra virgin or defective quality, reaffirming the preference of these consumers for defective oils and their rejection of extra virgin, higher quality oils.

Table 3 shows willingness to purchase scores for each of the two clusters identified.

Table 2. Mean overall liking scores evaluated on a 9-point hedonic scale according to cluster.

Sample	Cluster 1 (n = 51)		Cluster 2 (n = 48)	
	Without description	*With description*	*Without description*	*With description*
A	6.6^{bB}	7.0^{bB}	3.6^{aA}	4.3^{aA}
B	6.6^{bB}	6.6^{bB}	4.7^{aA}	4.5^{aA}
C	4.5^{aA}	4.3^{aA}	7.0^{bB}	6.4^{bB}
D	4.8^{a,bA}	4.4^{aA}	6.8^{cB}	5.8^{bB}

Different lower case superscripts indicate significant differences within one row according to the Tukey test ($p \leq 0.05$). Different upper case superscripts indicate significant differences within one column ($p \leq 0.05$).

Table 3. Mean willingness to purchase scores evaluated on a 9-point scale according to cluster.

Sample	Cluster 1 (n = 51)		Cluster 2 (n = 48)	
	Without description	*With description*	*Without description*	*With description*
A	6.2^{bB}	6.7^{bB}	3.3^{aA}	4.4^{aA}
B	6.6^{bB}	6.6^{bB}	4.2^{aA}	4.6^{aA}
C	4.3^{aA}	3.9^{aA}	6.9^{bB}	6.3^{bB}
D	4.6^{a,bA}	3.9^{aA}	6.5^{cB}	5.4^{bB}

Different lower case superscripts indicate significant differences within one row according to the Tukey test ($p \leq 0.05$). Different upper case superscripts indicate significant differences within one column.

It can be noted from **Table 3** that willingness to purchase scores followed a similar trend to that of overall liking. Willingness to purchase changed significantly between Sessions 1 and 2 only for sample D; yet its willingness to purchase remained significantly above that of the extra virgin oils (samples A and B).

Clusters 1 and 2 were did not differ significantly (p < 0.05) for any of the socioeconomic variables surveyed. There was a significant difference (p = 0.031) in the brand of olive oil habitually consumed by either cluster. In Cluster 1, there was a large proportion (28%) of habitual consumers of extra virgin olive oil brands of recognized quality, compared to 10% of consumers of these brands in Cluster 2. This highlights the effect of consumption habits on food product acceptability and willingness to purchase, as Cluster 2 consumers, most of whom were not habitual consumers of good quality olive oil brands, preferred defective olive oils and appeared reluctant to change this preference based on newly available knowledge that the tasted oils had several sensory defects and were not of extra virgin quality.

3.4. Check-All-That-Apply-Question

Tables 4 and **5** show the frequency of mention for each of the terms of the CATA question according to consumer cluster.

With both clusters, the most frequently used terms were mild-flavored, aromatic, good quality, tasty and strong-flavored, whereas delicious, sweet, defective, bad-tasting and rancid had the lowest frequencies of mention.

Frequencies of mention provided by Cluster 1 to 13 of the 18 terms included in the CATA question differed significantly according to oil type. No significant differences were found for the terms bitter, fruity and sweet, although these attributes had been rated differently according to oil by the tasting panel (see **Table 1**). This shows that these consumers were unable to evaluate and discriminate olive oil quality based on these three attributes. No significant differences were found among the frequencies of mention of the terms delicious and strange-tasting. The low frequency of mention of delicious among Cluster 1 consumers may indicate that olive oil in itself is not associated with pleasurable sensations,

Table 4. Results of the check-all-that-apply question for Cluster 1: frequency of mention by attribute and by sample.

Attribute	Samples			
	A (extra virgin)	B (extra virgin)	C (ordinary virgin)	C (ordinary virgin)
Tasty[*]	45	33	29	20
Aromátic[**]	65	47	37	27
Bitter[ns]	20	27	14	14
Pungent[***]	45	41	14	24
Fruity[ns]	20	10	14	12
Leafy notes[***]	37	45	10	8
Strange-tasting[ns]	18	14	16	25
Fresh[**]	31	31	12	10
Bad-tasting[***]	2	0	22	8
Strong-flavored[*]	45	27	25	20
Mild-flavored[*]	31	47	41	59
Sweet[ns]	2	8	6	10
Delicious[ns]	6	4	8	4
Rancid[***]	2	0	16	18
Defective[***]	4	0	18	6
Good quality[***]	47	45	16	24
Poor quality[***]	6	0	25	22
Cooked olives[***]	8	2	31	20

[*]p < 0.05; [**]p < 0.01; [***]p < 0.001. ns: No significant differences (p > 0.05) according to Cochran's Q test.

Table 5. Results of the check-all-that-apply question for Cluster 2: frequency of mention by attribute and by sample.

Attribute	Samples			
	A (extra virgin)	B (extra virgin)	C (ordinary virgin)	C (ordinary virgin)
Tasty[***]	15	27	60	46
Aromatic[***]	25	29	67	46
Bitter[***]	33	54	10	10
Pungent[***]	50	35	4	13
Fruity[ns]	17	17	13	17
Leafy notes[***]	50	44	10	2
Strange-tasting[***]	44	25	2	2
Fresh[***]	6	21	25	48
Bad-tasting[***]	23	17	2	2
Strong-flavored[*]	38	50	27	27
Mild-flavored[**]	27	23	52	60
Sweet[ns]	6	8	15	10
Delicious[**]	2	0	13	15
Rancid[**]	21	15	6	0
Defective[*]	8	10	0	2
Good quality[***]	10	17	50	44
Poor quality[***]	31	21	4	0
Cooked olives[*]	8	10	23	19

[*]p < 0.05; [**]p < 0.01; [***]p < 0.001. ns: No significant differences (p > 0.05) according to Cochran's Q test.

a possibility also consistent with the overall liking scores assigned to the oils, which were in no case greater than 7 on the 9-point hedonic scale (**Table 2**).

As shown in **Table 4**, Cluster 1 consumers described the extra virgin oils (A and B) in terms of aromatic, tasty, pungent, leafy notes and good quality. In particular, oil A was described as strong-flavored, in contrast with the mild-flavored attribute used by these consumers to describe oil B.

The two ordinary virgin oils (C and D) were described by Cluster 1 mainly in terms of bad-tasting, mild-flavored, rancid, defective, cooked-olive smell and poor quality. This shows that members of this cluster were readily able to identify those oils that had defects and relate such defects to overall product quality.

In Cluster 2, the frequencies of mention of 16 out of the 18 terms in the CATA questions differed significantly according to oil, showing that these consumers resorted to a greater number of terms to differentiate the oils. As was also the case with Cluster 1 consumers, no significant differences among the different oils were identified by Cluster 2 consumers in terms of fruity and

sweet, presumably due to the lack of the sensory training required to be able to rate these attributes.

There was considerable ignorance of olive oil quality facts among Cluster 2 consumers, who made negative appraisals of the positive attributes characteristic of extra virgin olive oil—i.e., these consumers described the bitter, pungent, and green grassy/leafy attributes of extra virgin olive oils A and B in terms of defective, bad-tasting, strange-tasting, poor quality and rancid. Conversely, they associated the cooked-olive smell typical of the fusty/muddy sediment defect of ordinary olive oils with good quality, tasty, sweet and aromatic attributes, which, in addition to mild-flavored, delicious and fresh, were used to describe oils C and D.

Among the factors key to the acceptability extra virgin olive oil are its bitterness and pungency (or spiciness). Australian olive oil producers consider these two attributes to be major determinants of an olive oil's profile [19] and Italian consumers perceive bitterness and pungency as the most appropriate sensory descriptors of an olive oil's extra virgin quality [20]. On the other hand, several authors have reported that the greater an olive oil's bitterness, the less consumers in emerging markets appear to like it [21-23], suggesting that such consumers have not been exposed to the product in sufficient degree to have developed an appreciation of the bitterness and pungency attributes of quality virgin olive oils [24]. In contrast with the above findings, a considerable proportion of the consumer sample (Cluster 1) were capable of appreciating the attributes of quality virgin olive oil by assigning high overall liking scores to oils A and B.

MFA was carried out based on CATA counts for both consumer clusters. The first two dimensions of the MFA accounted for 84.3% of the variance of the experimental data, representing 53.7% and 30.6% of the variance, respectively.

As shown in **Figure 1(a)**, there was a strong correlation between the evaluations made by the two clusters

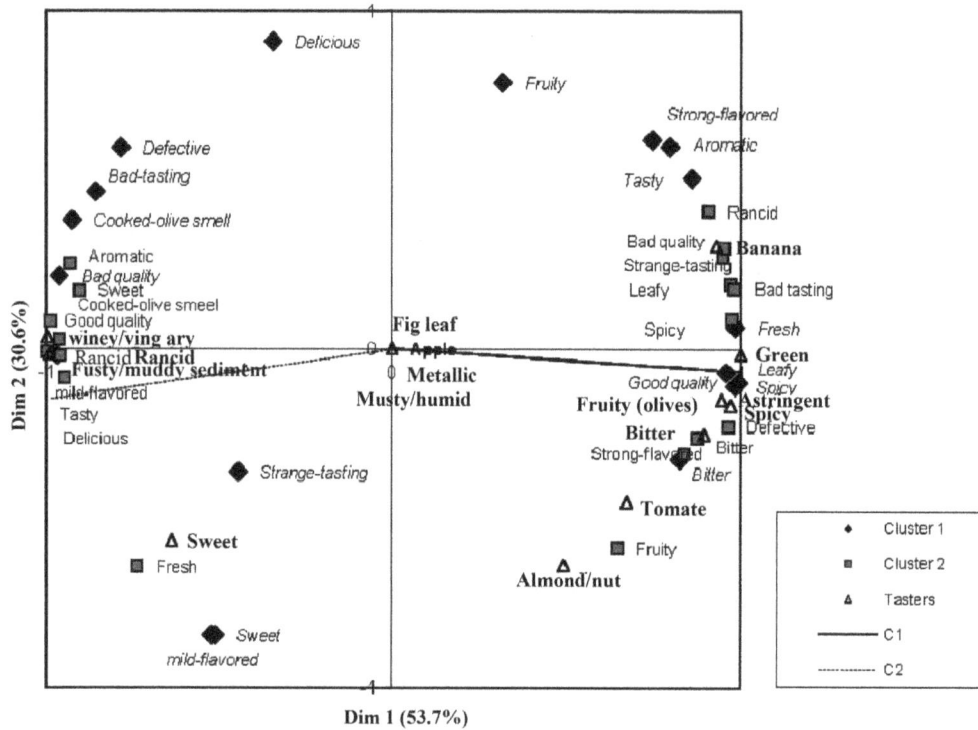

(a)

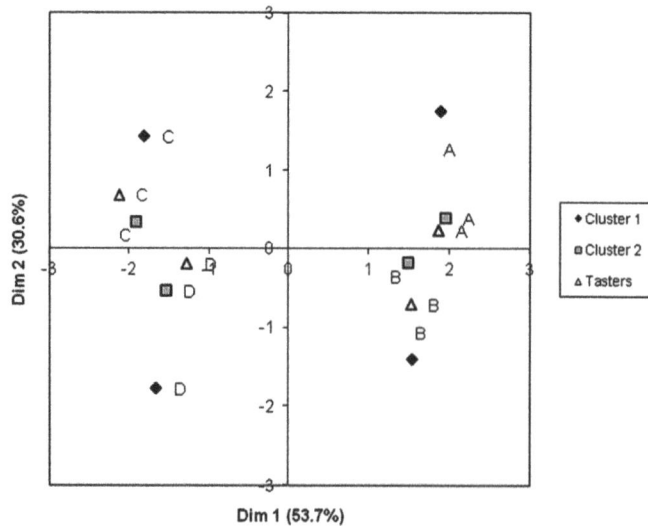

(b)

Figure 1. (a) Multiple factor analysis based on the check-all-that-apply question. Representation of attributes for Cluster 1 (black trapezoid and italic type), Cluster 2 (grey squares) and for the tasting panel (unfilled triangles and bold type; (b) Multiple factor analysis based on the check-all-that-apply question. Representation of oil samples.

with regard to most of the terms included in the CATA question, suggesting that the two clusters, despite their contrasting preference patterns, perceived the sensory attributes of the different oils similarly. Frequency of mention of pungency, bitter and cooked-olive smell correlated strongly between clusters, suggesting that the different oils were evaluated similarly by all consumers.

Also a strong correlation was found between otherwise contradictory appraisals—e.g., good quality and poor quality—made by consumers in either cluster, showing the different preconception of the sensory attributes of high quality olive oil; hence, the different preference patterns observed between consumer clusters.

The green (grassy/leafy notes) attribute assigned by the sensory tasters was associated with the terms fresh and leafy notes by Cluster 1 consumers, demonstrating the correct appreciation of the quality traits of extra virgin olive oil among members of this cluster. This is also confirmed by the strong correlation observed between the frequency of reference to good quality among Cluster 1 consumers and the ratings for the fruity, pungent, bitter and astringent attributes—i.e., those attributes typical of quality, extra virgin olive oil—provided by the sensory panel.

Likewise, the fusty/muddy sediment, rancid and winey defects identified by the tasting panel correlated strongly with the frequency of mention of rancid, poor quality, bad-tasting and cooked-olive smell among Cluster 1 consumers, and with that of good quality, sweet, aromatic, tasty, delicious and cooked-olive smell among Cluster 2 consumers.

The term cooked-olive smell is a sensory attribute associated with the fusty/muddy sediment defect, a characteristic flavor of oil obtained from olives that have been piled or stored in conditions such as to have undergone anaerobic fermentation to an advanced stage. It is worth noting that consumers were equally able to perceive this smell irrespective of cluster, although the smell was associated with poor quality in the case of Cluster 1 consumers and with good quality in the case of Cluster 2 consumers, who described the defective oils as aromatic. These results suggest that the two clusters had different drivers of liking, reflected in the difference in their preference patterns.

Figure 1(b) shows the representation of the studied oils in the MFA dimensions. It shows the coincidence in the spatial distribution of the oils among the panel of tasters and the two consumer Clusters. The first dimension separates the oils according to commercial quality, with the extra virgin oils on the right and the ordinary virgin oils on the left.

4. Conclusion

The results of this study indicate that a large proportion of consumers in Uruguay, an emerging olive-growing country, are unacquainted with the sensory traits of extra virgin olive oil, and clearly prefer defective, ordinary virgin olive oils, highlighting the need to sensitize consumers in this category.

REFERENCES

[1] URUGUAY XXI, Sector Olivícola en Uruguay, "Promoción de Inversiones y Exportaciones," Presidencia de la República Oriental del Uruguay, 2012, p. 5.

[2] A. Gámbaro, C. Dauber, G. Ares and A. C. Ellis, "Studying Uruguayan Consumers' Perception of Vegetables Oils Using Word Association," *Brazilian Journal of Food Technology*, Special Issue, 2011, pp. 131-139.

[3] M. Santosa, "Analysis of Sensory and Non-Sensory Factors Mitigating Consumer Behavior: A Case Study with Extra Virgin Olive Oil," Ph.D. Dissertation, Food Science: University of California, Davis, 2010.

[4] R. M. Duran, "Relationship between the Composition and Ripening of the Olive and the Quality of the Oil," *Acta-Horticulturae*, Vol. 286, 2010, pp. 441-451.

[5] F. Angerosa, "Influence of Volatile Compounds on Virgin Olive Oil Quality Evaluated by Analytical Approaches and Sensor Panels," *European Journal of Lipid Science and Technology*, Vol. 104, No. 9-10, 2002, pp. 639-660.

[6] R. Aparicio and J. Harwood, "Manual del Aceite de Oliva," AMV Ediciones, Mundiprensa, España, 2003.

[7] M. Uceda, M. P. Aguilera and I. Mazzuchelli, "Manual de Cata y Maridaje del Aceite de Oliva," Editorial Almuzara, Córdoba, 2010.

[8] A. Romero, J. Tous and L. Guerrero, "El Análisis Sensorial del Aceite de Oliva Virgen," In: J. Sancho, E. Bota and J. J. de Castro, Eds., *Introducción al Análisis Sensorial de los Alimentos*, Edicions Universitat de Barcelona, Barcelona, 1999, pp. 183-197.

[9] IOC, "Trade Standard Applying to Olive Oils and Olive-Pomace Oils," International Olive Council, COI/T.15/NC no.3/Rev.4, 2009.

[10] IOC, "Sensory Analysis of Olive Oil. Method for the Organoleptic Assessment of Virgin Olive Oil," International Olive Council, COI/T.20/Doc.No.15/Rev.4, 2011.

[11] S. Issanchou, "Sensory and Hedonic Consumer Expectations towards Typical Food Products," *European Conference on Sensory Science of Food and Beverages*, Florence, 26-29 September 2004.

[12] IOC, "Sensory Analysis of Olive Oil. Standard. Guide for the Selection, Training and Monitoring of Skilled Virgin Olive Oil Tasters," International Olive Council, COI/T.20/Doc.No.14/Rev.2, 2007.

[13] IOC, "Sensory Analysis of Olive Oil," Standard Glass for Oil Tasting, 1987.

[14] International Olive Council, COI/T.20/Doc.No.5.

[15] IOC, "Sensory Analysis of Olive Oil. Standard. Guide for the Installation of a Test Room," International Olive

Council, COI/T.20/Doc.No.6/Rev.1, 2007.

[16] ISO, "Sensory Analysis: General Guidance for the Design of Test Rooms, ISO 8589," International Organization for Standardization, Geneve, 1988.

[17] M. Bècue-Bertaut and J. Pagès, "Multiple Factor Analysis and Clustering of a Mixture of Quantitative, Categorical and Frequency Data," *Computational Statistics & Data Analysis*, Vol. 52, No. 6, 2008, pp. 3255-3268.

[18] M. Bécue-Bertaut, R. Alvarez-Esteban and J. Pagés, "Ratings of Products through Scores and Free-Text Assertions: Comparing and Combining Both," *Food Quality and Preference*, Vol. 19, 2008, pp. 122-134.

[19] A. M. Muñoz, V. G. Civille and B. T. Carr, "Sensory Evaluation in Quality Control," Van Mostrand Reinhold, New York, 1992.

[20] R. Gawel and D. A. G. Rogers, "The Relationship between Total Phenol Concentration and the Perceived Style of Extra Virgin Olive Oil," *Grasas y Aceites*, Vol. 60, No. 2, 2009, pp. 134-138.

[21] G. Caporale, S. Policastro, A. Carlucci and E. Monteleone, "Consumer Expectations for Sensory Properties in Virgin Olive Oils," *Food Quality and Preference*, Vol. 17, No. 1, 2006, pp. 116-125.

[22] G. Caporale, S. Policastro and E. Monteleone, "Bitterness Enhancement Induced by Cut Grass Odorant (cis-3-hexen-1-ol) in a Model Olive Oil," *Food Quality and Preference*, Vol. 15, No. 3, 2004, pp. 219-227.

[23] J. M. Garcıa, K. Yousfi, R. Mateos, M. Olmo and A. Cert, "Reduction of Oil Bitterness by Heating of Olive (*Olea europea*)," *Journal of Agricultural Food Chemistry*, Vol. 49, 2001, pp. 4231-4235.

[24] C. Delgado and J. X. Guinard, "How Do Consumer Hedonic Ratings for Extra Virgin Olive Oil Relate to Quality Ratings by Experts and Descriptive Analysis Ratings?" *Food Quality and Preference*, Vol. 22, No. 2, 2011, pp. 213-225.

[25] A. Recchia, E. Monteleone and H. Tuorila, "Responses to Extra Virgin Olive Oils in Consumers with Varying Commitment to Oils," *Food Quality and Preference*, Vol. 24, No. 1, 2012, pp. 153-161.

18

Impact of Preparation Process on the Protein Structure and on the Volatile Compounds in *Eisenia foetida* Protein Powders

Elias Bou-Maroun[1], Camille Loupiac[1], Aurélie Loison[1], Bernadette Rollin[1], Philippe Cayot[1], Nathalie Cayot[1*], Elil Marquez[2], Ana Luisa Medina[2]

[1]Unité Procédés Alimentaires et Microbiologiques, UMR A 02.102, AgroSup Dijon, Université de Bourgogne, Dijon, France; [2]Dpto Ciencia de Alimentos, Grupo Ecología y Nutrición, Facultad de Farmacia y Bioanalisis, Universidad de los Andes, Merida, Venezuela.

ABSTRACT

Protein powders from *Eisenia foetida* were prepared using different drying processes and fractionation. Differential scanning calorimetry was used to show that heat denaturation occurred during the drying process above 42°C. Protein solubility was also studied. The addition of dissociating reagents allowed concluding that solubility was decreased during oven drying due to thermo denaturation including hydrogen bonds. The volatile compounds of the different powders were extracted by solid phase micro-extraction and identified by mass spectrometry. Volatile compounds were related to lipid oxidation and Maillard reactions occurring during the preparation of the powders. High drying temperatures led to more volatile compounds resulting from Maillard reactions. In the protein powder preparation process, a fractionation step led to a "pulp fraction" and a "juice fraction" of earthworms. The "pulp fraction" contained less odorant volatile compounds resulting from Maillard reactions than the "juice fraction" did.

Keywords: *Eisenia foetida* Protein Powder; Drying Process; Protein Structure; Volatile Compounds; HS-SPME/GC-MS

1. Introduction

Some authors have pointed out the great potential of *Eisenia foetida* as a non-conventional protein source [1]. They determined that earthworms had a high protein content of about 62% dry weight [2] and that the content of essential amino acids, except for tyrosine [1], was higher than that recommended by the FAO. Earthworms could then have many different nutritional applications as animal feed and as an ingredient in food products for humans. They have already been used in an experimental diet for rainbow trout [3] and as chicken feed [4].

For practical reasons, earthworms were often dried and used as a powder in the different applications. We conducted former studies on oven-dried powders. Two main problems were encountered when these earthworm protein powders were used in food products. One was related to the odor properties, and the other to protein denaturation.

Concerning the odor properties, off-flavors can make the powder unacceptable to humans. In fact, the Solvent-Assisted-Flavor-Evaporation (SAFE) method was used in a previous study [5] to obtain the volatile fraction from the oven-dried protein powder. Gas chromatography coupled with mass spectrometry (GC-MS) was then used to analyze this volatile fraction and allow the identification of more than 70 volatile compounds. Among these volatile compounds, some might come from the raw material (*i.e.* the earthworm itself, its feed and its environment such as earth, water...); others might be due to microbiological modifications or to lipid peroxidation. Additionally, in the case of protein powder, drying is a crucial step because it may concentrate off-flavors and/or create new odorant compounds. The most abundant chemical groups found in the volatile fraction of the earthworm powder were ketones (29%) including undecan-2-one, alcohols (21%) including pentan-1-ol, and aldehydes (15%) including hexanal. Pyrazines (10%) were also a major chemical group and could be considered a product generated during drying as a result of thermally induced reactions. In fact, Maillard reactions are known to give alkylpyrazines, or benzene acetaldehyde among many other volatile compounds [6]. Some of the aldehydes and acids found in the volatile fraction may come from the degradation of lipids. In fact, the first step of lipid oxidation generates hydroperoxides, the decomposition of which leads to the formation of aldehydes, acids, esters and hydrocarbons. For example, model experiments of the autoxidation of oleic, linoleic and linolenic acids gave pentanal, hexanal, heptanal, E hex-2-enal [7]. This mixture of volatile compounds extracted from the earthworm powder gave it a strong animal odor, which was described by Cayot et al. [5] as "dried fish". This odor may limit the use of this protein powder in food products.

Nevertheless, dehydrated proteins offer a longer shelf life, and lower storage and transports costs. The traditional methods used to transform protein suspensions into dry powders are spray drying, solar drying, convective hot air drying and freeze drying. A number of effects are observed during the heat-drying (spray, solar and convective hot air drying) of protein-containing product, the main effect being the thermal denaturation of protein. Heat causes rapid thermal motion of the atoms leading to rupture of the hydrogen bonds that link them together. Under suitable conditions, thermal denaturation leads to coagulation. During the process, peptide chains are propagated and form agglomerates. The denatured protein is insoluble [8]. As a low temperature process, freeze-drying is a known method to ensure the long term stability of high-value or therapeutic proteins that are not stable enough in aqueous solutions and that are very sensitive to heat. However, some proteins are also sensitive to the stresses imposed by freeze drying and can be degraded or decomposed during the process [9]. The low temperature of freeze drying does not guarantee protein stability because some proteins also experience cold denaturation. Moreover, freeze drying is costly in terms of both time and energy, and can therefore only be used to dry high value proteins.

For these reasons, different processes to kill the earthworms, to crush them and to dry the different fractions obtained were tested in the present study. Two main types of drying process were used, oven drying (60°C during 4 h) and, for reference in terms of protein structure preservation, freeze drying. The volatile compounds were extracted from the protein powder by Headspace Solid Phase Micro Extraction HS-SPME, and were further analyzed by GC-MS. Protein denaturation was checked by Differential Scanning Calorimetry (DSC) and solubility measurements.

The aim of the study was to understand the impact of preparation process on volatile composition and on protein denaturation, in order to obtain fractions as odor-free as possible and with proteins as little denatured as possible.

2. Materials and Methods

2.1. Preparation of Protein Powder: From the Compost to the Dried Earthworm Powder

Samples were obtained from the species *Eisenia foetida* at the adult stage of development (3 months), with average length and weight of 8.5 cm and 0.45 g respectively (earthworm cultures from "Luis Ruiz Terán" Herbarium at the Faculty of Pharmacy, University of the Andes, Merida-Venezuela). The earthworms were fed on a diet of organic waste compost, obtained from a university canteen in the region. In order to guarantee optimum growth conditions, the temperature, moisture and pH of the compost were kept under control. A single batch of earthworm powder was prepared as follows and used throughout the study. Different preparation processes were then applied as summarized in **Figure 1** and the dried protein-rich powders obtained were stored until further analyses. The data we referred to in the introduction part were corresponding to ODEW sample of the present study.

2.2. Composition of the *Eisenia foetida* Powders

2.2.1. Water Content
The moisture content of each powder was determined in triplicate using volumetric Karl Fisher titrimetry. This chemical method allowed the determination of the water content with a reproducibility of 0.5%. The titration was checked with a water standard at 1% (Epura, Merk). A Karl Fisher titration apparatus from Mettler Toledo (DL38, Inc., Columbus, OH, USA) was used.

2.2.2. Protein Content
The Kjeldahl procedure [10] was used to determine the nitrogen content in earthworm powders (0.1 g), or suspensions (5 mL). Samples were weighed and transferred into the Kjeldahl digestion flask containing 10 mL of concentrated H_2SO_4 and a catalyst. After 4 h of digestion in a unit with electrical heat and fume removal (Labonco, Kansas City, MO, USA) and cooling to room temperature, 5 mL of distilled water was added to each flask. By distillation, ammonia was trapped as ammonium borate in a 3% boric acid solution (3 g of boric acid in 100 mL deionized water (w/v)). Total nitrogen was determined by titration with standardized HCl to recover

Sample label	FDEW	FDJ	ODJ	FDP	ODP	ODEW
Preparation of fresh earthworms	earthworms fed on a diet of organic waste compost separated from the compost washed and stored for 12 hours in an air insufflated water container					
Method to "kill" earthworms	liquid nitrogen for 5 min	boiling water for 1 min				
Fractionation	grinded using a household blender	Cut into pieces using a knife				
		Extraction using pressing				
		Juice		Pulp		
Drying	Freeze drying	Freeze drying	Oven 60 °C – 4 h	Freeze drying	Oven 60 °C – 4 h	Oven 60 °C – 4 h
Fractionation						ground with a mechanical crusher
Storage	closed plastic bags - at 4 °C until further analyses					

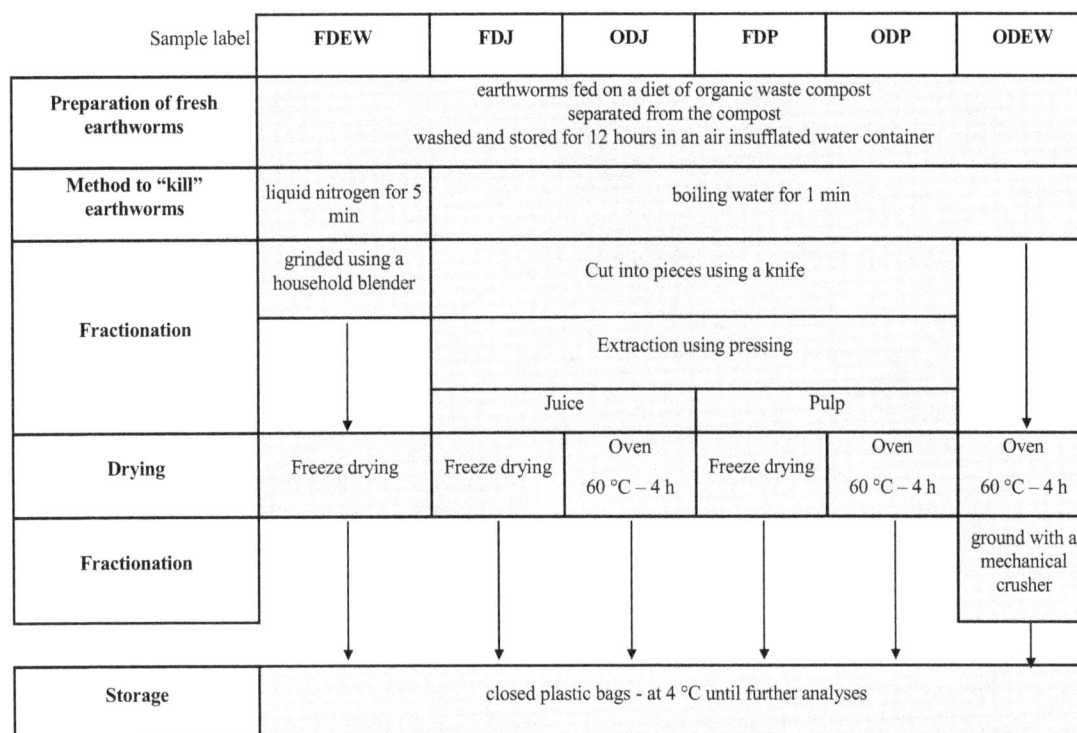

Figure 1. Process for the preparation of earthworm protein powders.

the initial boric acid pH (pH 4.6). The conversion factor from nitrogen content to protein content was 6.25.

2.3. Evaluation of Protein Denaturation

2.3.1. Protein Solubility

To determine protein solubility, protein powders were suspended in water at 40 $g \cdot L^{-1}$ and stirred overnight at 4°C. The pH of the suspension was adjusted to pH 6 at the end of the stirring step. The suspensions were centrifuged at 700 g for 30 minutes and the Kjeldahl procedure was used to analyze the protein content as previously described. Some samples of ODEW were suspended at 40 $g \cdot L^{-1}$ but in the presence of different reagents: 1% w/v of sodium dodecyl sulfate (SDS), 1 $mol \cdot L^{-1}$ of dithiotreitol (DTT), or 6 $mol \cdot L^{-1}$ of urea were added before centrifugation. Biochemical grade urea, SDS and DTT were purchased from Sigma.

2.3.2. Differential Scanning Calorimetry

A microcalorimeter DSC III (Setaram Instrumentation, France) was used to determine the denaturation temperature of the protein suspension. The range of temperature chosen for this study was 30°C to 110°C, with a temperature gradient of 0.5°C·min^{-1}, and a sample weight of approximately 600 mg (exact weighting) placed in a hermetically closed cell "batch". Water was used as the reference. The temperature profile of the thermogram

was done using the software of the apparatus (Setsoft, Setaram Instrumentation, France).

2.4. Extraction of Volatile Compounds by Headspace Solid Phase Micro-Extraction (HS-SPME)

For the HS-SPME analysis, 1 g of each protein powder was transferred into a 10 mL vial, which was immediately sealed with a Teflon-lined septum and a screw cap. The equilibration time was 24 h at room temperature. The powder was incubated at 30°C for 90 min with an agitation of 250 rpm at intervals of 20 s using an auto sampler (Gerstel MPS2). After incubation, the headspace volatiles were extracted using an SPME fiber (2 cm-50/30 μm DVB/Carboxen/PDMS/StableFlex, Supelco, USA) for 60 min at 30°C. Finally, for the injection, the fiber was inserted into the GC injection port and desorbed for 3 min in splitless mode. The fiber was cleaned at 240°C for 12 min after each injection. The relative standard deviation of this method was calculated by analyzing the ODEW sample five times and was equal to 2.17%. The other samples were analyzed twice.

2.5. Gas Chromatography Mass Spectrometry Analysis (GC-MS)

Volatile compounds were analyzed on a 5973 gas chromatograph (Hewlett-Packard, Palo Alto, CA, USA) equipped with a fused-silica capillary column (30 m × 0.32 mm ID, 0.5 μm film thickness) coated with a DB-Wax stationary phase (J & W Scientific, USA). The instrument was equipped with an injection port operating at 240°C. Helium was used as the carrier gas and the chromatographic temperature was programmed from 40°C to 240°C at a rate of 4°C/min, with a final isotherm of 10 min. Mass spectrometry was taken in the electron ionization mode at 70 eV and the scan range between 29 and 350 amu. The ion source was set at 230°C and the transfer line at 240°C. Compounds were identified by comparison with mass spectra libraries (WILLEY138, NIST, and INRA database) and by calculating and comparing the GC retention index of a series of alkanes (C8 to C30) with the retention index from published data calculated under the same conditions.

2.6. Statistical Data Analysis

The absolute area values of the four compounds chosen as related with heating process (hexanal, 2-pentyl furan, 2,6-dimethylpyrazine and benzaldehyde) were first centred and reduced in order to attribute the same importance to the four variables (four compounds). Standardized data were submitted to a hierarchical clustering analysis using Euclidean distances and Ward's aggregation method in order to identify homogeneous subgroups of powders. Then, the subgroups were verified by k-means clustering, calculated on the standardized data and using the un-scaled squared Euclidean distances. In fact, three k-means algorithms were made in parallel in order to verify the stability of the subgroups. Data analysis was done using the Statistica software (version 10).

3. Results and Discussion

3.1. Impact of the Preparation Process on the Protein Structure and Solubility

Differential scanning calorimetry was used to determine the impact of the preparation process on protein structure for the traditional powder (ODEW) and for a powder prepared without any heating step, neither during "killing" of earthworms, nor during drying (FDEW). **Figure 2** presents typical thermograms obtained by DSC on suspensions of freeze dried (FDEW) and oven dried (ODEW) protein powders. It was not possible to observe any endothermic peak on samples that had been dried in the oven at 60°C as ODEW. The thermogram obtained for samples which had been freeze dried (FDEW) exhibited an endothermic peak at 42°C, indicating that a denaturation process of non-denatured proteins had occurred during DSC at this temperature. As a consequence, FDEW can be considered a reference for earthworm powder as non-denatured protein and the powders that

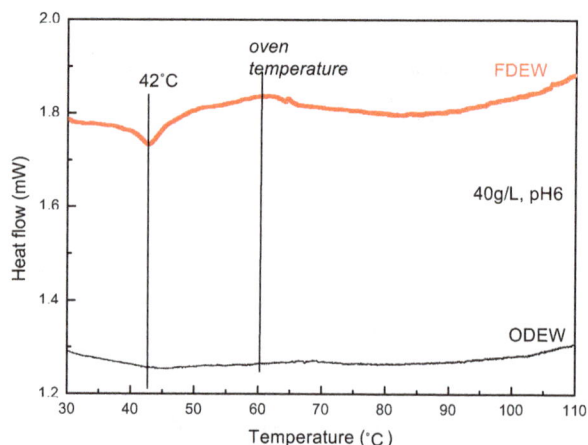

Figure 2. Thermograms obtained by differential scanning calorimetry of oven dried earthworm powder (ODEW, black line) and freeze dried earthworm powder (FDEW, bold line) (exo up).

had been prepared at temperatures above 42°C thus presented flat thermograms.

Preliminary studies on solubility were done from pH 2 to pH 8 on FDEW and on ODEW, and showed that maximum solubility was obtained for pH 6 and above (data not shown). As a pH value compatible with food products, pH 6 was selected to continue the study. Solubility at this pH value was 43% ± 22% for ODEW and 117% ± 27% for FDEW (value above 100% was due to the precision of the method). FDEW was thus less denatured than ODEW but also far more soluble.

It is well established that the functionality of proteins is strongly influenced by the processing conditions. The lack of solubility can be due to protein aggregation via covalent or non-covalent bonding. For example, heating whey proteins leads to denaturation of the whey proteins and beta-lactoglobulin, which modifies the availability of some free thiol groups. These free thiol groups can react with disulphide bonds, or can react with another thiol group to form a disulphide bond [11]. Non-covalent bonding can include the modification of dipole-dipole interactions, hydrogen bonds, or hydrophobic interactions.

To study which phenomena were involved in the loss of solubility of earthworm proteins, the solubility of ODEW at pH 6 was measured in the presence of different dissociating compounds, such as DTT, SDS and urea. In fact, hydrogen bonds and hydrophobic interactions in proteins can be destabilized by urea [12]. Urea denatures a protein molecule through preferential adsorption with charged protein solutes. This mechanism dehydrates the molecules and causes repulsion between proteins, which in turn stabilizes the unfolded form [13]. SDS allows the breaking of hydrophobic bonds [14]. DTT is often used to reduce the disulphide bonds of proteins and to prevent intra- and intermolecular disulphide bonds from forming between cysteine residues [15].

In our experiments, the addition of DTT did not change the solubility of ODEW (36% ± 14%). In the presence of SDS, the solubility increased to 64% ± 19%, and finally, the addition of urea led to 100% ± 19% solubility. It can then be concluded that the loss of solubility of earthworm protein due to heat denaturation involved hydrogen bonds and probably also hydrophobic interactions.

From these experiments on protein structure and solu-bility, FDEW rather than ODEW should be considered the reference for the preparation of earthworm powder. It was thus important to check the impact of the preparation process also on the volatile composition of these two powders.

The volatile compounds were identified for the whole powders, for two fractions of earthworms and for the two drying processes and were reported in **Table 1**. Among the identified compounds, some compounds are strongly

Table 1. Volatile compounds identified by GC-MS in an HS-SPME extract from protein powders and their corresponding peak area.

RI[a]	Identification[b]	Name	CAS number	FDEW	ODEW	FDJ	ODJ	FDP	ODP
1083	A	Hexanal ($m/z = 82$)	66-25-1	435603	792038	510838	995617	620397	919452
1165	A	Pent-1-en-3-ol	616-25-1	0	16862807	1571370	0	0	0
1188	A	Heptan-1-al	111-71-7	2957276	4237536	2182076	2973887	2127609	2548139
1200	A	Dodecane	112-40-3	0	0	10271953	6095898	0	0
1202	A	p-mentha-1,8-diene (limonene)	95327-98-3	0	3086859	0	0	4993981	5098878
1235	A	2-pentyl furan	3777-69-3	387000	3492850	306153	1457416	206160	377380
1241	A	6-methyl-heptan-2-one	928-68-7	0	1050938	0	0	0	0
1258	A	Pentan-1-ol	71-41-0	1334302	3499926	0	1879380	0	0
1273	A	1-methyl-4-(1-methylethyl)benzene	99-87-6	1846764	278404	0	0	0	1152006
1289	A	Octan-2-one	111-13-7	263751	1436805	0	0	0	286278
1307	A	Oct-1-en-3-one	4312-99-6	1098471	726976	769269	1252966	590119	652081
1327	A	2,6-dimethylpyrazine	108-50-9	0	3059196	0	4220157	0	1920495
1340	A	Oct-3-en-2-one	18402-82-9	255951	2942391	2465463	248874	68187	689294
1343	A	6-methyl-hept-5-en-2-one	110-93-0	467307	249658	509444	0	0	189492
1362	A	Hexan-1-ol	111-27-3	1004912	2627738	2649907	2068947	623320	1485772
1395	A	Nonan-2-one	821-55-6	0	120005	0	0	0	0
1398	A	Nonan-1-al	124-19-6	528165	446024	1186327	2088480	0	459443
1420	B	5-ethyl-1-formylcyclopentene	-	372893	294482	413625	843813	212376	543645
1434	A	Cyclohex-2-en-1-one	930-68-7	0	2610028	0	3483881	0	1950115
1451	A	Acetic acid	9035-69-2	760807	8190101	753608	32515159	0	641846
1458	A	Oct-1-en-3-ol	3391-86-4	8132508	8788512	6007789	5404839	3453999	5900573
1492	B	3,5,5-trimethyl-hex-2-en	26456-76-8	592147	857455	324731	498399	0	0
1498	A	Decan-2-one	693-54-9	0	426681	0	0	0	0
1521	A	Octa-3,5-dien-2-one ($m/z = 95$)	30086-02-3	2925190	843142	2027823	962378	1241743	973075
1524	A	Benzaldehyde ($m/z = 105 + 106$)	100-52-7	1031804	3058667	767255	3816239	484894	1318509
1540	A	Propanoic acid	79-09-4	4535554	1958294	2535796	9546045	661820	1475021
1568	A	Octan-1-ol	111-87-5	0	414594	114375	617447	0	0
1573	B	Octa-3,5-dien-2-one (ui)[*]	-	3625640	2923317	2142919	1064940	1383211	1294623
1604	A	Undecan-2-one ($m/z = 170$)	112-12-9	0	27665	0	0	0	0
1605	B	2,6-dimethyl Cyclohexanol ($m/z = 128$)	5337-72-4	0	51268	42104	0	0	34043
1624	A	Oct-2-en-1-ol (E)	18409-18-2	0	780186	592421	1118664	0	0
1630	A	Butanoic acid	107-92-6	0	3092595	39101	0	0	0
1649	B	Undecan-6-ol	23708-56-7	0	0	1392786	1230098	1203617	920051
1665	A	Furfuryl alcohol	98-00-0	0	0	0	3542250	0	0
1710	A	Dodecan-2-one	6175-49-1	0	419726	0	459043	0	381307
1805	A	4-methyl pentanoic acid	646-07-1	0	579506	0	0	0	0
1815	A	Tridecan-2-one	593-08-8	83529	298119	0	241804	0	0
1848	A	Hexanoic acid	142-62-1	706807	1657848	523068	5309713	301330	585619
1914	A	2-phenyl ethanol	60-12-8	581848	917356	1026107	152704	0	0
1943	A	β-Ionone	14901-07-6	63746	99855	117107	0	0	128966
1955	A	Heptanoic acid	111-14-8	1131334	864757	1376250	647697	0	1494944
		Sum of obsolute areas of identified compounds multiplied by 10^{-7}		3.51	8.41	4.26	9.47	1.82	3.34
		Sum of absolute areas of HPIC multiplied by 10^{-6}		1.85	10.40	1.58	14.03	1.31	4.54
		Sum of absolute areas of Maillard compounds multiplied by 10^{-6}		1.42	9.61	1.07	13.04	0.69	3.62

[*]ui = Unidentified isomer. [a]Retention index calculated with a DB-Wax stationary phase using a series of alkanes between C8 and C30. [b]A, compounds identi-fied by MS and GC retention index as compared with those published in the literature using a similar stationary phase [16-18]; B, tentative identification by MS. In bold: Products obtained from a heat induced reaction such as: Maillard reactions, lipid oxidation and lipid degradation. n.d. = not detected.

dependent on thermal treatments: 1) Benzaldehyde comes from Strecker degradation and is known to be a Maillard reaction product [19,20]. 2) Alkylpyrazines generates from two accepted mechanisms: the Strecker degradation and the ammonia/acyloin reaction. In both mechanisms, the temperature is a determinant factor. 2,6-dimethylpyrazine may come from the thermal degradation of serine or threonine [21]. 3) 2-pentyl furan comes from the thermal degradation of linoleic acid [22]. Strecker degradation produced by some lipid oxidation products such as 4,5(E)-epoxy-2(E)-decenal yields 2-pentyl furan [23]. 4) Hexanal is often found to be a major compound in the volatile profile of meat products and is often chosen as an indicator of oxidation in meats, especially during the early oxidative changes [24].

These four compounds were further used to compare the different powders. The term "heat process implicated compounds" (HPIC) will be used here to name these compounds.

3.2. Impact of the Preparation Process on the Volatile Composition of ODEW and FDEW

A global comparison of ODEW and FDEW volatile composition was done by comparing the sum of identified compounds areas. This comparison shows that ODEW contains 2.4 times more volatiles compounds than FDEW (**Table 1**). Compounds from various chemical classes such as aldehydes, alcohols, benzene derivatives, or acids were encountered in both powders. As previously reported [25], volatile compounds are originated from lipid oxidation, Maillard reactions, and animal feed. Amino acid catabolism, environmental pollutants and the growth of microorganisms have also been reported as possible origins of volatile compounds detected in earthworm powders [7].

The amount of Maillard reactions products and HPIC was higher in ODEW than in FDEW. 2,6-dimethylpyrazine was found in ODEW but not in FDEW. Killing earthworms in boiling water and drying them in a traditional oven favor Maillard reactions due to the higher temperature during the process and, consequently, lead to the production of Maillard compounds in larger quantities.

As many parameters were different in the preparation processes of these two powders, a deeper study of the impact of the drying step and the fractionation step on volatile compounds production was necessary.

3.3. Impact of the Drying Step on the Volatile Composition of the Powders

Each protein fraction was dried using either a traditional oven (ODJ and ODP) or using freeze drying (FDJ and FDP). In order to study the impact of the drying on the volatile composition of the powders, we compared powders that differ only by the drying step, *i.e.* the two couples FDJ/ODJ and FDP/ODP.

Benzaldehyde, 2,6-dimethylpyrazine and 2-pentyl-furan were chosen as indicators of Maillard reactions and their relative quantities were reported in **Figure 3**.

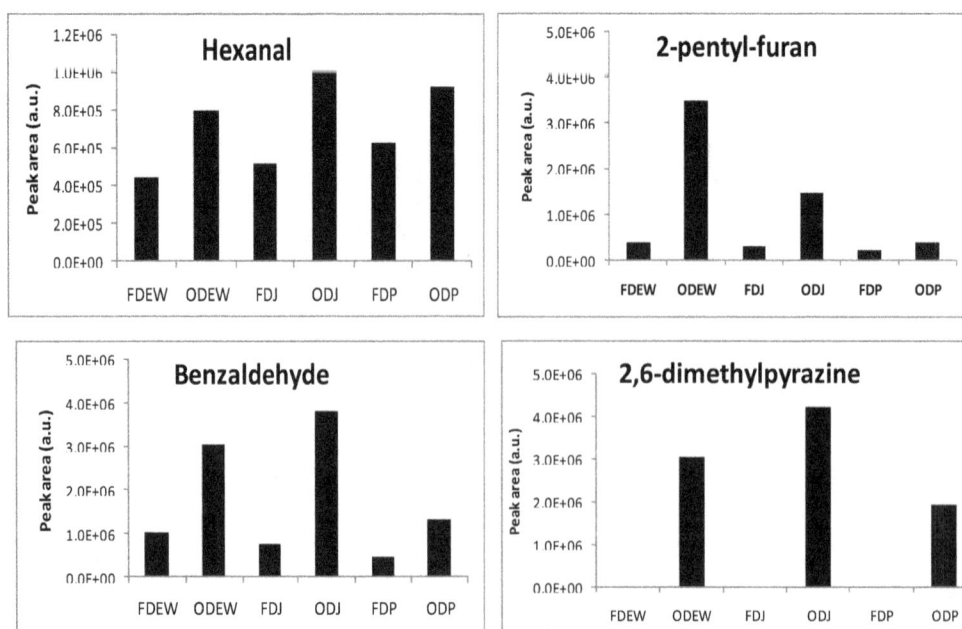

Figure 3. Absolute peak areas in the different powders obtained by HS-SPME-GC-MS. For 2-pentyl-furan and 2,6 dimethylpyrazine, areas were calculated from the TIC (Total Ion Chromatogram). For hexanal, areas were calculated from single ion $m/z = 82$ and for benzaldehyde, areas were calculated from single ions $m/z = 105 + 106$.

Oven dried juice contained 12.2 more Maillard compounds than freeze dried one and oven dried pulp contained 5.2 more Maillard compounds than the freeze dried one. This is clearly linked with the temperature of the drying process, which is higher in the oven than in the freeze dryer. It is noticeable that 2,6-dimethylpyrazine was found in all oven dried powders but not in freeze dried powders.

Hexanal was chosen as an indicator for lipid oxidation. It occurred in all the powders and its relative quantities found in the different powders were reported in **Figure 3**. Here again, oven dried powders contained higher quantities of hexanal than the corresponding freeze dried powders, which is consistent with the fact that lipid oxidation was accelerated by temperature increase. The larger amount of hexanal in oven dried powders in comparison with the freeze dried powders can be related to protein denaturation. Indeed, the oven dried proteins were more denatured than the freeze dried ones. This increases the availability of heme groups which are more efficient catalysts for lipid peroxidation than metal ions are. Belitz *et al.* [7] have already suggested that the activity of a heme protein towards hydroperoxides is influenced by its steric accessibility to fatty acid hydroperoxides.

Water content was higher in FDJ than in ODJ and it was higher in FDP than in ODP (**Table 2**). The Maillard compounds were higher in ODJ than in FDJ and higher in ODP than in FDP. The water content was therefore inversely proportional to the amount of Maillard compounds in these powders. Lu *et al.* [26] have already shown that volatile compounds generated from Maillard reactions increases as water content decreases.

3.4. Impact of the Fractionation Step on the Volatile Composition of the Powders

As detailed in **Figure 1**, different earthworm powders were prepared. ODEW was obtained using traditional drying in an oven. FDEW was our reference for less denatured protein and was done using freeze drying. The soluble fraction (juice) of earthworms was separated from the solid fraction (pulp). Protein content was higher in the pulp fraction than in the juice fraction (**Table 2**). Conversely, sugars are supposed to be mainly contained in the soluble fraction. As a consequence, Maillard reactions should be reduced in the pulp fraction.

In order to study the impact of the fractionation on the volatile composition of the powders, we compared the powders that differed only by the fractionation step, *i.e.* the two couples FDJ/FDP and ODJ/ODP. Even though the separation of the two fractions (juice and pulp) was not perfect, and some reducing sugar might remain in the pulp fraction, the quantity of Maillard reactions compounds (**Table 1**) (Benzaldehyde, 2,6-dimethylpyrazine and 2-pentyl-furan) was found to be 1.6 times lower in the pulp than in the juice for freeze dried powders and 3.6 times lower in the pulp than in the juice for the oven dried powders. This confirmed that reducing sugar was mainly in the juice. Additionally, as the proteins that were oven dried were denatured, NH_2 groups of amino acids might have been more available for reactions with reducing sugar, leading to more Maillard volatile compounds.

Juice and pulp fractions both contained larger quantities of hexanal than did the whole fractions. Fraction separation might favor oxygen contact with the intracellular medium and membranes and thus might have increased lipid oxidation. Oxygen contact could also have an impact on Maillard products. In fact, Fong Lam [27] showed that molecular oxygen can influence bond cleavage, which may lead to greater amounts of benzaldehyde. This can explain the high amount of benzaldehyde found in ODJ powder.

A hierarchical clustering analysis and k-means clusterings were done on the standardized areas of the "heat process implicated compounds" (HPIC) extracted from the different powders. The resulting dendrogram (**Figure 4**) shows three subgroups, which were verified by k-means clusterings: a first subgroup containing FDEW, FDJ and FDP, a second subgroup containing ODEW and ODJ and a final subgroup containing ODP alone. This dendrogram indicates that the three freeze-dried powders FDEW, FDJ and FDP are close together with a distance of aggregation below 13% and far from the oven dried powders. This result shows, as cited above, the importance of reducing the temperature in the preparation process of the powders in order to obtain earthworm powders containing low amounts of HPIC. On the dendrogram, the FDJ powder is closer to FDEW than FDP and ODJ is closer to ODEW than ODP. It clearly appears that the pulp fractions are different from the other powders. In agreement with this observation, the pulp fractions contain, for each thermal treatment, less volatile

Table 2. Water and protein content of the earthworm powders.

Sample	ODEW	FDEW	FDJ	ODJ	FDP	ODP
Water content (%)	5.24 ± 0.60	6.05 ± 0.15	5.38 ± 0.14	1.14 ± 0.51	5.04 ± 0.23	3.99 ± 0.24
Protein content (% w/w wb)	63.78 ± 3.90	66.16 ± 1.50	53.63 ± 1.01	53.73 ± 1.14	71.30 ± 0.42	61.68 ± 1.01
Protein content (% w/w db)	67.31 ± 3.88	70.42 ± 1.50	56.68 ± 1.01	54.35 ± 1.13	75.08 ± 0.42	64.24 ± 1.01

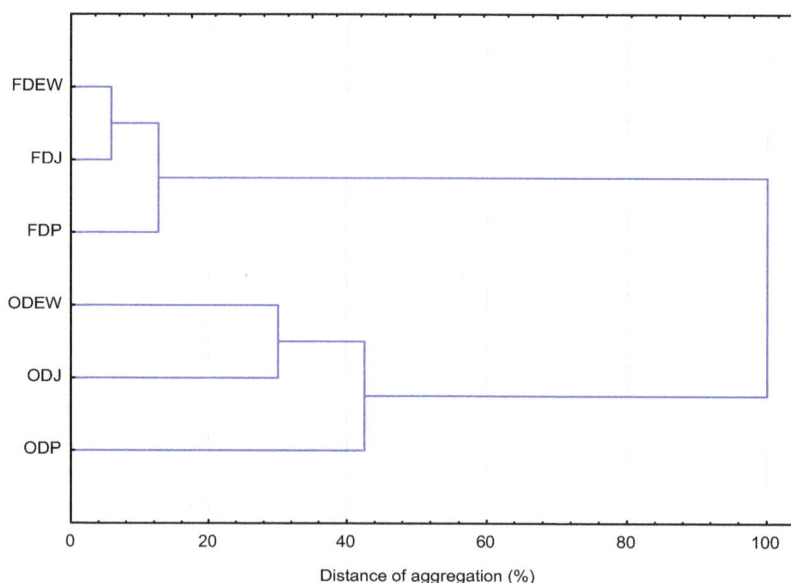

Figure 4. Dendrogram obtained from the hierarchical cluster analysis of the "heat process implicated compounds" HPIC analysed by HS-SPME-GC-MS in earthworm powder samples.

compounds than the other powders (**Table 1**). The difference between the pulp fractions and the other powders is mainly due to the volatile compounds from Maillard reactions than those from lipid oxidation (**Figure 3**).

In this work, differential scanning calorimetry showed that a drying process using temperature above 42°C led to protein denaturation. This thermo-denaturation led to a drastic decrease in solubility. This loss of solubility could be due to protein denaturation involving changes in hydrogen bonding and hydrophobic interactions. Finally, the FDEW powder was 3 times more soluble than the ODEW powder. Moreover, the odor properties were also modified by the drying process, and these were assessed through the analysis of volatile compounds. The amount of Maillard reactions compounds and lipid oxidation compounds was higher in powders that were obtained using oven drying rather than freeze drying. Fractionation was a determinant step in the preparation process of the earthworm powders. For the same heat treatment, the pulp fraction contained less volatile compounds resulting from Maillard reactions than the juice fraction did. It was possible to produce a more neutral protein by using fractionation and freeze drying: freeze dried pulp (FDP).

4. Acknowledgements

We are grateful to the Conseil Régional de Bourgogne for funding this research program.

REFERENCES

[1]   R. Vielma, D. J. F. Ovalles, L. A. Leon and A. L. Medina, "Valor Nutritivo de la Harina de Lombriz (*Eisenia foetida*) Como Fuente de Aminoacidos y su Estimacion Cuantitativa Mediante Cromatografia en Fase Reversa (HPLC) y Derivatizacion Precolumna con o-Ftalaldehido (OPA)," *ARS Pharmaceutica*, Vol. 44, 2003, pp. 43-58.

[2]   A. L. Medina, J. A. Cova, R. A. Vielma, P. Pujic, M. P. Carlos and J. V. Torres, "Immunological and Chemical Analysis of Proteins from *Eisenia foetida* Earthworm," *Food and Agricultural Immunology*, Vol. 15, No. 3-4, 2003, pp. 255-263.

[3]   H. Bastardo, A. L. Medina and S. Sofia, "Sustitución Total de Harina de Pescado por Harina de Lombriz en Dietas para Iniciador de Trucha Arcoíris," *Agroalimentación y Desarrollo Sustentable*, Vol. 17, 2005, pp. 1-6.

[4]   L. Taboga, "The Nutritional Value of Earthworms for Chickens," *British Poultry Science*, Vol. 21, No. 5, 1980, pp. 405-410.

[5]   N. Cayot, P. Cayot, E. Bou-Maroun, H. Laboure, B. Abad-Romero, K. Pernin, N. Seller-Alvarez, A. V. Hernández, E. Marquez and A. L. Medina, "Physico-Chemical Characterisation of a Non-Conventional Food Protein Source from Earthworms and Sensory Impact in Arepas," *International Journal of Food Science and Technology*, Vol. 44, No. 11, 2009, pp. 2303-2313.

[6]   D. S. Mottram, "The Effect of Cooking Conditions on the Formation of Volatile Heterocyclic Compounds in Pork," *Journal of the Science of Food and Agriculture*, Vol. 36, No. 5, 1985, pp. 377-382.

[7]   H. D. Belitz, W. Grosch and P. Schieberle, "Food Chemistry," Springer, Berlin, 2004.

[8]   C. Strumillo, W. Kaminski and I. Zbicinski, "Some Aspects of the Drying of Protein Products," *The Chemical Engineering Journal*, Vol. 58, No. 2, 1995, pp. 197-204.

[9]   T. Arakawa, S. J. Prestrelski, W. C. Kenney and J. F. Carpenter, "Factors Affecting Short-Term and Long-Term Stabilities of Proteins," *Advanced Drug Delivery Reviews*, Vol. 46, No. 1-3, 2001, pp. 307-326.

[10] Association of Official Analytical Chemists, "Crude Protein in Meat: Block Digestion Method. (981.10)," Official Methods of Analysis, 1990.

[11] R. Floris, I. Bodnar, F. Weinbreck and A. C. Alting, "Dynamic Rearrangement of Disulfide Bridges Influences Solubility of Whey Protein Coatings," *International Dairy Journal*, Vol. 18, No. 5, 2008, pp. 566-573.

[12] J. A. Gordon and W. P. Jencks, "The Relationship of Structure to the Effectiveness of Denaturing Agents for Proteins," *Biochemistry*, Vol. 2, No. 1, 1963, pp. 47-57.

[13] A. Wallqvist, D. G. Covell and D. Thirumalai, "Hydrophobic Interactions in Aqueous Urea Solutions with Implications for the Mechanism of Protein Denaturation," *Journal of American Chemist Society*, Vol. 120, No. 2, 1998, pp. 427-428.

[14] A. V. Moroni, S. Ametti, F. Bonomi, E. K. Arendt and F. Dal Bello, "Efficient Buffers for the Extraction of Total Protein in Gluten Free Cereals," *Food Chemistry*, Vol. 121, No. 4, 2010, pp. 1225-1230.

[15] U. T. Rüegg, J. Rudinger, C. H. W. Hirs and S. N. Timasheff, "Reductive Cleavage of Cystine Disulfides with Tributylphosphine," In: C. H. W. Hirs and S. N. Timasheff, Eds., *Methods in Enzymology*, Vol. 47, Academic Press, New York, 1977, pp. 111-116.

[16] P. Farkas, J. L. Le Quéré, H. Maarse and M. Kovác, "The Standard GC Retention Index Library of Flavour Compounds," In: H. Maarse and D. G. Van Der Heij, Eds., *Trends in Flavour Research*, Elsevier Science B.V., Amsterdam, 1994, pp. 145-149.

[17] F. Bianchi, M. Careri, A. Mangia and M. Musci, "Retention Indices in the Analysis of Food Aroma Volatile Compounds in Temperature-Programmed Gas Chromatography: Database Creation and Evaluation of Precision and Robustness," *Journal of Separation Science*, Vol. 30, No. 4, 2007, pp. 563-572.

[18] A. M. El-Sayed, "The Pherobase: Database of Pheromones and Semiochemicals," 2008. http://www.pherobase.com

[19] J. adamiec, J. Rössner, J. Velíšek, K. Cejpek and J. Šavel, "Minor Strecker Degradation Products of Phenylalanine and Phenylglycine," *European Food Research and Technology*, Vol. 212, No. 2, 2001, pp. 135-140.

[20] N. A. Mancilla-Margalli and M. G. López, "Generation of Maillard Compounds from Inulin during the Thermal Processing of *Agave tequilana* Weber Var. azul," *Journal of Agricultural and Food Chemistry*, Vol. 50, No. 4, 2002, pp. 806-812.

[21] C.-K. Shu, "Pyrazine Formation from Serine and Threonine," *Journal of Agricultural and Food Chemistry*, Vol. 47, No. 10, 1999, pp. 4332-4335.

[22] O. Mandin, S. C. Duckham and J. M. Ames, "Volatile Compounds from Potato-Like Model Systems," *Journal of Agricultural and Food Chemistry*, Vol. 47, No. 6, 1999, pp. 2355-2359.

[23] F. J. Hidalgo and R. Zamora, "Strecker-Type Degradation Produced by the Lipid Oxidation Products 4,5-Epoxy-2-Alkenals," *Journal of Agricultural and Food Chemistry*, Vol. 52, No. 23, 2004, pp. 7126-7131.

[24] F. Shahidi, "Assessment of Lipid Oxidation and Off-Flavour Development in Meat and Meat Products," In: F. Shahidi, Ed., *Flavour of Meat and Meat Products*, Chapman and Hall, Glasgow, 1994, pp. 247-266.

[25] E. Bou-Maroun and N. Cayot, "Odour-Active Compounds of an *Eisenia foetida* Protein Powder. Identification and Effect of Delipidation on the Odour Profile," *Food Chemistry*, Vol. 124, No. 3, 2011, pp. 889-894.

[26] C.-Y. Lu, Z. Hao, R. Payne and C.-T. Ho, "Effects of Water Content on Volatile Generation and Peptide Degradation in the Maillard Reaction of Glycine, Diglycine, and Triglycine," *Journal of Agricultural and Food Chemistry*, Vol. 53, No. 16, 2005, pp. 6443-6447.

[27] C. Fong Lam, "Elucidation of Selected Maillard Reaction Pathways in Alanine and Phenylalanine Model Systems through Isotope Labelling and Pyrolysis-GC/MS Based Techniques," Ph.D. Dissertation, McGill University, Montreal, 2009.

19

Determination of Rennet Clotting Time by Texture Analysis Method

Zerrin Yuksel

Department of Food Technology, Bayramiç Vocational Collage, Çanakkale Onsekiz Mart University, Çanakkale, Turkey.

ABSTRACT

In this study, texture analysis method was used for the determination of rennet flocculation time (t_{floc}) and rennet clotting time (t_{clot}) of rennet-induced reconstitued milk samples with different $CaCl_2$ concentrations. The rennet flocculation time (RFT) and rennet clotting time (RCT) were also determined by using the Berridge test and sensory evaluation. The hardness value versus renneting time curves derived from texture analysis gave a good modified exponential relationship for each $CaCl_2$ concentration and the curves were used to calculate flocculation time and clotting time parameters. It was found that the parameters (t_{floc} and t_{clot}) appeared strongly correlated with RFT and RCT, respectively. Texture analysis was proved as a suitable method to control the rennet-induced coagulation and determine the rennet clotting time. It was also determined that enrichment of milk with $CaCl_2$ leaded to a decrease in flocculation and clotting times and an increase in rate of clotting and gel hardness.

Keywords: Rennet Flocculation Time; Rennet Clotting Time; Texture Analysis; Hardness; Rennet Gel; $CaCl_2$

1. Introduction

The first stage of cheese manufacture is the conversion of liquid milk to cheese curd. After the addition of chy mosin to the milk, there is little apparent reaction for some time and then the milk coagulates rapidly. During this *lag* phase, the enzyme hydrolyses the κ-casein which stabilizes the casein micelles. This phenomenon, which is the first step of cheesemaking, results from two stages. The enzymatic proteolysis forms the first or primary phase. In this phase, milk-clotting enzymes spli κ-casein at the junction between the para-κ-casein and macropeptide moieties, *i.e.* in bovine κ-casein, at the Phe_{105} - Met_{106} bond. When sufficient amount of κ-casein has been hydrolyzed, the destabilized micelles begin to aggregate and this eventually leads to a three-dimensional cheese curd [1,2]. In casein micelles, proteolysis of a small number of κ-casein molecules will have much less effect on the aggregation properties, since the micelles contain many hundreds or even thousands of such molecules. The para-κ-casein produced in the micelles by renneting can only aggregate when the whole micelle is capable of aggregating, and it is this that causes the *lag* phase before aggregation is observed. Renneted micelles

appear to be incapable of aggregating until about 60% - 80% of their κ-casein has been destroyed, after which the concentration of micelles capable of aggregating increases rapidly. This behavior can be explained either by the loss of surface charge during renneting or by loss of steric stabilization [1].

The aggregation rate of renneted micelles is unaffected by the concentration of rennet or by the size of the micelles. However, it is very sensitive to the concentration of ionic calcium. It is thought that the ionic calcium does not directly affect the enzymatic phase, although addition of $CaCl_2$ does reduce milk pH, which accelerates the hydrolysis reaction. However, addition of calcium reduces the rennet coagulation time, even at constant milk pH, and flocculation occurs at a lower degree of κ-casein hydrolysis. Addition of calcium also increases the rate of firming of rennet-induced milk gels and firmness of the gel. This effect of ionic calcium could be explained by the masking of charged groups and the hydrophobicity increase [3].

The most easily detected outcome of chymosin proteolysis and rennet clotting is the visible observation of the presence of flocs in a milk sample in a rotating tube. The time taken for their appearance is defined as the rennet coagulation time, and for the cheese producers interested in the activity of an enzyme preparation, this may be the only quantity of interest [1]. Numerous devices have been developed to study and control of phenomena occurred during the rennet-induced coagulation due to the importance of the curd cutting time on the final cheese quality [4]. Various laboratory techniques have also been described for measuring visco-elastic properties of milk gels as a reference for cheese making; the most widely used are Formagraph and low-amplitude dynamic shear measurements [5]. Payne *et al.* [6] used a diffuse reflectance technique to predict optimal cutting time as measured by Formagraph on composite milks prepared from varying proportions of cream, skim milk and condensed skim milk. They found that the inflection point of the sigmoidal phase of the diffuse reflactance curve was well correlated with the Formagraph measure of the rennet clotting time. It was compared that the performance of a NIR transmission probe with the Formagraph used skimmed and whole milk and it has found that there was a good correlation between the time to inflection point of the transmission signal and the rennet coagulation time by the Formagraph.

Sharma *et al.* [7] evaluated on-line measurements of coagulation time and coagulation firmness of renneted milk using a torsional vibration technique, namely a Nametre viscometer. It was defined as the point where complex viscosity became higher than the initial value at rennet addition using a coagulation time parameter.

Turbidity or light scattering methods were used for following early aggregation phase while the gel formation and development are most monitored by rheometer. Each technique suffers from limitations. Light scattering requires a dilute dispersion of particles so that only singly scaretted photons are collected at the detector. Studies using light scattering are thus limited to initial stage of aggregation, where growth of molecular weight or degree of polimerization is obtained as function of reaction time. Rheological measurements suffer from the opposite failing. There, the limitation is instrument-sensitivity and a detectable signal is realized only after the reaction has progressed to a significant extent [1].

Some of the techniques mentioned above have also been used to determine the influence of various factors such as temperature, pH, milk composition and $CaCl_2$ concentration on the rennet coagulation process.

In most cases, studies were carried out with using techniques mentioned above. There is no research on determination of both of the rennet flocculation and clotting times with using the texture analysis. The objective of this study is to predict both of the flocculation and clotting times with using results of hardness measurements at different $CaCl_2$ concentrations during the rennetting.

2. Materials and Methods

2.1. Reconstitued Milk Sample

Reconstitued milk sample was prepared by reconstituting low-heat skimmed milk powder (purchased from ENKA Dairy Co., Konya, Turkey) at ratio of 12 g/100ml deionized [8]. $CaCl_2$ (Merck, Germany) was added to obtained concentrations of 0.02% (Sample A), 0.04% (Sample B), 0.06% (Sample C), 0.08% (Sample D), 0.10% (Sample E), 0.12% (w/v) (Sample F). Total protein content of the reconstitued milk was 3.2%, w/w.

2.2. Rennet Coagulation Experiment

Maxiren 600® (DSM Foods, Delft, The Netherlands) with a declared activity of 160 IMCUmL^{-1} was used for renneting of the reconstitued milk. A solution of rennet was prepared daily by diluting 1 mL of Maxiren 600® in 100 mL of deionized water. Rennet solution was added to the reconstitued milk at 32°C to a final concentration of 0.01% (v/v). Samples were divided into several assays (200 ml each) for preparing renneting time-series (0 - 70 min).

2.3. Measurement of Gel Hardness

The hardness (N) of viscoelastic rennet gels was determined by means of texture analysis. A texturometer model TA-PLUS (Lloyd Instruments, Fareham, UK) with a 10 N cross-head was used to measure the hardness of rennet gels. The measuring probe consisted of an acrylic cone (diameter 4 cm and 60° angle) was thrust into gels in cylindrical containers (200 ml in volume) by the 80% compression depth from surface. A single compression method (one cycle) was applied at a constant crosshead velocity of 2 mm·s^{-1}. The hardness value of the gels was calculated on line by using the software, Nexygen® 2.0 software (Lloyd Instruments, Fareham, UK). Exponential relationships between rennetting time (t, min) and hardness value (*H*, N) have been determined in the rennet gels at different $CaCl_2$ concentrations. It was found that the curves fitted "modified exponential model" (correlation coefficient, r ≥ 0.98; standard error, SE < 0.006 were found). According to this model, equation derived from the curves was given as such;

$$H = H_o \cdot \exp(a/t) \qquad (1)$$

where *H* is hardness value (N), t is the time after the addition of rennet (min), H_o is the initial hardness value (N), *a* is the regression coefficient.

Four different parameters were extracted from the hardness changes according to renneting time; rennet clotting time (t_{clot}) corresponding to the time when the hardness value was 0.12 N (maximum hardness value for Sample A), flocculation time (t_{floc}) corresponding to the time when the hardness value became greater than the initial value, rate of clotting (r_{clot}) corresponding to the rate of hardness increase, and final hardness (H_{70}) corresponding to the hardness value at 70 min after the addition of rennet. Curve Expert 1.3 (a comprehensive curve fitting system for Windows) was used to evaluate Hardness-renneting time curves.

2.4. Determination of Rennet Flocculation and Clotting Times

The rennet floculation time (RFT), the time from addition of the rennet to the first formation of visible floccules, was determined according to the method by Berridge [8]. The rennet clotting time (RCT) was determined with sensory evaluation by using a spatula by means of traditionally detection of curd-cutting time.

2.5. Total Protein Content

The total protein content of reconstituted milk was determined by Bradford method [9].

2.6. Calcium Ion Activity

Calcium ion activity of the reconstitued milk samples was determined with an Ion Analyzer (Orion, Model 407 A, USA) and an ion selective electrode (Sartek, UK).

2.7. pH Measurements

pH values of the reconstitued milk samples were determined with using a pH meter (Sartorius PB-20, Germany).

2.8. Statistical Analysis

Each experiment was carried out in triplicate. The data were analyzed statistically by using SPSS for Windows® software (SPSS Inc., Chicago, IL, USA).

3. Results

The hardness values of the samples measured during 70 min after the addition of rennet were shown from **Figure 1**. It was observed that the hardness was increased to a maximum and it was then kept almost constant for all $CaCl_2$ concentration during the renneting. There was also a *lag* phase observed at the samples added $CaCl_2$ at below of 0.10%. The differences between slopes of the curves and length of the *lag* phase obtained by different $CaCl_2$ concentrations, allowed to the determination of the rennet flocculation and clotting times and the explanation of the effect of $CaCl_2$ on the renneting with using texture analysis method.

Plotting the hardness value as a function of the renneting time gave a good modified exponential relationship (correlation coefficient ≥ 0.98; standard error < 0.006) for each $CaCl_2$ concentrations. Rennet clotting

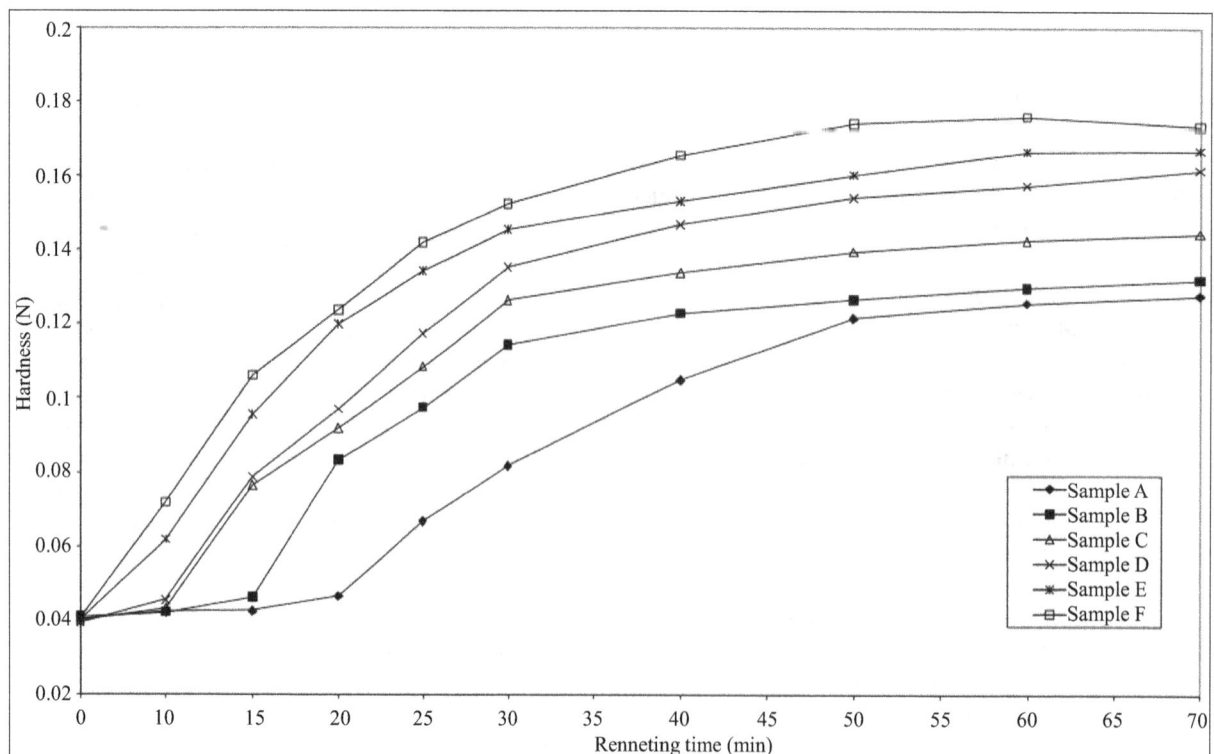

Figure 1. Hardness values of the samples during the renneting.

time (t_{clot}), flocculation time (t_{floc}) and Hardness value at 70 min after the addition of rennet (H_{70}) were determined by using the equation $H = H_o \cdot \exp(a/t)$ derived from the curves, where H was hardness value (N), t was the time after the addition of rennet (min) (**Table 1**). For all $CaCl_2$ concentrations, t_{floc} values were calculated for the hardness value (0.05 N) that was greater than the initial value as an observed *lag* phase on the diagrams. And t_{clot} values of the all samples were calculated for 0.12 N, which was the maximum hardness value of the lowest, $CaCl_2$-induced sample A. The renneting time for hardness value of 0.12 N was assumed as curd-clotting time of sample A. On the other hand, the sample A showed the longest rennet clotting time. Because the sample A reached to the curd-clotting maturity at this hardness, the time needed for to reach to this hardness value (0.12 N) of the other samples were used for calculation and comparison of rennet flocculation and clotting times of the samples. According to these assumptions, it was used the follow-ing equations;

$$t_{floc} = -a/(2.99 + \ln H_o) \quad (2)$$

$$t_{clot} = -a/(2.12 + \ln H_o) \quad (3)$$

Rate of clotting values (r_{clot}, N/min) were calculated from the slope of the plot of hardness versus renneting time. r_{clot} values corresponding to the rate of hardness increase as a function of renneting time were presented in **Table 1**.

It was also shown from **Table 2** the rennet flocculation time (RFT) and rennet clotting time (RCT) determined by Berridge method and sensory evaluation, respectively. It was found that the time parameters (t_{floc} and t_{clot}) derived from the hardness changes according to the renneting time which were in correlation with the RFT and RCT (determination coefficients, $R^2 = 0.993$ for t_{floc}-RFT line and $R^2 = 0.999$ for t_{clot}-RCT line). The relationship was validated on the samples at different $CaCl_2$ concentrations (0.02% - 0.12%) (**Figure 2**).

Table 1. Rennet coagulation parameters derived from hardness-renneting time curves (t_{floc}, flocculation time; t_{clot}, clotting time; H_{70}, hardness value at 70 min after the addition of rennet; r_{clot}, rate of clotting).

	Sample A	Sample B	Sample C	Sample D	Sample E	Sample F
t_{floc} (min)	18.80 ± 1.0	12.23 ± 0.7	10.19 ± 1.2	9.34 ± 0.6	7.30 ± 0.8	6.75 ± 0.5
t_{clot} (min)	54.21 ± 1.5	42.31 ± 1.6	31.88 ± 2.2	26.00 ± 1.8	20.46 ± 2.0	17.80 ± 1.5
H_{70} (N)	0.133 ± 0.015	0.138 ± 0.02	0.150 ± 0.01	0.163 ± 0.01	0.169 ± 0.012	0.179 ± 0.019
r_{clot} (N/min)	2.44×10^{-3}	2.91×10^{-3}	3.96×10^{-3}	4.37×10^{-3}	4.92×10^{-3}	5.08×10^{-3}

Table 2. Rennet flocculation and clotting times (RFT and RCT) of the samples determined by the Berridge method at 32°C.

	Sample A	Sample B	Sample C	Sample D	Sample E	Sample F
pH	6.76 ± 0.05	6.68 ± 0.05	6.62 ± 0.05	6.58 ± 0.05	6.52 ± 0.05	6.47 ± 0.05
$[Ca^{2+}]$ (ppm)	61 ± 1.2	72 ± 1.0	90 ± 1.5	110 ± 1.5	135 ± 2.0	162 ± 2.2
RFT (min)	19.17 ± 1.2	13.30 ± 1.0	11.10 ± 0.8	9.20 ± 0.5	7.75 ± 0.5	6.50 ± 0.7
RCT (min)	55.50 ± 1.8	42.75 ± 2.0	33.75 ± 2.4	27.17 ± 1.5	21.92 ± 1.8	18.60 ± 2.0

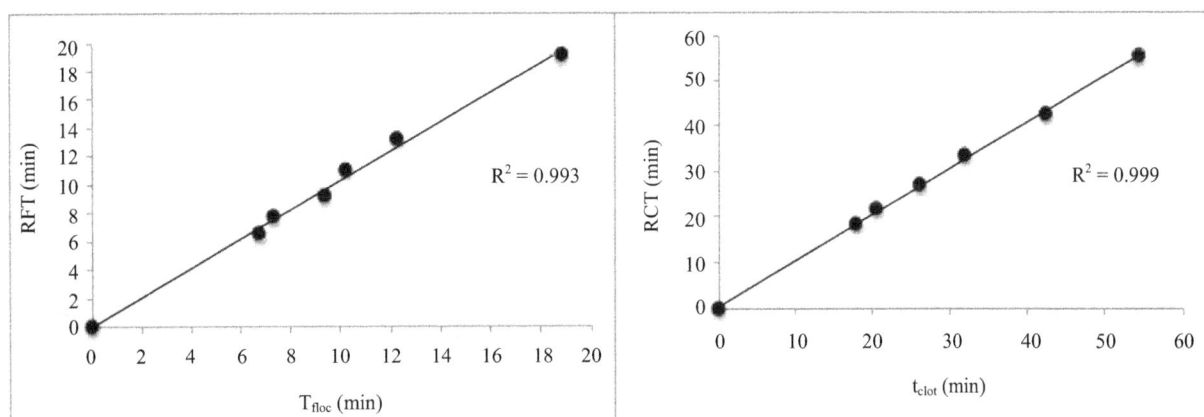

Figure 2. Rennet flocculation time (RFT) determined by Berridge method versus rennet flocculation time (t_{floc}) derived from texture analysis; rennet clotting time (RCT) determined with sensory evaluation versus rennet clotting time (t_{clot}) derived from texture analysis at different $CaCl_2$ concentrations. R^2, determination coefficient.

Calcium ion activity and pH values of the samples at different CaCl₂ concentrations are presented in **Table 2**. pH of the samples decreased with increasing CaCl₂ concentration of the samples because of the "Lewis acid role of the calcium". The addition of CaCl₂ not only increased the calcium concentration but also reduced the pH of milk, resulting in an increased aggregation rate [10]. It was also reported that the percentage of κ-casein proteolyzed was dependent on milk pH and ionic calcium content [1].

As it is shown from in **Figure 1**, decreasing of CaCl₂ concentration leads to prolongation of the lag phase of the renneting that represents the preliminary of flocculation stage of para-κ-casein micelles. Both of t_{floc}-t_{clot} and RFT-RCT decreased with increasing CaCl₂ concentration (**Table 1** and **2**). CaCl₂ has also leaded to an increase in the gel hardness during the coagulation process. The hardness value at 70 min after the addition of rennet (H_{70}) increased with increasing CaCl₂ concentration. Rate of clotting (r_{clot}) also increased with increasing CaCl₂ concentration (**Table 1**).

4. Discussion

In this study, the milk samples were characterized by different rennet coagulation properties due to CaCl₂ added at different concentration to the reconstituted milk. Different CaCl₂ concentration of rennet gels allowed evaluation of the method on the samples characterized by different rennet flocculation and clotting times. Two different methods were used and compared to determine rennet flocculation and clotting times of the rennet gels. It was showed that the textural analysis allowed determining both of the rennet flocculation and clotting times. There are many techniques in the literature provided to determine or predict individually early aggregation phase and gel formation-development. Initial aggregation phase of rennetting was generally followed by some methods such as using turbidity or light scaretting whereas the gel formation and development was monitored by another method such as rheometer[1]. However, in this study it was found that the hardness measurement (texture analysis) allowed monitoring the both of early phase and curding cutting step during the rennetting via evaluation of the time parameters (t_{floc} and t_{clot}).

Castillo *et al.* [11] studied on prediction of both gelation time and cutting time in Cottage cheese-type gels and monitored the coagulation process simultaneously using a light backscatter sensor, a rheometer and a pH-meter. It was developed the models for the prediction of the rheologically defined gelation time using light backscatter and for the prediction of cutting time using light backscatter. It was determined gelation time by using an equation and cutting time by using another equation.

However, in this study one method based on texture analysis during the renneting was provided determination of rennet flocculation and clotting times as a function of hardness. This is confirmed that the texture analysis method was sensitive to all stages of renneting. It was showed that determination of rennet flocculation and clotting times with this method is possible for all given CaCl₂ concentrations. It was exposed that flocculation and clotting times can be determined by using the same equation $H = H_o \cdot \exp(a/t)$, where H have taken values of 0.05 and 0.12 N for calculation of t_{floc} and t_{clot}, respectively while constants of H_o and a have taken different values for different CaCl₂ concentrations.

Besides the determination of the time parameters, this method contributed to the observation of the hardness changes during the renneting and evaluation of influence of CaCl₂ on rennet flocculation and coagulation phases. κ-casein molecules hydrolyse to para-κ-casein and macropeptide during the lag phase of renneting and aggregation phase starts after the sufficient amount of κ-casein hydrolyses [1]. The length of lag phase decreases with increasing calcium ion activity. Above of the 0.08% addition of CaCl₂, there was no lag phase observed. This result indicates that the CaCl₂ concentration affects the primary phase of renneting. It possibly means that flocculation may occur at a lower degree of κ-casein hydrolysis at higher ionic calcium concentrations.

It is known that the rate of formation and final strength of rennet gels are influenced greatly by calcium [12]. While not completely understood, it has been shown that ionic calcium influence coagulation and gel formation [13,14]. In this study, it was observed that the maximum hardness value of the gels increased and the time needed to reach to the maximum hardness decreased with increasing CaCl₂ concentration (**Figure 1** and **Table 2**). Furthermore, an increase of rate of clotting with increasing CaCl₂ concentration could be explained mainly by charge neutralization of the negatively charged groups on the micelle surface and possibly by the formation of calcium bridges (**Table 1**). It is reported that enrichment of milk with calcium leads to an increase in casein aggregation rate during the coagulation process, consequently decreased the flocculation time, and an increase in firmness of rennet gel [4]. It was shown from **Table 1** and **Table 2**, rennet flocculation and clotting times decreased with increasing CaCl₂ concentration. Sample F (0.12% CaCl₂) has reached maximum H_{70} values while sample A (0.02% CaCl₂) has minimum H_{70} value during the renneting (**Table 1**). In this study, it was shown that the CaCl₂ affects on the both of the primary and secondary phases of renneting with using the texture analysis method.

5. Conclusion

In this study, it was indicated that the texture analysis

could be used accurately for determination of rennet clotting time during the renneting of milk. The obtained hardness values were plotted against the renneting time and evaluated for calculation of rennet flocculation and clotting times. It was found to be important of texture analysis method (hardness measurement) at the rennet-induced samples with different $CaCl_2$ concentrations in terms of prediction both of the flocculation time and clotting time. Furthermore, it was shown that $CaCl_2$ concentration which is the one of the main parameters in the renneting process was affected on the primary and secondary phases of renneting with using the texture analysis method.

6. Acknowledgements

The author expresses her gratitude to Prof. Dr. Yaşar Kemal Erdem for his permition for using the texturometer and to Miss Elif Avcı for her helpings.

REFERENCES

[1] F. Fox, P. L. H. McSweeney, T. M. Cohan and T. P. Guinee, "Cheese Chemistry, Physics and Microbiology," Vol. 1, Elsevier Ltd., Oxford, 2004.

[2] P. Walstra, J. T. M. Wouters and T. J. Geurts, "Dairy Science and Technology," CRC Press, Taylor and Francis Group, LLC, 2006.

[3] A. Eck and J. C. Gillis, "Cheesemaking—From Science to Quality Assurance," Intercept Ltd., Andover, Hampshire, 2000.

[4] A. H. Klandar, A. Lagaude and D. Chevalier-Lucia, "Assessment of the Rennet Coagulation of Skim Milk: A Comparison Methods," International Dairy Journal, Vol. 17, No. 10, 2007, pp. 1151-1160.

[5] D. J. O'Callaghan, C. P. O'Donnell and F. A. Payne, "A Comparison of On-Line Techniques for Determination of Curd Setting Time Using Cheesemilk under Different Rates of Coagulation," Journal of Food Engineering, Vol. 41, No. 1, 1999, pp. 43-54.

[6] F. A. Payne, C. L. Hicks and P. S. Shen, "Predicting Optimal Cutting Time of Coagulating Milk Using Diffuse Reflectance," Journal of Dairy Science, Vol. 76, No. 1, 1993, pp. 48-61.

[7] S. K. Sharma, A. R. Hill and G. S. Mittal, "Evaluation of Methods to Measure Coagulation Time of Ultrafiltered Milk," Milchwissenschaft, Vol. 47, No. 11, 1992, pp. 701-704.

[8] N. J. Berridge, "Some Observation on the Determination of the Activity of Rennet," The Analyst, Vol. 77, No. 911, 1952, pp. 57-62.

[9] M. M. Bradford, "A Rapid and Sensitive Method for Quantitation of Microgram Quantities of Protein Utilizing the Principle of Protein-Dye Binding," Analytical Biochemistry, Vol. 72, No. 1-2, 1976, pp. 248-254.

[10] A. F. Wolfschoon-Pompo, "Influence of Calcium Chloride Addition to Milk on the Cheese Yield," International Dairy Journal, Vol. 7, No. 4, 1997, pp. 249-254.

[11] M. Castillo, F. A. Payne, T. Wang and J. A. Lucey, "Effect of Temperature and Inoclulum Concentration on Predicition of Both Gelation Time and Cutting Time. Cottage Cheese-Type Gels," International Dairy Journal, Vol. 16, No. 2, 2006, pp. 147-152.

[12] J. A. Lucey and P. F. Fox, "Importance of Calcium and Phosphate in Cheese Manufacture," Journal of Dairy Science, Vol. 76, No. 6, 1993, pp. 1714-1724.

[13] S. I. Shalabi and P. F. Fox, "Influence of pH on the Rennet Coagulation of Milk," Journal of Dairy Research, Vol. 49, No. 1, 1982, pp. 153-157.

[14] G. J. O. Martin, R. P. W. Williams, C. Choong, B. Lee and D. E. Dunstan, "Comparison of Rennet Gelation Using Raw and Reconstituted Skim Milk," International Dairy Journal, Vol. 18, No. 10-11, 2008, pp. 1077-1080.

20

Development of Child-Friendly Fish Dishes to Increase Young Children's Acceptance and Consumption of Fish[*]

Lyndsey R. Huss[1], Sean D. McCabe[2], Jennifer Dobbs-Oates[3], John Burgess[1], Carl Behnke[4], Charles R. Santerre[1], Sibylle Kranz[1#]

[1]Department of Nutrition Science, Purdue University, West Lafayette, USA; [2]Department of Statistics, Purdue University, West Lafayette, USA; [3]Department of Human Development and Family Studies, Purdue University, West Lafayette, USA; [4]Department of Hospitality and Tourism Management, Purdue University, West Lafayette, USA.

ABSTRACT

Background: The Dietary Guidelines for Americans 2010 recommend that Americans age two years and older consume seafood, especially fish high in omega-3 polyunsaturated fatty acids, at least twice a week. Although fish is of particular importance during childhood to support proper brain and eye development, it is under-consumed in the US pediatric population. This study examined if substituting salmon for chicken would increase preschooler's fish consumption. **Methods:** Two-to-five years old children (n = 45) were served eight lunches (four pairs of comparable chicken versus salmon dishes) twice, totaling sixteen lunches over a period of three months to test the hypothesis that children will consume fish at least once a week, thus increasing docosahexaenoic acid (DHA) intake. The plate waste method was used to collect intake data and consumption of total energy and DHA intake in the chicken and the fish dishes were compared using contrasts within a mixed effect ANOVA (significance at P < 0.05). **Results:** Dietary intake estimates showed that there were no significant differences in energy intake when the chicken and fish dishes looked similar (macaroni-and-cheese and wraps), but when the fish dishes looked new (nuggets and dumplings), energy intake on fish days was lower than on the chicken day. DHA intake increased significantly on all days the fish was served. **Conclusions:** This pilot study indicates that fish intake can meet recommendations if salmon is incorporated into familiar dishes such as salad wraps or macaroni-and-cheese, in the childcare setting. Although fish is more expensive, childcare centers may serve this highly nutritious protein once a week without experiencing undue amounts of food wastes if incorporated into well-accepted main dishes. Further studies in larger and more diverse samples of children, different experimental dishes, and longer exposure periods may elucidate additional venues to increase children's diet quality by increasing consumption of fatty fish.

Keywords: Fish Consumption; Salmon; Diet Quality; Young Children; Acceptance; Omega-3 PUFAs; DHA

1. Introduction

The polyunsaturated fatty acid (PUFA) known to be critical for brain and eye development, docosahexaenoic acid (DHA, 22:6n-3), supports nervous tissue growth and function, such as learning and memory [1,2]. The parent 18-carbon fatty acid, α-linolenic acid (ALA), can be converted to various long-chain omega-3 PUFAs, including eicosapentaenoic acid (EPA) and DHA. ALA is found in plant-based foods, with higher levels in soybean, canola, and flax seed oils. Another PUFA, the 18-carbon omega-6 fatty acid linoleic acid (LA) is the precursor of omega-6 fatty acids and is abundant in modern food supplies, as it contributes more than 50% of all the fatty acids in soybean, corn, safflower, and sunflower oils.

The conversion of ALA to DHA is very inefficient. Using compartment models, the conversion rate has been estimated to be less than 1% [3,4]. Also, measurement of peak or area-under-the-curve plasma contents of previously labeled fatty acids showed that less than 8% of ALA is converted into EPA and less than 4% to DHA in men [5-8]; in women, less than 21% of ALA is converted to EPA and less than 9% to DHA [9]. Thus, consumption of dietary sources of DHA during the years of brain de-

[*]This work was funded by the US Department of Agriculture, National Institute of Food and Agriculture IND0-2010-01295.
[#]Corresponding author.

velopment (up to age 25) [10] is critical for children's cognitive functioning.

Once obtained from the diet, ALA can be metabolized to EPA, while LA is metabolized using the same enzymes to the 20-carbon chain omega-6 fatty acid, arachidonic acid [11]. The pathway generally accepted for metabolism of EPA to DHA involves elongation of the fatty acid carbon chain to yield DHA [12]. In animals, an intake of 1% energy from LA, with an LA to ALA ratio of 2:1 or lower, supports high levels of DHA in the developing brain [13,14]. However, high intakes of LA inhibit desaturation of ALA to EPA and DHA and reduce accretion of DHA in the brain, retina, and other organs [13-17]. This means that dietary omega-3 PUFAs, the amount and type of omega-3 PUFA (ALA and DHA), and high LA determine DHA accretion in the developing brain and retina. When omega-3 PUFA intake is inadequate, DHA decreases and omega-6 fatty acids increase in the brain. Human milk and infant formulas now provide more than 3% energy from LA, which suggests circulating levels and tissue levels of DHA will be low unless DHA is provided [2].

DHA is only found in animal tissue lipids, with fatty fish being the best dietary source. It is not present in plant sources (such as vegetable fats and oils, grains, nuts, and seeds), although those may provide other omega-3 PUFAs [2]. Because humans lack specific desaturases, they are unable to form omega-3 or omega-6 PUFAs de novo and must obtain these fatty acids from their diet [11]. DHA is the major omega-3 PUFA esterified in the glycerophospholipids that form the structural matrix of brain grey matter and retinal membranes [18,19]. Therefore, DHA accumulation in the brain and retina as well as in other organs depends on the amount and types of omega-3 PUFAs consumed in the diet. In addition, dietary intake of omega-6 fatty acids plays a role, as omega-6 PUFA interact and compete with omega-3 PUFAs in the fatty acid metabolic pathway [13,15,20-24].

The *in vivo* conversion of ALA to DHA is not very efficient, and the Dietary Guidelines for Americans 2010 recommend consuming more foods rich in EPA and DHA by eating seafood at least twice a week. Health-conscious parents strive to offer food sources of DHA to incorporate into their children's diets but in most US regions fish consumption, especially intake of fatty fish, are very low and children may not accept fish into their diet easily.

Prenatal DHA availability is determined by maternal dietary DHA intake during pregnancy and has been shown to have a significant effect on quality of movement in seven-year-old children [25]. Prenatal DHA availability is critical for brain development in utero but also later in life. For that reason, DHA supplementation in infant formula is recommended as it has been shown

to lead to visual acuity and IQ maturation similar to that of breast-fed infants [26]. Even past the usual age of weaning, the brain is not fully developed. The frontal lobes of the brain, which are responsible for executive functions, develop in spurts between birth-to-two years of age, from seven-to-nine years of age, and in the mid-teenage years—up to the age of 25 years old [10]. Hence, adequate intake of DHA prior to adulthood is imperative to support healthy brain development and consumption of DHA-rich foods, such as fatty fish, is recommended for all Americans.

According to the Institute of Medicine report [27], it is recommended that 2 - 5 years old children consume two age-appropriate servings (1-to 2-ounce) of seafood per week. However, previous research has shown that fish consumption is less than adequate, especially when compared to pasta, a dish that is common in the Western diet. In a previous study [28], when pasta was served to 23 two-to-five year old children, intake was 81% greater than when fish was served. This is the case even for immigrants from Asia that would normally have a high fish intake [29], but have been submersed into Western culture. When assessing the dietary patterns of Korean adolescents, fish dishes such as kimchi, fish cake soup, and fish cutlets comprised the majority of meals, whereas Korean-American adolescents consumed a more typical American diet of milk, soda, hamburgers, etc. [29].

Despite the benefits of DHA in the diet, Western diets are low in omega-3 PUFAs, especially ALA found in plant oils and DHA found in fish [2]. According to the Institute of Medicine and National Academy of Sciences [30], the median intake of EPA and DHA in adults is 0.05% of dietary energy. Contrary to popular belief, even an individual meeting the estimated intake recommendations for omega-3 PUFA intake may consume less than optimal amounts of DHA, if the majority of the omega-3 PUFAs are from plants. Consequently, this becomes an issue for the developing brain. Western diets low in omega-3 PUFAs and high in omega-6 fatty acids contribute to poor brain development and function [2]. By adding fatty fish to children's diets, DHA intake would increase, and aid in the prevention of poor brain development. Dalton *et al.* [10] showed that when seven-to-nine year old children's diets were supplemented with a fish-flour spread rich in omega-3 PUFAs, verbal learning ability and memory were improved. Thus, effective ways to encourage habitual consumption of a diet high in DHA are warranted.

For children, the general requirements of total fat and fatty acids have not yet been adequately established [31]. Currently, recommended DHA intake levels are based on values calculated on a per kilogram (kg) body weight basis. However, according to Koletzko *et al.* [32], EPA and DHA intake recommendations using the estimation based on body weight may result in suboptimal DHA amounts in 2 - 12 years old children because their DHA needs are higher relative to their body weight. In adulthood, DHA needs are lower compared to the time from birth to age 25.

Once the importance of EPA and DHA was established, numerous alternative sources of omega-3 PUFAs were developed. The only natural sources of DHA are marine food sources, such as fish and seaweed, and their products, such as fish and algal oil. However, food companies have taken the initiative to incorporate DHA and other omega-3 PUFAs into foods such as breads, pastas, milk, eggs, processed meats, salad dressings, margarines, mayonnaise, peanut butters, pizzas, nutrition bars, cereals, yogurts, and juices [33]. Although the fortification with DHA for these processed foods is based on marine sources, such as fish or algal oils, there are many issues that should be considered. The bioavailability of the omega-3 PUFAs in synthetic form is understudied. Furthermore, foods with added DHA usually only contain 12 - 50 mg EPA and DHA combined. The average DHA amount consumed per one serving of these foods is 32 mg, which is equivalent to less than a teaspoon of salmon. With product claims such as "good source of DHA," parents may become misled and confused. While the population seems to have accepted these synthetic sources of DHA and other omega-3 PUFAs, skepticism remains in the field of nutrition.

The Dietary Guidelines for Americans 2010 recommend consuming seafood at least twice a week. Based on the estimated intake of DHA in Americans, especially American children, the development of venues to increase fish consumption by offering child-friendly fish dishes is a critical public health concern. The present pilot study was designed to explore two modes of offering fatty fish to 2 - 5 years old children—by incorporating it in well-accepted dishes and by offering novel dishes. The specific aims of this research were to 1) substitute salmon for chicken without significantly affecting 2 - 5 years old children's energy intake; and 2) increase children's intake of DHA by offering salmon for lunch. We hypothesized that offering canned or cooked salmon to preschoolers during lunch at the childcare center would lead to children's consumption of fish at least once a week.

2. Methods

2.1. Study Participants

Recruitment of participants was based upon attendance at the Ben and Maxine Miller Child Development Laboratory School, a childcare center located at Purdue University (West Lafayette, Indiana). As the unit of analysis was group-level average consumption, no individual

child was identified and all data were recorded and analyzed for the group of participating children. Eligibility was restricted to children between the ages of 2 - 5 years; exclusion criteria included the presence of food restrictions, food allergies, or digestive diseases, such as Crohn's Disease or Cystic Fibrosis. Forty-five children from three different classrooms participated in the study. This study was approved by the Institutional Review Board of Purdue University. As no individual data was collected, consent procedures consisted of approval by the director and teachers of the childcare in addition to verbal assent by each participating child prior to collecting data on each study day. Children's refusal to participate in the study, such as not wanting to rate the foods, was honored.

2.2. Study Design

This study functioned as a pilot study with one within-subject factor (meal). Over a period of three months, children were twice served the four regularly scheduled chicken main dishes and the study foods: salmon nuggets, salmon dumplings, salmon salad wraps, and salmon macaroni and cheese. The chicken main dishes were regularly scheduled once a week, and each main dish was served once a month. The salmon main dishes substituted for regularly scheduled items over the duration of the study, with a salmon main dish served once a week and each salmon main dish served once a month. During each study lunch, children rated the liking of the dish (appearance, taste, texture, smell, and overall liking). To calculate food intake and total energy consumption (in kilocalories), the plate waste method was employed. The plate waste method and scale to rate liking of foods are described below.

2.3. Dietary Assessment Methods

Children's liking of the chicken or fish dish was measured using a three-point Likert-type scale using smiley faces. First, the main dish was presented to the child on his or her plate with the rest of the lunch items and the child was asked to take a bite to taste the dish and then categorize the food as "yummy", "yucky" or "just okay". The children were asked to provide five ratings for each main dish in regards to the appearance, taste, texture, smell and the overall liking of each main dish. The children's responses were recorded by the researcher and entered into excel sheets for analysis.

The plate-waste method was used to measure children's consumption of lunch and to determine percent waste, macronutrient consumption, and actual weight and caloric density of foods consumed and discarded. In each classroom, each food served at lunch was weighed in their respective serving bowls prior to being distributed to the children's lunch plates. The weight (in grams) of each food was recorded. Once lunchtime was over, the waste of each food was collected, combined with the respective leftover food, and weighed to determine how much of that particular food was not consumed. From there, the waste was subtracted from the initial weight to determine how much food the children consumed in grams to calculate energy consumption in kilocalories.

2.4. Experimental Meals

The four novel fish main dishes were designed to be similar to the regularly served chicken main dishes, which were already incorporated into the childcare center's 8-week menu rotation (**Tables A1-A8**). The portion sizes of the meals were based on the United States Department of Agriculture (USDA) Food and Nutrition Service Child and Adult Care Food Program (CACFP) Child Meal Pattern for Lunch for 1 - 2 years old and 3 - 5 years old [34]. The comparison of total energy (kcal/100g) and DHA (mg/100g) content provided by each main dish is reflected in **Table 1**. The fish-based dishes were designed to be of equivalent energy density as the chicken-based dishes.

The chicken-based dishes and the comparable fish-based dishes were a chicken patty versus salmon nuggets, chicken macaroni and cheese versus salmon macaroni and cheese, chicken salad wrap versus salmon salad wrap, and chicken stir-fry versus salmon dumplings. Due to the texture of salmon, a fish stir-fry would not have been acceptable to most children, thus, pot-sticker dumplings were served instead. Although canned pink salmon was directly substituted for canned chicken in the salad wrap and macaroni and cheese recipes, the other recipes required further development. To develop the salmon nuggets and salmon dumplings, cooked Atlantic salmon was pureed with canned Great Northern Beans to help provide a soft and smooth textured protein base for the salmon nuggets and dumplings. The salmon-bean mix-

ture was prepared in two different ways: 1) lightly coated in breadcrumbs and baked to form salmon nuggets and 2) portioned into Wonton wrappers and steamed to form salmon dumplings. The four experimental dishes were developed and taste-tested in a preschool-age population. Based on the taste-test responses, the recipes were modified until at least 80% of the children liked the test foods.

2.5. Procedures

On each study day, teachers in participating classrooms were instructed to follow standard mealtime procedures for lunch. In each classroom the children would sit at a table together and were served lunch by a research assistant. Children were not encouraged to eat more or less than usual and were instructed not to share food. All food liking results and plate-waste measurements were entered into excel sheets for additional analysis. Children's intake was entered into the Nutrition Data System for Research (NDSR) 2012. Food intake was recorded as grams of food consumed and total energy for each food component at lunch and for the whole meal (kcal) was calculated as well as intake of DHA (mg). For a more accurate analysis of how much DHA was provided by the two types of salmon used in the study (canned and cooked), gas chromatography with a flame ionization detector was used.

The brand of canned salmon (3 lots) and cooked Atlantic salmon (1 fillet) were purchased from local stores around Lafayette, Indiana in 2013. From each lot, the total contents were combined and ground in a food processor to obtain a composite sample. The following methods have been replicated from a previous study by Shim, Dorworth, Lasrado, and Santerre [35]. For determination of total fat, two composite samples were randomly chosen from each lot, thawed, and mixed well. A modified Folch method [36] was used to determine total fat concentration. Five grams of composite tissue was mixed with 100 mL of chloroform/methanol (2:1, v/v,

Table 1. Energy density of each main dish (kcal/100g) and DHA provided (mg/100g).

Main dish	Energy density (kcal/100g)	DHA (mg/100g)
Chicken stir-fry, fresh boneless chicken thighs, stir-fried	236.3	38.8
Chicken macaroni & cheese, fully cooked, diced, dark and white blend	205.0	1.5
Chicken breast patty, breaded, baked	263.2	18.4
Chicken salad wrap, fully cooked, diced, dark and white blend	280.4	3.6
Salmon dumplings, cooked Atlantic salmon pureed with Great Northern Beans, steamed	197.1	584.5
Salmon macaroni & cheese, canned pink salmon	198.3	453.9
Salmon nuggets, cooked Atlantic salmon pureed with Great Northern Beans, baked	215.4	584.5
Salmon salad wrap, canned pink salmon	229.5	453.9

HPLC grade for chloroform, pesticide grade for methanol, Fisher Scientific, Fair Lawn, NJ, USA) for 2 h to extract the fat. The mixture was filtered (Whatman filter paper nr 1, 150-mm dia, Whatman Intl. Ltd. Maidstone, England) and 50 mL of 0.88% potassium chloride (ACS reagent, Sigma, St. Louis, Mo., USA) was added to the filtrate. After removing the aqueous layer (upper), the solvent (lower) was reduced by evaporation using a Turbo Vap® (Zymark Corp., Hopkinton, Mass, USA). The extract was transferred to a pre-weighed flask and placed in a desiccator overnight. Duplicated blanks were included in each run during the fat extraction. Ninety-five percent recovery of total fat was determined using a Standard Reference Material (SRM) (Lake Superior fish tissue 1946, Natl. Inst. of Standards and Technology, Gaithersburg, Md., USA). Determination of fatty acids was carried out using the AOAC method 991.39 (AOAC 2000). Polyunsaturated fatty acids, including LA, ALA, stearidonic acid (SDA), ARA, EPA, docosapentaenoic acid (DPA), and DHA were quantified by gas chromatography with a flame ionization detector (GC/FID, Varian 3900 GC, CP-8400 auto sampler, CP-8410 auto injector, Varian Analytical Instruments, Walnut Creek, Calif., USA). Operating conditions were as follows: injection port temperature, 240°C; detector temperature, 300°C; oven programmed from 175°C for 4 min to final hold temperature of 240°C for 5 min with an increase of 3°C /min; helium carrier gas (99.999% pure, Inweld, Inc., Lafayette, Ind., USA); and wall coated open tubular (WCOT) fused silica capillary column, 30 m × 0.32 mm, coated with Chrompack (CP) wax 52CB, DF 0.25 mm (CP 8843, Varian).

2.6. Statistical Analysis

All statistical analyses were conducted using the Statistical Analysis Software (version 9.3, 2010, SAS Institute Inc., Cary, NC). For the three-point Likert scale ratings, participants' responses were recorded and coded: "–1" for "yucky", "0" for "okay", and "+1" for "yummy". The values were entered for each participant, for each main dish (chicken and salmon), and for each category (appearance, taste, texture, smell, and overall liking). Average responses were calculated. A two independent sample t-test was conducted to determine statistical differences between each of the food categories. Kilocalorie consumption per child and DHA consumption per child were analyzed using a mixed model analysis of variance. Factors included in the model were classroom (3 levels), main dish type (4 levels), food type (chicken and salmon), and the interaction between food type and main dish type. Contrasts were then used to compare food types for each main dish. Statistical significance was defined as $P < 0.05$.

3. Results

Complete intake data were obtained for 45 children. Data were aggregated by classroom, and a total of 48 eating occasions (3 classrooms × 16 meals) were collected. The analysis of variance indicated that the interaction between main dish and food type was statistically significant ($P < 0.0001$). To investigate this, contrasts were used to compare food types for each main dish. Means and standard deviations of energy and DHA intake of the main dishes at lunchtime are provided in **Table 2**. Energy intake decreased by 83% for salmon nuggets compared to the intake of the chicken patty (43 versus 256 kcal, $P < 0.0001$). However, DHA intake increased by 550% (117 versus 18 mg, $P = 0.0024$). Energy intake decreased by 54% for the salmon dumplings compared to the chicken stir-fry (50 versus 108 kcal, $P = 0.0120$) but DHA intake increased by 722% (148 versus 18 mg, $P = 0.0001$). No significant difference for energy intake was observed for the substitution of chicken in the wrap as energy intake decreased by 28% when the salmon salad

Table 2. Comparison of children's energy and DHA intake of main dish (chicken versus salmon) (mean ± SD).

Meal	Main dish energy (kcal)	P-value	DHA (mg)	P-value
Chicken Patty	256 ± 40	<0.0001*	18 ± 3	0.0024*
Salmon Nuggets	43 ± 30		117 ± 83	
Chicken Macaroni & Cheese	152 ± 58	0.8640	1 ± 0	<0.0001*
Salmon Macaroni & Cheese	148 ± 50		339 ± 115	
Chicken Stir Fry	108 ± 26	0.0120*	18 ± 4	<0.0001*
Salmon Dumplings	50 ± 16		148 ± 46	
Chicken Salad Wrap	108 ± 40	0.1916	1 ± 1	0.0001*
Salmon Salad Wrap	78 ± 32		155 ± 62	

*Statistical significance at $P < 0.05$.

wrap was served instead of the chicken salad wrap (78 versus 108 kcal, P = 0.1916) but DHA intake increased by 15400% (155 versus 1 mg, P < 0.0001). Likewise energy intake only decreased by 3% when the main course was salmon macaroni and cheese as compared to chicken macaroni and cheese (152 versus 148 kcal, P = 0.8640) and DHA intake increased by 33800% (339 versus 1 mg, P < 0.0001). Thus, in two of the substitutions, total energy intake decreased significantly but DHA intake increased despite the reduction in total food intake.

The results of the Likert scale ratings used to determine liking of the main dishes are provided in **Table 3**. No significant differences were observed between the chicken and salmon macaroni and cheeses, the chicken and salmon salad wraps, or the chicken patty and salmon nuggets. As for the difference between the chicken stir-fry and salmon dumplings, only overall liking of the dumplings was significantly lower (P = 0.001). Overall, the Likert scale ratings corresponded to energy consumed by participants. There were no significant differences in the ratings of the chicken and salmon macaroni and cheeses or the chicken and salmon salad wraps, and both of these pairs of main dishes did not have significant differences in energy intake. Conversely, there was a significant difference in overall liking and energy intake for the chicken stir-fry in comparison to the salmon dumplings.

4. Discussion

DHA supports healthy brain and eye development and certain fish are the best dietary sources of DHA. Therefore, changing children's consumption patterns to include fatty fish is a critical public health issue. This study was designed to explore the feasibility of increasing preschooler's fish consumption to help meet the dietary guideline for seafood consumption and to increase DHA intake. Results showed that although children's intake of the main dish decreased when some of the salmon-based foods were served, DHA intake was significantly higher than when the regularly scheduled chicken-based dishes were served.

In the two instances where the salmon main dishes consisted of a mixture of cooked beans and Atlantic salmon and did not resemble the chicken main dishes in appearance (and were unfamiliar to the children), main dish intake significantly decreased (salmon nuggets and salmon dumplings) (**Table 2**). This finding was expected as it is well documented that young children are resistant to accepting new foods into the diet and consumption of novel foods only increases with repeated exposure. In this particular study, only two exposures were provided and it is probable that consumption would have increased if the children had increased exposures to the two novel fish dishes.

However, in the two instances where the salmon main dishes incorporated canned pink salmon and did resemble the chicken main dishes in appearance (and were therefore familiar to the children), main dish intake did not significantly decrease (salmon salad wrap and salmon macaroni and cheese) (**Table 2**). In addition, there were no significant differences between the children's ratings on appearance, taste, texture, smell, and overall liking of the chicken and salmon versions of the salad wraps and macaroni and cheese main dishes (**Table 3**). Since fatty fish is the best dietary source of DHA [2], our results indicate that modifying main dishes to incorporate salmon can prove to be an effective approach to increase DHA intake at meals.

With the exception of the overall liking of the dumplings, children did not report any difference in the liking of the five characteristics of the four different salmon main dishes versus the chicken main dishes studied (**Table 3**). Therefore, the results of this study support our hypothesis that 2 - 5 years old children will consume fish at least once a week and therefore increase DHA intake. It must be noted that these four fish dishes were created to be similar to the regularly served chicken main dishes. By children rating the salmon and chicken dishes similarly on the itemized list of food characteristics, this study demonstrates that it is possible to improve young children's diet quality by serving fish in childcare centers.

Study results support the premise that parents and caretakers of children should introduce children to fish at a young age. Through repeated exposure to certain foods, children develop a liking for the food's characteristics

Table 3. Differences between chicken main dish scores and salmon main dish scores.

	Appearance	Taste	Texture	Smell	Overall
Chicken Patty versus Salmon Nuggets	0.187	0.006	0.038	0.180	0.015
Chicken versus Salmon Macaroni and Cheese	0.671	0.291	0.162	0.331	0.345
Chicken Stir-Fry versus Salmon Dumplings	0.051	0.128	0.141	0.196	0.001*
Chicken versus Salmon Salad Wrap	0.708	0.865	0.115	0.155	0.305

*Statistical significance at P < 0.0025 when using the Bonferroni adjustment.

(appearance, smell, taste, and texture) [37]. If fatty-fish were offered to children as a protein source at least twice a week, children would predictably choose to eat more fish, thereby increasing their DHA intake. Although the main dishes tested in this study were only representative of a small portion of the main dishes children usually consume, providing fish at least once a week can have an additive effect on children's average fish consumption and DHA intake. Whether fish is served on its own (such as a salmon filet) or incorporated into mixed dishes (such as salmon macaroni and cheese or a salmon salad wrap), children's average daily intake of DHA will increase.

The present study had several strengths and limitations. This study was highly innovative as we are not aware of any other studies on the development of child-friendly fish dishes in an effort to increase children's fatty fish consumption. Due to the results of this study, especially the high acceptance of the salmon macaroni and cheese and the salmon salad wrap, the daycare menu was revised to incorporate these two dishes into the childcare center's 8-week menu rotation (**Tables A1-A8**). Since data were collected on the group level, individual changes in intake or liking of the foods were not identified. The study was based on a university population (higher parent education, more international diversity, less domestic ethnic diversity, etc.) and was therefore not representative of the US pediatric population. The results of this study indicate that children's intake in a childcare setting can be modified to improve overall diet quality and support children's growth and development.

Future research on this topic should be based on larger and more diverse samples of children and include individual data, provide more exposures, and a larger variety of fish-based foods. Despite the limitations of this study, the findings strongly indicate that it is feasible and advisable for childcare centers to include offering high-fat fish once a week. Although some children may not consume as much of the fish-based lunch initially, this change in the menu may help children adopt a healthier diet for life as well as provide essential omega-3 PUFAs, specifically DHA, which are critical in the development of the nervous system.

5. Acknowledgements

The authors thank the parents and teachers of the participating childcare center. Without your help and tolerance, our research would not have been possible. We also thank Lauren Suttie (Undergraduate Independent Researcher), Elizabeth Schlesinger-Devlin (Child Care Director), the staff of the Ben and Maxine Miller Child Development Laboratory School, Jamie Wittenberg (Lab Manager), Dennis Cladis (Graduate Student), Mary Brauchla (PhD Candidate), Selena Baker (MS Student), and other volunteers in the Kranz Lab for their contributions to this research project.

REFERENCES

[1] D. Cao, K. Kevala, J. Kim, H.-S. Moon, S. B. Jun, D. Lovinger and H.-Y. Kim, "Docosahexaenoic Acid Promotes Hippocampal Neuronal Development and Synaptic Function," *Journal of Neurochemistry*, Vol. 111, No. 2, 2009, pp. 510-521.

[2] S. M. Innis, "Dietary Omega 3 Fatty Acids and the Developing Brain," *Brain Research*, Vol. 1237, 2008, pp. 35-43.

[3] R. J. Pawlosky, J. R. Hibblen, Y. Lin, *et al.*, "Effects of Beef- and Fish-Based Diet on the Kinets of n-3 Fatty Acid Metabolism in Human Subjects," *American Journal of Clinical Nutrition*, Vol. 77, No. 3, 2003, pp. 565-572.

[4] P. L. Goyens, M. E. Spilker, P. L. Zock, M. B. Katan and R. P. Mensink, "Compartmental Modeling to Quantify Alpha-Linolenic Acid Conversion after Longer-Term Intake of Multiple Tracer Boluses," *Journal of Lipid Research*, Vol. 46, 2005, pp. 1474-1483.

[5] G. C. Burdge, Y. E. Finnegan, A. M. Minihane, C. M. Williams and S. A. Wootton, "Effect of Altered Dietary n-3 Fatty Acid Intake upon Plasma Lipid Fatty Acid Composition, Converstion of [^{13}C] Alpha-Linolenic Acid to Longer-Chain Fatty Acids and Partitioning towards Beta-Oxidation in Older Men," *British Journal of Nutrition*, Vol. 90, No. 2, 2003, pp. 311-321.

[6] G. C. Burdge, A. E. Jones and S. A. Wootton, "Eicosapentaenoic and Docosapentaenoic Acids Are the Principal Products of Alpha-Linolenic Acid Metabolism in Young Men," *British Journal of Nutrition*, Vol. 88, No. 4, 2002, pp. 355-364.

[7] E. A. Emken, R. O. Adlof and R. M. Gulley, "Dietary Linoleic Acid Influences Desaturation and Acylation of Deuterium-Labeled Linoleic and Linolenic Acids in Young Adult Males," *Biochimica et Biophysica Acta*, Vol. 1213, No. 3, 1994, pp. 277-288.

[8] N. Hussein, E. Ah-Sing, P. Wilkinson, C. Leach, B. A. Griffin and D. J. Millward, "Long-Chain Conversion of [^{13}C] Linoleic Acid and Alpha-Linolenic Acid in Response to Marked Changes in Their Dietary Intake in Men," *Journal of Lipid Research*, Vol. 46, 2005, pp. 269-280.

[9] G. C. Burdge and S. A. Wootton, "Conversion of Alpha-Linolenic Acid to Eicosapentaenoic, Docosapentaenoic and Docosahexaenoic Acids in Young Women," *British Journal of Nutrition*, Vol. 88, No. 4, 2002, pp. 411-421.

[10] A. Dalton, P. Wolmarans, R. C. Witthuhn, M. E. van Stuijvenberg, S. A. Swanevelder and C. M. Smuts, "A Randomised Control Trial in Schoolchildren Showed Improvement in Cognitive Function after Consuming a Bread Spread, Containing Fish Flour from a Marine Source," *Prostaglandins Leukotrienes and Essential Fatty Acids*,

Vol. 80, No. 2-3, 2009, pp. 143-149.

[11] S. M. Innis, "Perinatal Biochemistry and Physiology of Long-Chain Polyunsaturated Fatty Acids," *Journal of Pediatrics*, Vol. 143, No. 4, 2003, pp. S1-S8.

[12] H. W. Sprecher, Q. Chen and F. Q. Yin, "Differences in the Regulation of Biosynthesis of 20- versus 22-Carbon Polyunsaturated Fatty Acids," *Prostaglandins, Leukotrienes and Essential Fatty Acids*, Vol. 52, No. 2-3, 1995, pp. 99-101.

[13] J. M. Bourre, M. Piciotti, O. Dumont, G. Pascal and G. Durand, "Dietary Linoleic Acid and Polyunsaturated Fatty Acids in Rat Brain and Other Organs. Minimal Requirements of Linoleic Acid," *Lipids*, Vol. 25, No. 8, 1990, pp. 465-472.

[14] W. E. Lands, A. Morris and B. Libelt, "Quantitative Effects of Dietary Polyunsaturated Fats on the Composition of Fatty Acids in Rat Tissues," *Lipids*, Vol. 25, No. 9, 1990, pp. 505-516.

[15] L. D. Arbuckle, M. J. MacKinnon and S. M. Innis, "Formula 18:2(n-6) and 18:3(n-3) Content and Ratio Influence Long-Chain Polyunsaturated Fatty Acids in the Developing Piglet Liver and Central Nervous System," *The Journal of Nutrition*, Vol. 124, No. 2, 1994, pp. 289-298.

[16] R. R. Brenner and R. O. Peluffo, "Effect of Saturated and Unsaturated Fatty Acids on the Desaturation *in Vitro* of Palmitic, Stearic, Oleic, Linoleic, and Linolenic Acids," *Journal of Biological Chemistry*, Vol. 241, No. 22, 1966, pp. 5213-5219.

[17] J. Rahm and R. T. Holman, "Effect of Linoleic Acid upon the Metabolism of Linolenic Acid," *The Journal of Nutrition*, Vol. 84, 1964, pp. 15-19.

[18] P. S. Sastry, "Lipids of Nervous Tissue: Composition and Metabolism," *Progress in Lipid Research*, Vol. 24, No. 2, 1985, pp. 69-176.

[19] N. M. Giusto, G. A. Salvador, P. I. Castagnet, S. J. Pasquare and M. G. Ilincheta de Boschero, "Age-Associated Changes in Central Nervous System Glycerolipid Composition and Metabolism," *Neurochemical Research*, Vol. 27, No. 11, 2002, pp. 1513-1523.

[20] J. M. Bourre, M. Francois, A. Youyou, O. Dumont, M. Piciotti, G. Pascal and G. Durand, "The Effects of Dietary Alpha-Linolenic Acid on the Composition of Nerve Membranes, Enzymatic Activity, Amplitude of Electrophysiological Parameters, Resistance to Poisons and Performance of Learning Tasks in Rats," *The Journal of Nutrition*, Vol. 119, No. 12, 1989, pp. 1880-1892.

[21] C. Galli, H. I. Trzeciak and R. Paoletti, "Differential Effects of Dietary Fatty Acids on the Accumulation of Arachidonic Acid and Its Metabolic Conversion through the Cyclooxygenase and Lipoxygenase in Platelets and Vascular Tissue," *Lipids*, Vol. 16, No. 3, 1981, pp. 165-172.

[22] N. Hrboticky, M. J. MacKinnon and S. M. Innis, "Effect of a Vegetable Oil Formula Rich in Linoleic Acid on Tissue Fatty Acid Accretion in the Brain, Liver, Plasma, and Erythrocytes of Infant Piglets," *American Journal of Clinical Nutrition*, Vol. 5, No. 2, 1990, pp. 173-182.

[23] N. Hrboticky, M. J. MacKinnon and S. M. Innis, "Retina Fatty Acid Composition of Piglets Fed from Birth with a Linoleic Acid-Rich Vegetable-Oil Formula for Infants," *American Journal of Clinical Nutrition*, Vol. 53, No. 2, 1991, pp. 483-490.

[24] S. M. Innis, "Essential Fatty Acids in Growth and Development," *Progress in Lipid Research*, Vol. 30, No. 1, 1991, pp. 39-103.

[25] E. C. Bakker, G. Hornstra, C. E. Blanco and J. S. H. Vles, "Relationship between Long-Chain Polyunsaturated Fatty Acids at Birth and Motor Function at 7 Years of Age," *European Journal of Clinical Nutrition*, Vol. 63, No. 4, 2009, pp. 499-504.

[26] E. E. Birch and D. R. Stager Sr., "Long-Term Motor and Sensory Outcomes after Early Surgery for Infantile Esotropia," *Journal of the American Association for Pediatric Ophthalmology and Strabismus*, Vol. 10, No. 5, 2006, pp. 409-413.

[27] Institute of Medicine, "Seafood Choices: Balancing Benefits and Risks," 2006. http://www.iom.edu/Reports/2006/Seafood-Choices-Balancing-Benefits-and-Risks.aspx

[28] L. R. Huss, S. Laurentz, J. O. Fisher, G. P. McCabe and S. Kranz, "Timing of Serving Dessert but Not Portion Size Affects Young Children's Intake at Lunchtime," *Appetite*, Vol. 68, 2013, pp. 158-163.

[29] S. Y. Park, H.-Y. Paik, J. D. Skinner, A. A. Spindler and H.-R. Park, "Nutrient Intake of Korean-American, Korean, and American Adolescents," *Journal of the American Dietetic Association*, Vol. 104, No. 2, 2004, pp. 242-245.

[30] Food and Nutrition Board, Institute of Medicine, "Dietary Reference Intakes for Energy, Carbohydrate, Fiber, Fat, Fatty Acids, Cholesterol, Protein, and Amino Acids (Macronutrients)," A Report of the Panel on Macronutrients, Subcommittees on Upper Reference Levels of Nutrients and Interpretation and Uses of Dietary Reference Intakes, and the Standing Committee on the Scientific Evaluation of Dietary Reference Intakes, National Academy Press, Washington DC, 2002.

[31] FAO/WHO, "Interim Summary of Conclusions and Dietary Recommendations on Total Fat & Fatty Acids," Joint FAO/WHO Expert Consultation on Fats and Fatty Acids in Human Nutrition, Geneva, 2008.

[32] B. Koletzko, R. Uauy, A. Palou, *et al.*, "Dietary Intake of Eicosapentaenoic Acid (EPA) and Docosahexaenoic Acid (DHA) in Children—A Workshop Report," *British Journal of Nutrition*, Vol. 103, No. 6, 2010, pp. 923-928.

[33] J. Whelan, L. Jahns and K. Kavanagh, "Docosahexaenoic Acid: Measurements in Food and Dietary Exposure," *Prostaglandins, Leukotrienes and Essential Fatty Acids*, Vol.

8, No. 2-3, 2009, pp. 133-136.

[34] E. Foland, L. Graves and C. Markle, "CACFP—What's in a Meal?—Healthy Hoosier Edition," 2008. http://www.doe.in.gov/student-services/nutrition/cacfp-whats-meal-healthy-hoosier-edition

[35] S. M. Shim, L. E. Dorworth, J. A. Lasrado and C. R. Santerre, "Mercury and Fatty Acids in Canned Tuna, Salmon, and Mackerel," *Journal of Food Science*, Vol. 69, 2004,

pp. 681-684.

[36] J. Folch, M. Lees and G. H. Sloane-Stanley, "A Simple Method for the Isolation and Purification of Total Lipids from Animal Tissues," *Canadian Journal of Biochemistry and Physiology*, Vol. 226, 1956, pp. 497-509.

[37] A. Drewnowski, "Taste Preferences and Food Intake," *Annual Review of Nutrition*, Vol. 17, 1997, pp. 237-253.

Appendix

See Tables A1-A8.

Table A1. Week 1 menu rotation.

	Monday	Tuesday	Wednesday	Thursday	Friday
Main dish	Chicken patty on whole wheat bun	Teriyaki chicken w/rice	Turkey meatloaf w/biscuits	Salmon macaroni & cheese	Cheese ravioli
Veggie	Salad w/tomatoes	Steamed veggies	Carrots	Zucchini	Green beans
Fruit	Apples	Peaches	Grapes	Orange wedges	Pears
Dairy	Milk	Milk	Milk	Milk	Milk
Dessert					Cookie

Table A2. Week 2 menu rotation.

DAY	Monday	Tuesday	Wednesday	Thursday	Friday
Main dish	Sloppy Joe on whole wheat bun	Chicken enchiladas	Pancakes and turkey sausage	Salmon nuggets	Chicken salad wrap
Veggie	Steamed peas	Green beans	Tater tots	Broccoli & Cauliflower	Carrots
Fruit	Watermelon	Grapes	Oranges	Apples	Pineapple
Dairy	Milk	Milk	Milk	Milk	Milk
Dessert					Cookie

Table A3. Week 3 menu rotation.

DAY	Monday	Tuesday	Wednesday	Thursday	Friday
Main dish	Chicken Pot Pie	Turkey Tacos	Chicken Macaroni & Cheese	Salmon Dumplings w/Wild Rice	Spaghetti and Meatballs
Veggie	Carrots	Lettuce/Tomatoes/Olives	Tomato/Cucumber	Cauliflower Blend	Zucchini/Squash
Fruit	Fresh Fruit	Fresh Fruit	Grapes	Fresh Fruit	Mandarin Oranges
Dairy	Milk	Milk	Milk	Milk	Milk
Dessert					Cookie

Table A4. Week 4 menu rotation.

DAY	Monday	Tuesday	Wednesday	Thursday	Friday
Main dish	4 × 6 Tony's pizza	Chicken stir-fry w/rice	Turkey sandwiches	Salmon salad wrap	Toasted cheese sandwich
Veggie	Salad w/carrot and cucumber	Stir-fry veggies	Peas & corn	Sweet potato wedges	Cherry tomatoes
Fruit	Diced pears	Fresh fruit	Fresh fruit	Blueberries	Canned fruit
Dairy	Milk	Milk	Milk	Milk	Milk
Dessert					Cookie

Table A5. Week 5 menu rotation.

DAY	Monday	Tuesday	Wednesday	Thursday	Friday
Main dish	Chicken and noodles w/dinner roll	Pasta primavera	Turkey meatloaf w/biscuits	Salmon macaroni & Cheese	Cheese pizza
Veggie	Diced carrots	Italian blend vegetables	Carrots	Steamed veggies	Peas
Fruit	Apple wedges	Fresh fruit	Grapes	Fresh fruit	Blueberries
Dairy	Milk	Milk	Milk	Milk	Milk
Dessert					Cookie

Table A6. Week 6 menu rotation.

DAY	Monday	Tuesday	Wednesday	Thursday	Friday
Main dish	Grilled chicken w/breadstick	Salmon salad wrap	Turkey sausage & Pancakes	Fish nuggets w/wheat roll	Tofu & Rice
Veggie	Salad	Salad	Cucumbers	Green beans	Stir-fry veggies
Fruit	Watermelon	Strawberries	Oranges	Fresh fruits	Pineapple
Dairy	Milk	Milk	Milk	Milk	Milk
Dessert					Cookie

Table A7. Week 7 menu rotation.

DAY	Monday	Tuesday	Wednesday	Thursday	Friday
Main dish	Beef & Vegetable stroganoff	Salmon dumplings w/white rice	Chicken macaroni & Cheese	Pork loin w/wheat roll	Chicken fajitas
Veggie	Stroganoff vegetables	Salad	Cucumbers	Mashed potatoes	Zucchini & squash
Fruit	Strawberries	Fresh fruit	Grapes	Orange wedges	Watermelon
Dairy	Milk	Milk	Milk	Milk	Milk
Dessert					Cookie

Table A8. Week 8 menu rotation.

DAY	Monday	Tuesday	Wednesday	Thursday	Friday
Main dish	Chicken pot pie	Ham & Cheese sandwich	Turkey w/whole wheat bread	Salmon nuggets w/wheat roll	Toasted cheese sandwich
Veggie	Peas & carrots	Mixed green salad	Corn	Spinach & corn	Tomato soup
Fruit	Diced pineapple	Fresh fruit	Fresh fruit	Apple wedges	Fresh fruit
Dairy	Milk	Milk	Milk	Milk	Milk
Dessert					Cookie

21

Antioxidants from Syrah Grapes (*Vitis vinifera L. cv. Syrah*). Extraction Process through Optimization by Response Surface Methodology

Youssef El Hajj[1,2], Espérance Debs[3], Catherine Nguyen[2], Richard G. Maroun[1*], Nicolas Louka[1]

[1]Unité Technologies et Valorisation Alimentaire, Centre d'Analyses et de Recherche, Faculté des Sciences, Université Saint-Joseph de Beyrouth, Beirut, Lebanon; [2]INSERM, U928, Technological Advances for Genomics and Clinics Laboratory, Marseille, France; [3]Department of Biology, Faculty of Sciences, University of Balamand, Tripoli, Lebanon.

ABSTRACT

In this work, optimization of phenolic compounds (PC) and monomeric anthocyanins (MA) extraction from Syrah (Sy) wine grapes (*Vitis vinifera L. cv. Syrah*) using response surface methodology was conducted. The comparisons between two extraction mixtures, Acetone/Water (A/W) as well as Methanol/Water (M/W) and the effects of three critical variables, Extraction Time (between 8 and 88 h), Extraction Temperature (between $1°C$ and $35°C$) and Solvent Content (between 63% and 97%), on Phenolic Compounds Yield (PCY), Monomeric Anthocyanins Yield (MAY) and the DPPH Free Radical Inhibition Potential (DFRIP) were studied. The highest PCY was obtained in 63% A/W solvent content after 88 h incubation at $35°C$. The highest MAY was acquired in 97% M/W solvent content after 8 h incubation at $17°C$. The highest DFRIP of the extract was attained using 97% A/W solvent content after 16 h incubation at $35°C$. The low cost of this process, on economic and environmental levels, could lead to interesting applications on an industrial scale. It could be used to obtain bioactive phytochemicals from direct material or byproducts for either therapeutic or nutritional purposes.

Keywords: Phenolic Compounds; Monomeric Anthocyanins; Antiradical Scavenging Potential; Extraction Optimization; Grape; Time; Solvent and Temperature

1. Introduction

Importance of natural antioxidants for medical and food sectors has been underlined by numerous works [1-15]. Nowadays, many studies report increasing interest in wine grapes such as Syrah (*Vitis vinifera L. cv. Syrah*) as source of powerful antioxidants, mainly phenolic compounds (PC).

The aromatic ring, bound to a hydroxyl group, is the basal structure common to all phenolic compounds [3]. This configuration allows radical scavenging potential and gives phenolic compounds a multitude of bioactive roles [4]. Furthermore, the high industrial value of these molecules is well proven, especially due to food lipid antioxidation [5]. Thus PC can be considered as added-

value phytochemicals of plant waste material, justifying their isolation from the plant matrix by extraction.

Recovery of PC is commonly performed through a solvent-extraction procedure but, currently, only ambiguous data on the methods and conditions for extraction are available and sometimes contradictory.

The aim of an extraction process should be to provide a possible maximum yield of substances of the highest quality (quantity of target compounds and antioxidant potential of the extract). Just few works, on antioxidant recovery from grapes, have targeted the optimization of some process parameters [6]. The variables mostly studied have been: the type of extraction solvent or solvent mixture, extraction time and extraction temperature [6].

Type of solvent has been the most investigated factor. Acetone and methanol were reported as two of the best

---

*Corresponding author.

solvents for extraction of PC and MA from plants [7-10]. Mixtures of solvents and water have revealed to be more efficient in extracting phenolic constituents than the corresponding mono-component solvent system [11,12].

Time and temperature of extraction are important parameters to be optimized especially in order to minimize energy cost of the process. Many authors agreed on the fact that an increase in the working temperature favors extraction and enhances both the solubility of the solute and the diffusion coefficient. However, phenolic compounds can be denatured beyond a certain range [2,11, 13]. More contradictories are the data available for incubation time during extraction: some authors chose quite short extraction times [11,13,14]; other long ones [2,9, 12,15,16].

After having optimized the extraction process of Total Phenolic Compounds (TPC) and MA from Cabernet Sauvignon (CS) grapes in a previous study [1], our objective in this work is to optimize the extraction process of TPC and MA from Sy grapes in addition to the determination of FRIP (Free Radical Inhibition Potential) of the extracts. This will provide us with a better understanding of the extraction process parameters impact on the quality of the extracted PC and MA. In order to achieve our goal, we used the response surface methodology (RSM) with a five-level, and three-variable central composite design. We have noticed that literature lacks optimization studies regarding the extraction process of TPC and MA from grapes as well as FRIP analysis of the obtained extracts. Therefore we determined the optimal parameters (solvent type, water concentration in the solvent system, extraction time and extraction temperature) needed to give the highest TPC, MA yields and FRIP from Sy grapes extracts. We draw a response surface plot of the extraction kinetics corresponding to these parameters and aimed to improve the extraction to a low cost and energy depending procedure. A solvent extraction method was proposed with simple, no complex machinery and without expensive pre-treatments of the starting material or excessive heating. We obtained a PC rich extracts with a high free radical scavenging potential. These extracts could be used as additives for food preservation, as well as in pharmaceutical, cosmetics and nutraceutical industries.

2. Material and Methods

2.1. Reagents

The solvents used for the extraction of the samples were pure water, acetone and methanol of analytical grade from Scharlau (Barcelona, Spain) same as Ethanol (Merck) used for DFRIP (DPPH Free Radical Inhibition Potential) determination. The Folin reagent (Sigma Che-

mical Co., St. Louis, MO, USA) and sodium carbonate (Fluka, Buchs, Switzerland) were used for the measurement of the total phenolic compounds concentrations using the Folin-Ciocalteu method, the calibration curve was constructed with gallic acid (Sigma Chemical Co., St. Louis, MO, USA). Potassium chloride (Fluka, Buchs, Switzerland) and sodium acetate (Scharlau, Barcelona, Spain) were used for total monomeric anthocyanin determination by the pH-differential method. Resveratrol (Sigma Chemical Co.) and DPPH (2,2-Diphenyl-picrylhydrazyl) reagent (Sigma Chemical Co.) were used for DFRIP determination.

2.2. Sample Preparation

Grapes (*Vitis vinifera L. cv. Syrah*) were collected from different crop areas located at different regions in the Lebanese Bekaa valley. Harvesting took place during summer/fall of 2010 (August till October). Grapes from different regions, at different maturity stages and from several localization on the vine were placed in a single container. All batch was crushed to obtain a fine grape paste (maximum particle size = 1 mm). The paste was frozen at $-80^{\circ}C$ until use. Each tube/experimental point was subjected to a different parametrical pattern (**Table 1**). Extracts were then centrifuged (6000 g) and filtered through RC membranes (0.2 μm). Samples were kept at $-80^{\circ}C$ ready to be analyzed.

2.3. Total Phenolic Compound Determination

Total phenolic compounds were determined according to the Folin-Ciocalteu reagent with the Micro method previously described by Andrew Waterhouse (Department of Viticulture and Enology, University of California, Davis, USA). The absorbance of each solution was determined at 765 nm against the blank (water). A calibration curve was created by plotting absorbance vs. concentration of the standards (solutions of different Gallic Acid concentrations) and the total phenols concentrations were determined in all samples. Phenolic Compound Yield (PCY) was calculated by transforming milligrams of Gallic Acid Equivalents (GAE) per liter (mg GAE/L) into grams of GAE per 100 g of grape paste or fresh weight (g GAE/100g) which is % GAE.

2.4. Total Monomeric Anthocyanin Determination

Monomeric anthocyanins were measured by the pH-differential method, which relies on the structural transformation of the anthocyanin chromophore as a function of pH, and can then be measured using optical spectroscopy [17]. Two dilutions of each sample were prepared using the appropriate, previously determined dilution factor:

Table 1. Central composite arrangement for independent variables and their responses for both extraction mixtures.

Run	Variables levels (coded/uncoded)			A/W			M/W		
	X_1^a	X_2^b	X_3^c	PCY (%GAE)	MAY (mg/100g)	DFRIP (%)	PCY (% GAE)	MAY (mg/100g)	DFRIP (%)
1	−1 (24)	−1 (8)	−1 (70)	0.73	99.36	63.35	0.54	104.87	54.61
2	1 (72)	−1 (8)	−1 (70)	0.85	82.72	45.91	0.58	94.48	45.68
3	−1 (24)	1 (28)	−1 (70)	0.67	77.39	60.92	0.52	96.41	46.84
4	1 (72)	1 (28)	−1 (70)	0.88	54.85	49.55	0.67	93.45	46.36
5	−1 (24)	−1 (8)	1 (90)	0.72	91.52	57.28	0.56	115.71	56.8
6	1 (72)	−1 (8)	1 (90)	0.89	82.78	40	0.61	105.43	42.95
7	−1 (24)	1 (28)	1 (90)	0.59	65.35	67.48	0.51	102.82	55.58
8	1 (72)	1 (28)	1 (90)	0.81	50.42	54.77	0.75	104.47	42.73
9	$-\alpha$ (7.63)	0 (18)	0 (80)	0.83	93.91	54.57	0.62	99.81	48.95
10	α (88.36)	0 (18)	0 (80)	0.83	56.78	65.81	0.66	92.15	44.96
11	0 (48)	$-\alpha$ (1.18)	0 (80)	0.78	90.84	67.59	0.59	50.36	50.69
12	0 (48)	α (34.81)	0 (80)	0.87	33.5	48.48	0.68	79.49	47.07
13	0 (48)	0 (18)	$-\alpha$ (63.18)	0.84	81.02	56.08	0.56	90.44	51.34
14	0 (48)	0 (18)	α (96.81)	0.84	80.22	76.7	0.64	111.39	47.42
15	0 (48)	0 (18)	0 (80)	0.8	82.78	54.02	0.64	102.88	49.07
16	0 (48)	0 (18)	0 (80)	0.78	80.85	56.49	0.64	103.22	49.48
17	0 (48)	0 (18)	0 (80)	0.85	79.03	54.64	0.65	102.76	46.8
18	0 (48)	0 (18)	0 (80)	0.82	81.93	50.93	0.65	104.24	48.45

PCY, Phenolic compounds Yields; MAY, Monomeric Anthocyanins Yields; DFRIP, DPPH Free Radical Inhibition Percentage; % GAE, Percentage Gallic Acid Equivalent; A/W, Acetone/Water; M/W, Methanol/Water. [a]Time (h); [b]Temperature (°C); [c]Solvent Content (%).

once with potassium chloride buffer at 0.025 M and pH 1.0 and the other with sodium acetate buffer at 0.4 M and pH 4.5. The dilutions were equilibrated for 15 min. The absorbance of each dilution was measured at the $\lambda_{vis-max}$ vis-max and at 700 nm against blank cell filled with distilled water. The absorbance (A) of the diluted sample was calculated as follows:

$$A = \left(A_{\lambda vis} - A_{700}\right)_{pH1} - \left(A_{\lambda vis-max} - A_{700}\right)_{pH4.5} \quad (1)$$

The monomeric anthocyanin pigment (MAP) concentration in the original sample was calculated using the following formula:

$$MAP_{(mg/L)} = \left(A \times MW \times DF \times 1000\right)/\left(molA \times L\right) \quad (2)$$

where MW and molA are the molecular weight and the molar absorptivity, respectively of the pigment cyaniding-3-glucoside used as reference; MW = 449.2 g/mole and molA = 26900 $mg^{-1} \cdot l^{-1} \cdot cm^{-1}$. DF is the dilution factor. Milligrams of MA (Monomeric Anthocyanin) per

liter of extract (mg/L) were then transformed into Monomeric Anthocyanin Yield (MAY) which is milligrams per 100 grams of grape paste or fresh weight (mg/100g).

2.5. DPPH Free Radical Inhibition Potential Determination

Antiradical potential of the extracts was assessed according to the DPPH assay, which is based on the ability of antioxidants to interact with the radical DPPH decreasing its absorbance at 517 nm. 1 mg of Resveratrol was dissolved in 1 ml of Ethanol (Merck) to form the 1 mg/ml positive control stock solutions. The grape sample extracts and the positive control solutions were diluted with Ethanol (Merck) to obtain a concentration for each sample and for the positive control at 100 μg/l. 3.9 mg of DPPH powder (Sigma Chemical Co.) were dissolved in 200 ml methanol to form a 0.1 mM DPPH solution wich can be stored at 4°C. 315.2 mg of Tris-HCl were dissolved in 40 ml of water and the pH was elevated to 7.4

Antioxidants from Syrah Grapes (Vitis vinifera L. cv. Syrah). Extraction Process through
Optimization by Response Surface Methodology

171

with NaOH solution (10 mM) to obtain a Tris-HCl (50 mM, pH 7.4) buffer.

A mix of 50 µl of the different sample and positive control dilutions, 450 µl of the Tris-HCl buffer and 1.5 ml of the DPPH solution were incubated for 30 minutes at room temperature then the absorbance was read at 517 nm.

DFRIP of the original sample was calculated using the following formula:

$$DFRIP_{(\%)} = \left(A_{res} - A_s\right)/\left(A_{res}\right) \times 100 \qquad (3)$$

where A_{res} is the absorbance of the solution containing DPPH after inhibition of its free radicals by resveratrol, and A_s is the absorbance of the solution containing DPPH after inhibition of its free radicals by the grape extract sample.

2.6. Experimental Design

In this response surface methodology study, a rotatable central composite design was used to evaluate the main effects of the factors: extraction time (X_1), extraction temperature (X_2), and solvent content (degree or percentage) (X_3) and their interaction on total phenolic compounds, monomeric anthocyannins yields and DPPH Free Radical Inhibition Potential obtained from grapes (*Vitis vinifera L. cv. Syrah*) using Acetone or Methanol separately as extraction solvents. Eighteen experiments (**Table 1**) were performed per extraction solvent mixture including four experiments as the repeatability of the measurements at the center of the experimental domain.

All the factors levels are reported in **Table 2**. Data pertaining to three independent, and two response, variable were analyzed to get a multiple regression equation:

$$Y = b_0 + \sum_{n=1}^{3} b_n X_n + \sum_{n=1}^{3} b_{nn} X_n^2 + \sum_{n=1}^{2} \sum_{m=n+1}^{3} b_{nm} X_n X_m \qquad (4)$$

where Y is the predicted response, X_n and X_m are the coded values of the factors, b_0 is the mean value of responses at the central point of the experiment; and b_n, b_{nn} and b_{nm} are the linear, quadratic and interaction coeffi-

cients, respectively. Analysis of the coefficients of regression models was carried out using an ANOVA table to find the significance of each coefficient (**Table 3**). This significance was illustrated using Pareto charts (**Figure 1**). The significance of Lack of fit for each extraction model is shown in **Table 4**. The process was optimized using response surface methodology for two independent variables at a time. The third parameter was fixed at zero level, the response surface graphs gave values of independent variables allowing the optimization of the process, considering that all the independent variable conditions can be identified for maximum PCY, MAY yields and DFRIP (**Figures 2** and **3**). The optimum experimental conditions were deduced from this study. **Table 5** shows the best time, temperature and solvent degree where the highest PCY, MAY and DFRIP are obtained using A/W or M/W.

Table 3. Test of significance effect for each independent variable, quadratic and interaction effect between variables. (A parameter is significant if Pi < 0.05).

	A/W			M/W		
	P1 (PCY)	P2 (MAY)	P3 (DFRIP)	P4 (PCY)	P5 (MAY)	P6 (DFRIP)
X_1^a	0.0463	0.0002	0.0185	0.0001	0.0008	0.0022
X_2^b	0.7792	0.0001	0.5357	0.0007	0.0019	0.0439
X_3^c	0.7049	0.0234	0.0274	0.0013	0.0001	0.674
X_{12}^d	0.7411	0.0617	0.3929	0.0134	0.0196	0.3028
$X_1 X_2^e$	0.5662	0.0775	0.2022	0.0004	0.002	0.0659
$X_1 X_3$	0.8041	0.0428	0.8683	0.0088	0.0891	0.0139
X_{22}	0.6869	0.0009	0.8709	0.0079	0.0000	0.476
$X_2 X_3$	0.4639	0.1537	0.0359	0.3081	0.1056	0.1887
X_{32}	0.8539	0.3216	0.0222	0.0008	0.0008	0.271

[a]Time, [b]Temperature, [c]Solvent Content, [d]Quadratic effect of time, [e]Interaction effect between Time and Temperature.

Table 4. Validation of the model showed by the responses lack of fit values.

		Lack of fit		
		Sum of squares	Df	f ratio	P value
A/W	PCY	0.052	5	11.68	0.0351
	MAY	135.83	5	10.38	0.0412
	DFRIP	930.545	5	34.82	0.0074
M/W	PCY	0.0119	5	72	0.0025
	MAY	1260.14	5	557.79	0.0001
	DFRIP	54.558	5	7.85	0.0601

Table 2. Independent variables and their levels used for central composite rotatable design.

Independent variables	Symbol	Coded variable levels				
		$-\alpha$	-1	0	$+1$	$+\alpha$
Extraction time (h)	X_1	7.64	24	48	72	88.36
Extraction temperature (°C)	X_2	1.18	8	18	28	34.82
Solvent degree (%)	X_3	63.18	70	80	90	96.82

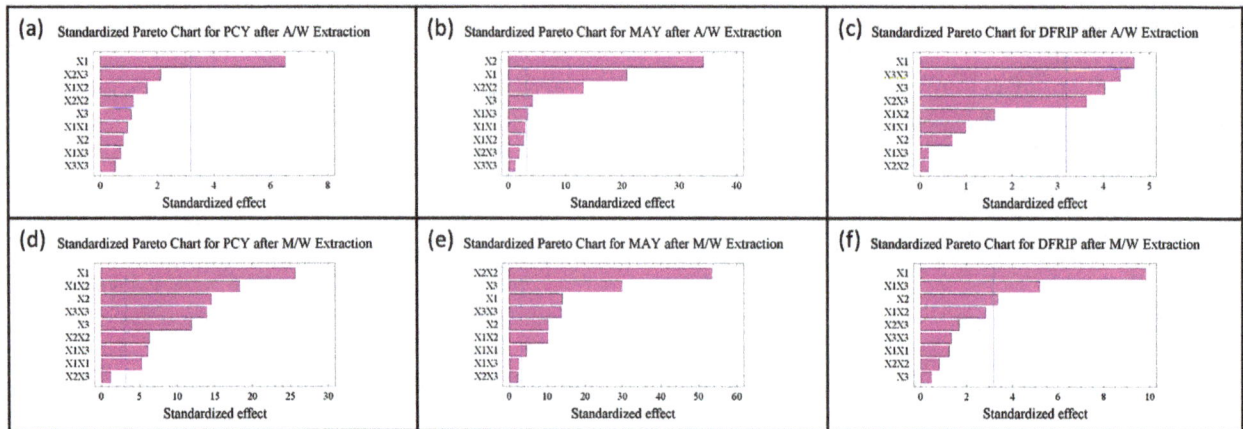

Figure 1. Standardized Pareto charts. Analysis shown for PCY (a); MAY (b) and DFRIP (c) using A/W and for PCY (d); MAY (e) and DFRIP (f) using M/W. The variables are: X_1, time; X_2, temperature; X_3, solvent content; X_1X_1, X_2X_2 and X_3X_3, quadratic effect of time, temperature and solvent content, respectively; X_1X_2, X_1X_3 and X_2X_3, interaction effect between time and temperature, time and solvent content and temperature and solvent content, respectively. The columns/parameters exceeding the vertical bar are statistically significant with more than 95% of confidence.

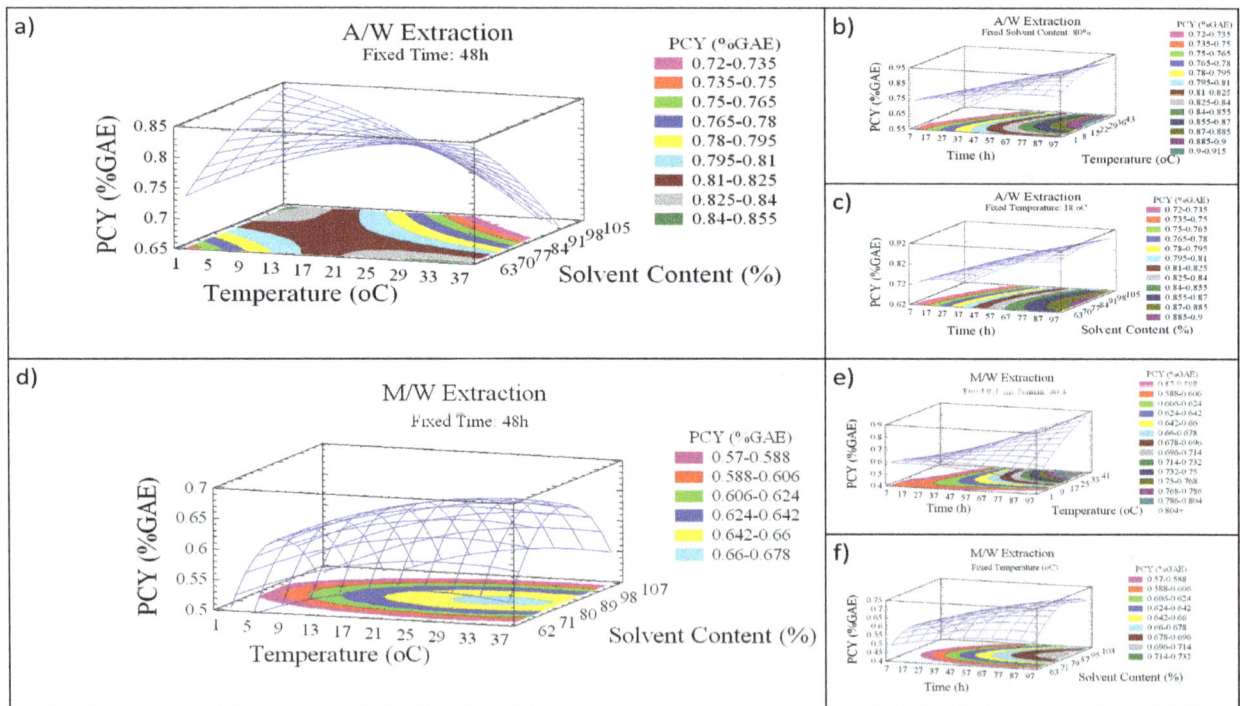

Figure 2. PCY response surface plots. Three-dimensional expressions by response surface plots of PCY, using A/W ((a), (b), and (c)) or M/W ((d), (e), and (f)) as extraction mixtures are shown. The three-dimensional graphs were plotted between two independent variables (temperature and Solvent Content; (a), and (d), time and temperature; (b), and (e), and Time and Solvent Content; (c), and (f)) while the remaining independent variable (time; (a), and (d), Solvent Content; (b), and (e), and Temperature; (c) and (f)) was kept at its zero level. The colored areas at the bottom of each graph indicate the iso-responses zones.

3. Results and Discussion

3.1. Optimal Yields and Antioxidant Activity

Response surface methodology was used to determine the parameters for optimal levels of PC and MA yields (PCY and MAY) and those for optimal antioxidant potential of the extracts (DFRIP) (Equations (1), (2) and (3), respectively). The corresponding independent variables and their levels are shown in (**Table 2**, Equation (4)).

Using RSM, **Table 1** gives the value of the three re-

Antioxidants from Syrah Grapes (Vitis vinifera L. cv. Syrah). Extraction Process through
Optimization by Response Surface Methodology

173

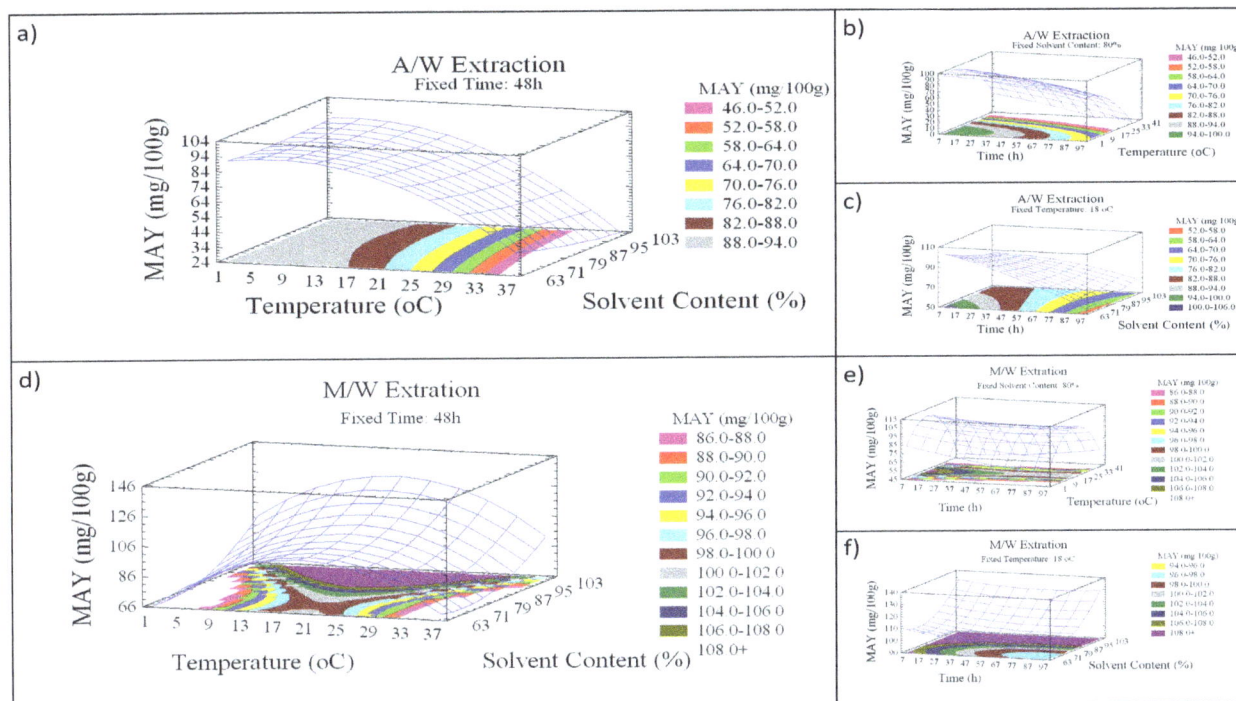

Figure 3. MAY response surface plots Three-dimensional expressions by response surface plots of MAY, using A/W ((a), (b), and (c)) or M/W ((d), (e), and (f)) as extraction mixtures are shown. The three-dimensional graphs were plotted between two independent variables (Temperature and Solvent Content; a, and d, Time and Temperature; (b), and (e), and Time and Solvent Content; (c), and (f)) while the remaining independent variable (Time; (a), and (d), Solvent Content; (b), and (e), and Temperature; (c) and (f)) was kept at its zero level. The colored areas at the bottom of each graph indicate the iso-responses zones.

Table 5. Optimum experimental conditions for maximal extraction yields and values of the corresponding responses.

| | Optimum condition | | | | | | Extraction yield | |
| | A/W as extraction mixture | | | M/W as extraction mixture | | | | |
	Time (h)	Temperature (°C)	Solvent content (%)	Time (h)	Temperature (°C)	Solvent content (%)	A/W	M/W
PCY	88.36	34.8	63.18	88.36	34.81	94.88	0.94% GAE	0.82% GAE
MAY	7.69	9.13	63.18	7.64	16.51	96.5	107.53 mg/100g	123.94 mg/100g
DFRIP	16.06	34.82	96.82	7.64	5.72	96.82	76.37%	62.36%

sponses: PCY, MAY and DFRIP in A/W and M/W mixtures. The optimized levels of the parameters were as follows: for PCY, 63.18% acetone in water, 88.36h of extraction incubation and at 34°C; for MAY, 96.5% methanol in water, 7.64 h of extraction incubation and at 16.51°C; for DFRIP, 96.82% acetone in water, 16.06 h of extraction length and at 34.82°C (**Table 5**).

In many works similar to this one, optimizations of the PC extraction parameters (extraction time [12,15-18], extraction temperature [2,19] and extraction solvent [20]) were done using different plant matrices. We looked to compare response optimums as an attempt to relate our study to previous ones, even though a complete correlation will not be possible due to differences in the starting

material, the diversity of the PC in each plant as well as the antioxidant profile of the extract and the changeability of the extraction process. Our work showed relatively high PCY of 0.94% GAE of fresh weight (grape paste), fair yields of MA: 123.94 mg/100g and our extract showed high antioxidant potential translated in a higher optimum level of DFRIP: 76.37% (**Table 5**). In literature, total PC yields were of lower levels compared to our study, as Spigno and De Faveri, (2007) [2] obtained a PCY of 0.27% GAE from powdered red grape pomace and of 0.33% GAE from red grape stems, Revilla et al. (1998) [8] obtained 0.51% GAE of PCY from fresh grapes and 0.25% GAE from fresh red grape skins, Cruz et al. (2004) [19] got 0.22% GAE of PCY from distilled grape pomace and Benvenuti et al., (2004) [21] yielded 0.88% GAE from black berries. MAY in literature were very close to ours; Revilla et al. (1998) [8] obtained 111 mg/100g from entire fresh grapes, and Fan et al. (2008) [22] had 152 mg/100g from dried sweet potato. In a previous work, antioxidant potential of extracts obtained (by classic or carbon maceration) did not exceed 50% [16].

3.2. Response Surface Model Design

The experimental values of PCY, MAY and DFRIP obtained in A/W or M/W are shown in **Table 1**.

The values for the coefficient of determination (R^2) were 46.83%, 97.1%, 29.71%, 81.51%, 64.06% and 78.11% for the experimental design of PCY, MAY and DFRIP in A/W mixture and of the same constituents in M/W mixtures, respectively. The value of R^2 for MAY (0.971) extracted by A/W mixture, is very close to 1, and indicates a high degree of correlation between the observed and predicted values. Whereas the values of R^2 for PCY and DFRIP in M/W mixture are reasonably close to 1, indicating reasonable agreement of the corresponding models with the experimental results.

A significant lack of fit ($P < 0.05$) was found in all models corresponding to PCY and MAY by A/W and M/W extractions and to DFRIP in A/W extraction condition. This shows no fit of all five models to real conditions, which means that the manipulator errors are negligible compared to the errors induced by the model (calculated from the repetitions at the field center), in contrast no significant lack of fit ($P > 0.05$) was in the model corresponding to DFRIP in M/W extraction condition. From all the above we can assume that the used model is considered as valid (**Table 4**).

3.3. Yields and Antioxidant Activity Are Affected by Parametrical Variation

Table 3 shows the significance of each parameter when using the ANOVA test for the analysis of the regression models coefficients. The effect of a parameter is considered as statistically significant when histograms cross the vertical line, translating the threshold of significance of 95%. According to **Figure 1** and **Table 3**, and in the field of variation of the process parameters, the results showed that Time had a significant linear (X_1) effect ($P < 0.05$) on PCY, MAY and DFRIP after extraction by both A/W and M/W mixtures. On another side, Temperature had a significant linear (X_2) effect on PCY, MAY and DFRIP in the presence of Acetone in the extraction mixture and on MAY in presence of A/W. Solvent content (X_3) linear effect ($P < 0.05$) was significant on MAY extracted by both A/W and M/W mixtures, on PCY extracted by M/W

and on DFRIP extracted by A/W.

The levels of independent variables for optimal extraction conditions of PC, MA and for DFRIP in A/W or M/W extraction mixtures were expressed in three dimensions using response surface graphs plotted between two independent variables while the remaining third independent variable was kept at zero level (**Figures 2-4**).

3.3.1. Total Phenolic Compounds

Response surface plots show the effect of each parameter on the responses (**Figures 2-4**). From the shape of the surface plot it can be noticed that Time affected significantly PCY, in both Acetone (A/W) (**Figures 2(b)** and **(c)**) and methanol (M/W) (**Figures 2(d)**, **(e)**, and **(f)**) extraction systems. A significant increase of PCY with the extraction length is translated into a clear steepness in the inclination of the plot ascent in **Figures 2(b)**, **(c)** and **(f)**. Thus, the extraction time parameter has a positive and significant effect on PCY. Some studies in literature were in accordance with what we found. Pekic et al. (1998) [15] noticed an increase in the yield of a group of the PC, the proanthocyanidins with the increase of extraction time to 24 h after undergoing an extraction from dried seeds of grape pomace, the same increasing effect in Spigno et al. (2007) [6] was also observed on powdered grape pomace total phenolics after 24 h of extraction process. Lapornik et al. (2005) [16] observed an increase in total phenolics with the extraction time from grape pomace obtained after classic maceration using just water or 70% ethanol as extractants, but also (and in disagreement with what we found) a decrease in total phenolics from this same grape material using 70% methanol as extractant was noticed.

Throughout literature, temperature is shown to be one of the most critical variables to be affecting the release of phenolic compounds from grape matrix [2,6,9,19], due to an increase in the coefficient of diffusion and solubility. In accordance most authors found an increase in the amount of total extracted phenols [2,13] while heating. In contrast, in the range of our study measurements a significant effect ($P < 0.05$) of temperature variation on PCY was shown in M/W extraction mixtures (**Figures 2(d)** and **(e)**), but not in A/W (**Figures 2(a)** and **(b)**). The corresponding plots (**Figures 2(a)** and **(d)**) are dome shaped, showing that at the fixed time level (48 h) a maximal value of PCY has been reached in the range of our measurements, and that an increase in the extraction temperature will increase PCY until the value of 0.85% GAE and 0.65% GAE, in acetonic (**Figure 2(a)**) and methanolic extracts (**Figure 2(d)**).

The effect of solvent content in water on PCY showed to be statistically significant ($P < 0.05$) in A/W (**Figures 1(a)** and **2(b)**) and M/W (**Figures 1(d)** and **2(e)**). In ac-

Antioxidants from Syrah Grapes (Vitis vinifera L. cv. Syrah). Extraction Process through
Optimization by Response Surface Methodology

175

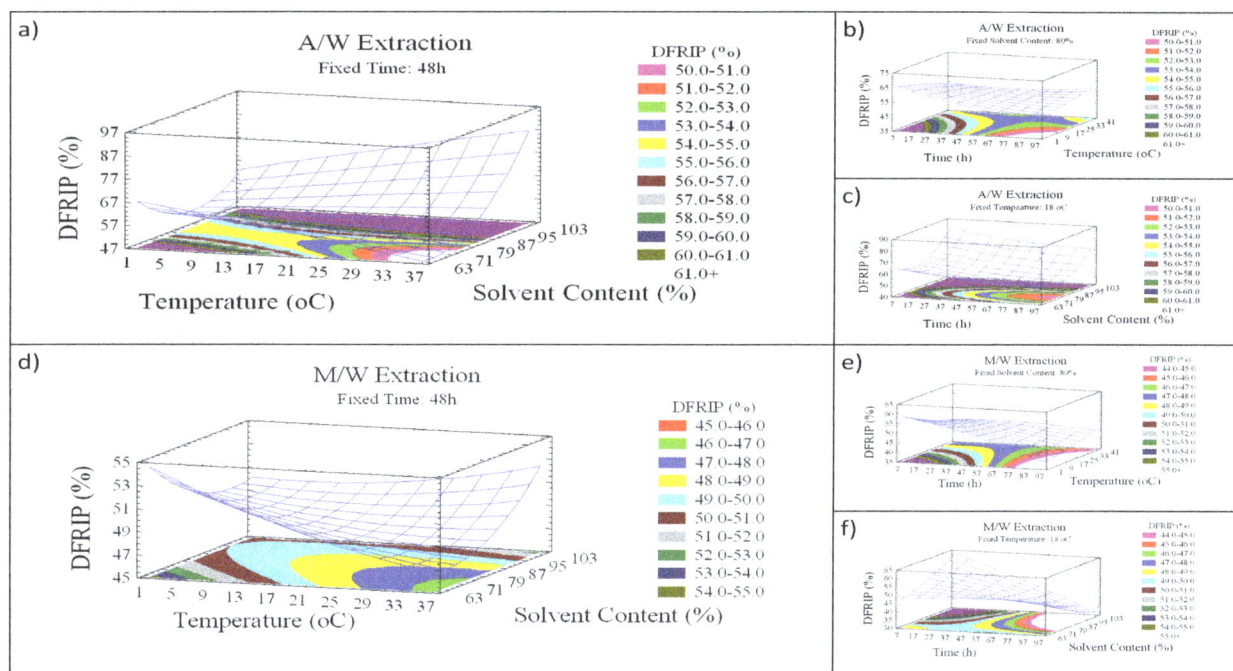

Figure 4. DFRIP response surface plots. Three-dimensional expressions by response surface plots of DFRIP, using A/W ((a), (b), and (c)) or M/W ((d), (e), and (f)) as extraction mixtures are shown. The three-dimensional graphs were plotted between two independent variables (Temperature and Solvent Content; (a), and (d), Time and Temperature; (b), and (e), and Time and Solvent Content; (c), and (f)) while the remaining independent variable (Time; a, and d, Solvent Content; (b), and (e), and Temperature; (c) and (f) was kept at its zero level. The colored areas at the bottom of each graph indicate the iso-responses zones.

cordance with literature, our results showed that the optimal PCY were obtained using A/W and not M/W as the extraction mixture.

Literature studies on extractions from grape materials showed that aqueous acetone was a better mixture for extracting PC (which had an overall unpolar character) than aqueous methanol [23].

3.3.2. Monomeric Anthocyanins

Figures 3(b), (c), (e) and (f) show plots with a clear steepness in the inclination of their profile, which can be translated into a significant decrease of MAY with the extraction time. Thus this parameter had a negative significant effect on MAY.

Lapornik *et al.* (2005) [16] showed a decrease in total anthocyanins with water extractions after a long extraction time similarly to what appeared in our study, but an increase in anthocyanins was noted using 70% ethanol and methanol as extractants.

In accordance to what was said previously, we noticed a statistical ($P < 0.05$) and negative influence of temperature on MA yields in both A/W (**Figures 3(a)** and **(b)**) and M/W (**Figures 3(d)** and **(e)**) mixtures; in fact, the corresponding plots showed a significant tendency in MAY towards higher temperatures ($35°C$ - $37°C$). Our samples subjected to extraction temperature between $1°C$

and $17°C$ showed higher MAY than samples extracted at temperatures around $35°C$. This could be explained by conformational changes or degradation of monomeric anthocyanins at higher temperatures or by color change and co-pigmentation, which is an interaction and coupling of the anthocyanins with other components making them trapped and undetectable by usual tests.

We chose in our study to consider the temperature margins in which were extracted, at the same time, both grape compound groups (PC and MA) without subjecting them to degradation. In the same context, literature showed temperature to be one of the major degradation factors of the anthocyanins along with oxygen and photo degradation [24]. In addition to this and according to Vatai and Skerget, (2009) [25] a relatively low extraction temperatures ($20°C$) were more suitable than high temperature ($60°C$) for extracting higher MA yields from Cabernet and Merlot grapes.

Zhou and Yu [23], showed that methanol was better than acetone or water for MAY extraction [8,16]. In accordance with this observation, our results showed that the optimal MAY was obtained using M/W and not A/W.

MAY was affected significantly ($P < 0.05$) by both the acetone and methanol contents in water (**Figures 1(b)** and **(e), Figures 3(a)** and **(d)**). According to literature, MA are better extracted using more polar solvents like methanol than by other organic solvents. It is also well described that methanol and alcoholic solvents are better extractants than water for anthocyanins. Our study showed a negative effect of acetone content in the A/W mixture on the MA yields (**Figures 3(a)**, and **(c)**), meaning that MA are more affine to the water part of the mixture, in accordance to the previous assumptions. The same comment could be given for the positive effect of methanol content on MAY, in the M/W mixture (**Figures 3(d)** and **(f)**). We can conclude that the affinity of solvents to MAY is as follows: methanol is the best, followed by water then acetone.

3.3.3. Antioxidant Activity

Figures 4(b), **(c)**, **(e)** and **(f)** show plots with a clear steepness in the inclination of their profile, which can be translated into a significant decrease of DFRIP with the extraction time. Thus this parameter had a negative significant effect on the antioxidant activity.

In addition to that and in contrast to our study other authors described an increase in the antioxidant potential of PC extracts [16]. In those works these increases were described after undergoing extraction not longer than 2 to 3 h. The short times of extraction might explain this behavior. The very long extraction time (up to 88 h) used in our study, might be responsible for the loss in the antioxidant potential translated by the decrease in DFRIP. It is well known that at this level we can observe, in function of time, a competition between two phenomena; extraction v/s oxidation.

Additionally, temperature showed a significant and negative effect on the antioxidant potential (DFRIP) of the extract in the presence of methanol as extraction mixture (M/W) (**Figures 4(d)** and **(e)**). This loss of antioxidant potential could be explained by a degradation process of the phenolics taking place in the presence of methanol at higher extraction temperatures. In literature and in disagreement with what we found, some studies showed an increase in the antioxidant potential of the extract with higher extraction temperatures [13], while others showed the exact decreasing effect [16]. This dichotomy in the results could be explained by a double effect that temperature could produce on PC; first it could promote a better extraction of antioxidants from the matrix, and second could cause degradation of those antioxidants, decreasing by this the overall antiradical potential.

Some authors show that antioxidant potential of the extract was better preserved in the presence of unpolar solvents like acetone [26,27]. Higher values were noticed as compared to those obtained by Rajha *et al.* (2013) [28] for phenolic compounds water extracts. In accordance with this, our results showed that DFRIP optimal level was obtained with Acetone (**Table 5**).

3.4. Simultaneous Response Optimization

In the previous section of this work we designated the parameters in order to extract optimum yields of PC, MA and the best DFRIP of the extract. In this part we show simultaneously, by the desirability function, how the three responses could be affected by the parameters (**Figure 5**). It can be seen that PCY, MAY and DFRIP concentric circles, converge mostly towards different regions in the superposition plots. Opposite localization of the optimum PCY, MAY and DFRIP are observed on most of the plots of **Figure 5**. This emphasizes that PC need long extraction time to reach a maximum yield while on the contrary, MA are extracted in an optimal way in the first hours of the extraction process. Plots **(a)**, **(d)**, **(c)**, and **(f)** also show how PCY are maximized at a high range extraction temperature, while MAY and DFRIP showed best values at low extraction temperatures. High methanol content in the solvent mixture (near 100%) gave the best values for all three responses (**Figures 5(d)-(f)**). This was the case as well for DFRIP in A/W but low Acetone percentages in water (near 63%) were better for PCY and MAY (**Figures 5(a)-(c)**). This divergence in optimal parameters for each response shows that PCY, MAY and DFRIP cannot be maximized in the same extract. Nevertheless, we can guide the extraction process to obtain the best yield or the best antioxidant potential as the plots show (**Figure 5**, green marks). Thus, the parameters will be compromised between PCY, MAY and DFRIP. In some specific cases, these parameters could be favored towards PCY, MAY or even DFRIP depending on the final application of the extracted compounds.

4. Conclusions

Hereby, we attempted to explore the field of antioxidants and their extraction from grapes. Keeping in mind the variability of the techniques and the grape matrices due to soil and season climate, we proposed here a model that could be applied for industrial purposes. Our major findings lead us to suggest that extracting or optimizing the extraction of phenolic compounds and antioxidants like anthocyanins from Sy grapes can be done easily, without heavy or expensive machinery and could be environmentally friendly. We propose as well that multiple response optimizations could be used to define the optimum area which can lead to choose the convenient ratio between PCY and MAY in addition to favoring between yields of antioxidants and the antiradical potential of the extract. We showed that aqueous Acetone is better than Methanol and/or water in extracting total PC and in preserving the antioxidant potential of the extract, but Methanol seemed to be more suitable in the extraction of MA followed by water than Acetone. In parallel, the ex-

Antioxidants from Syrah Grapes (Vitis vinifera L. cv. Syrah). Extraction Process through
Optimization by Response Surface Methodology

177

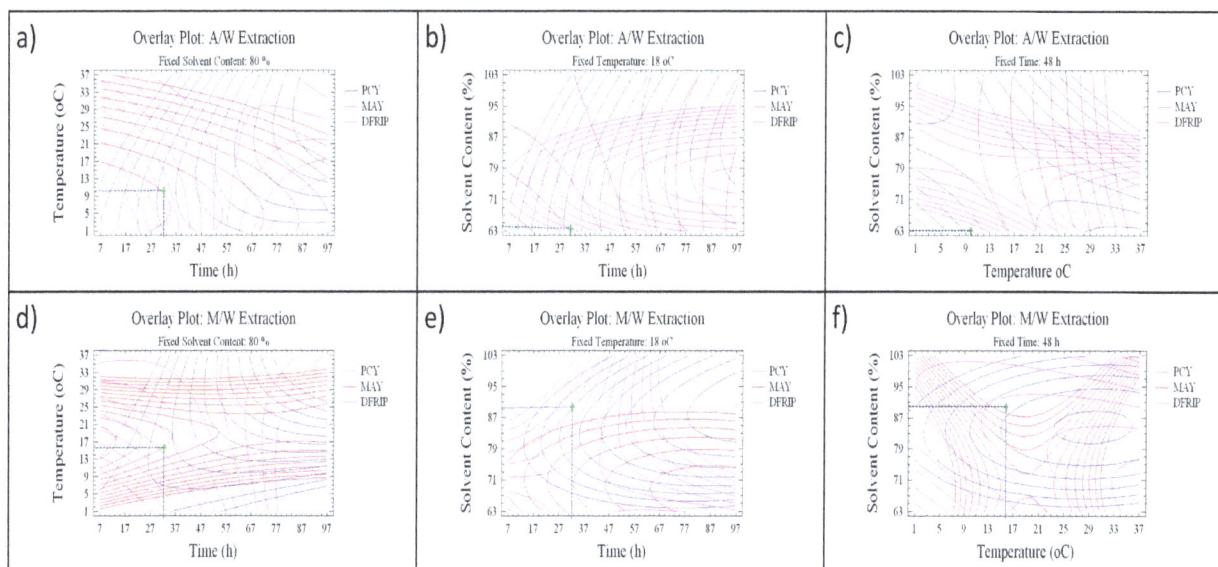

Figure 5. Desirability analysis. Superposition plots, showing the best experimental parameters (Time, Temperature, and Solvent Content) that maximizes PCY, MAY and DFRIP at the same time are shown. In A/W ((a), (b), and (c)) and M/W ((d), (e), and (f)) extraction mixtures, the contours graphs were plotted between two independent variables (Temperature and time; (a), and (d), Solvent Content and Time; (b), and (e), and Solvent Content and Temperature; (c), and (f)) while the remaining independent variable (Solvent Content; (a), and (d), Temperature; (b), and (e), and Time; (c), and (f)) was at its zero level. The green mark shows the "middle way" parameters that compromise between PCY and MAY.

traction time effect showed to be significant on the grape phenolic yields; MA were fast to extract from Sy grapes, and 8 h as extraction time gave higher yields than longer time, while extraction yielded the maximum of PC after 88 h. Longer extraction times contributed in decreasing the antioxidant properties of the extract. We demonstrated as well that the extraction temperature had a significantly negative effect on the MA extraction and on the antioxidant potential of the extract and low extraction temperature ($10°C$) yielded more MA and gave more antioxidant potential to the extract than high extraction temperature ($35°C$).

Finally, complementary studies are needed in order to solve the problem concerning preservation of the antioxidant capacity of a grape extracts, and to be able to fully understand the effect of parameters such as extraction time, temperature and solvent content on the extraction process. Future work should take into consideration larger intervals of the extraction parameters, in order to be able to enhance the quantity and the quality of the extracted compounds. All the combined data obtained through an optimal extraction strategy could lead to the production of pure natural antioxidant molecules. These could be used for the quality improvement of several industrial products such as cosmetics, pharmaceutics and agrofood.

5. Acknowledgements

This project was funded by the Research Council of Saint-Joseph University-Lebanon (Project FS20). The authors are grateful to Joseph Yaghi and Nada El Darra for technical assistance.

REFERENCES

[1] Y. El Hajj, N. Louka, C. Nguyen and R. Maroun, "Low Cost Process for Phenolic Compounds Extraction from Cabernet Sauvignon Grapes (*Vitis vinifera* L. cv. Cabernet Sauvignon). Optimization by Response Surface Methodology," *Food and Nutrition Sciences*, Vol. 3, No. 1, 2012, pp. 89-103.

[2] G. Spigno and D. M. De Faveri, "Antioxidants from Grape Stalks and Marc: Influence of Extraction Procedure on Yield, Purity and Antioxidant Power of the Extracts," *Journal of Food Engineering*, Vol. 78, No. 3, 2007, pp. 793-801.

[3] C. Flanzy, "Oenologie: Fondements Scientifiques et Technologiques," Tec & doc-Lavoisier, 1998.

[4] M. F. Molina, I. Sanchez-Reus, I. Iglesias and J. Benedi, "Quercetin, a Flavonoid Antioxidant, Prevents and Protects against Ethanol-Induced Oxidative Stress in Mouse Liver," *Biological & Pharmaceutical Bulletin*, Vol. 26, No. 10, 2003, pp. 1398-1402.

[5] C. Negro, L. Tommasi and A. Miceli, "Phenolic Compounds and Antioxidant Activity from Red Grape Marc Extracts," *Bioresource Technology*, Vol. 87, No. 1, 2003, pp. 41-44.

[6] G. Spigno, L. Tramelli and D. M. De Faveri, "Effects of Extraction Time, Temperature and Solvent on Concentration and Antioxidant Activity of Grape Marc Phenolics," *Journal of Food Engineering*, Vol. 8, No. 1, 2007, pp. 200-208.

[7] R. Liorach, F. A. Tomas-Barberan and F. Ferreres, "Lettuce and Chicory Byproducts as a Source of Antioxidant Phenolic Extracts," *Journal of Agricultural and Food Chemistry*, Vol. 52, No. 16, 2004, pp. 5109-5116.

[8] E. Revilla, J. M. Ryan and G. Martin-Ortega, "Comparison of Several Procedures Used for the Extraction of Anthocyanins from Red Grapes," *Journal of Agricultural and Food Chemistry*, Vol. 46, No. 11, 1998, pp. 4592-4597.

[9] G. K. Jayaprakasha, R. P. Singh and K. K. Sakariah, "Antioxidant Activity of Grape Seed (*Vitis vinifera*) Extracts on Peroxidation Models *in Vitro*," *Food Chemistry*, Vol. 73, No. 3, 2001, pp. 285-290.

[10] N. G. Baydar, G. Özkan and O. Sagdiç, "Total Phenolic Contents and Antibacterial Activities of Grape (*Vitis vinifera* L.) Extracts," *Food Control*, Vol. 15, No. 5, 2004, pp. 335-339.

[11] Y. Yilmaz and R. T. Toledo, "Oxygen Radical Absorbance Capacities of Grape/Wine Industry Byproducts and Effect of Solvent Type on Extraction of Grape Seed Polyphenols," *Journal of Food Composition and Analysis*, Vol. 19, No. 1, 2006, pp. 41-48.

[12] M. Pinelo, P. Del Fabbro, L. Manzocco, M. J. Nuñez and M. C. Vicoli, "Optimization of Continuous Phenol Extraction from *Vitis vinifera* Byproducts," *Food Chemistry*, Vol. 92, No. 1, 2005, pp. 109-117.

[13] M. Pinelo, M. Rubilar, M. Jerez, J. Sineiro and M. J. Nuñez, "Effect of Solvent, Temperature, and Solvent-to-Solid Ratio on the Total Phenolic Content and Antiradical Activity of Extracts from Different Components of Grape Pomace," *Journal of Agricultural and Food Chemistry*, Vol. 53, No. 6, 2005, pp. 2111-2117.

[14] F. Bonilla, M. Mayen, J. Merida and M. Medina, "Extraction of Phenolic Compounds from Red Grape Marc for Use as Food Lipid Antioxidants," *Food Chemistry*, Vol. 66, No. 2, 1999, pp. 209-215.

[15] B. Pekic, V. Kovac, E. Alonso and E. Revilla, "Study of the Extraction of Proanthocyanidins from Grape Seeds," *Food Chemistry*, Vol. 61, No.1-2, 1998, pp. 201-206.

[16] B. Lapornik, M. Prošek and A. G. Wondra, "Comparison of Extracts Prepared from Plant By-Products Using Different Solvents and Extraction Time," *Journal of Food Engineering*, Vol. 71, No. 2, 2005, pp. 214-222.

[17] M. M. Giusti and R. E. Wrolstad, "Characterization and Measurement of Anthocyanins by UV-Visible Spectroscopy," *Current Protocols in Food Analytical Chemistry*, John Wiley & Sons, Inc. 2001.

[18] E. Alonso, M. Bourzeix and E. Revilla, "Suitability of Water/Ethanol Mixtures for the Extraction of Catechins and Proanthocyanidins from *Vitis vinifera* Seeds Contained in a Winery By-Product," *Seed Science and Technology*, Vol. 19, No. 3, 1991, pp. 542-552.

[19] J. M. Cruz, H. Dominguez and J. C. Parajo, "Assessment of the Production of Antioxidants from Winemaking Waste Solids," *Journal of Agricultural and Food Chemistry*, Vol. 52, No. 18, 2004, pp. 5612-5620.

[20] N. G. Baydar, G. Özkan and O. Sagdiç, "Total Phenolic Contents and Antibacterial Activities of Grape (*Vitis vinifera* L.) Extracts," *Food Control*, Vol. 15, No. 5, 2004, pp. 335-339.

[21] S. Benvenuti, F. Pellati, M. Melegari and D. Bertelli, "Polyphenols, Anthocyanins, Ascorbic Acid, and Radical Scavenging Activity of Rubus, Ribes, and Aronia," *Journal of Food Science*, Vol. 69, No. 3, 2004, pp. 164-169.

[22] G. Fan, Y. Han, Z. Gu and D. Chen, "Optimizing Conditions for Anthocyanins Extraction from Purple Sweet Potato Using Response Surface Methodology (RSM)," *LWT-Food Science and Technology*, Vol. 41, No. 1, 2008, pp. 155-160.

[23] K. Zhou and L. Yu, "Effects of Extraction Solvent on Wheat Bran Antioxidant Activity Estimation," *Lebensmittel-Wissenschaft und-Technologie*, Vol. 37, No. 7, 2004, pp. 717-721.

[24] R. L. Jackman, R. Y. Yada, M. A. Tung and R. A. Speers, "Anthocyaninsas Food Colorants—A Review," *Journal of Food Biochemistry*, Vol. 11, No. 3, 1987, pp. 201-247.

[25] T. Vatai and M. Skerget, "Extraction of Phenolic Compounds from Elder Berry and Different Grape Marc Varieties Using Organic Solvents and/or Supercritical Carbon Dioxide," *Journal of Food Engineering*, Vol. 90, No. 2, 2009, pp. 246-254.

[26] J. Tabart, C. Kevers, A. Cipel, J. Pincemail, J. O. DEfraigne and J. Dommes, "Optimisation of Extraction of Phenolics and Antioxidants from Black Currant Leaves and Buds and of Stability during Storage," *Food Chemistry*, Vol. 105, No. 3, 2007, pp. 1268-1275.

[27] N. El Darra, J. Tannous, P. BouMouncef, J. Palge, J. Yaghi, E. Vorobiev, N. Louka and R. G. Maroun, "A Comparative Study on Antiradical and Antimicrobial Properties of Red Grapes Extracts Obtained from Different *Vitis vinifera* Varieties," *Food and Nutrition Sciences*, Vol. 3, No. 10, 2012, pp. 1420-1432.

[28] H. N. Rajha, N. El Darra, E. Vorobiev, N. Louka and R. G. Maroun, "An Environment Friendly, Low-Cost Ex-

Antioxidants from Syrah Grapes (Vitis vinifera L. cv. Syrah). Extraction Process through
Optimization by Response Surface Methodology

179

traction Process of Phenolic Compounds from Grape Byproducts. Optimization by Multi-Response Surface Methodology," *Food and Nutrition Sciences*, Vol. 4, No. 6,
2013, pp. 650-659.

22

Anti-Ulcerogenic Activity of the Pomegranate Peel (*Punica granatum*) Methanol Extract

Ghazaleh Moghaddam[1], Mohammad Sharifzadeh[2], Gholamreza Hassanzadeh[3], Mahnaz Khanavi[4,5], Mannan Hajimahmoodi[1,5*]

[1]Department of Drug and Food Control, Faculty of Pharmacy, Tehran University of Medical Sciences, Tehran, Iran; [2]Department of Pharmacology and Toxicology, Pharmaceutical Research Center, Faculty of Pharmacy, Tehran University of Medical Sciences, Tehran, Iran; [3]Department of Anatomy, School of Medicine, Tehran University of Medical Sciences, Tehran, Iran; [4]Department of Pharmacognosy and Traditional Iranian Pharmacy Research Center, Faculty of Pharmacy, Tehran University of Medical Sciences, Tehran, Iran; [5]Department of Traditional Pharmacy, Faculty of Traditional Medicine, Tehran University of Medical Sciences, Tehran, Iran.

ABSTRACT

Pomegranate (*Punica granatum* L.) belongs to genera *Punica* and family Punicaceae. It is a herbal preparation that has been suggested as useful in the treatment of gastrointestinal disorders. However, to our knowledge, no study has been conducted to evaluate this therapeutic property. In the present study the antiulcerogenic effects of pomegranate peel methanol extract, was tested on male Wistar albino rats. Oral pretreatment with peel extracts (25, 50 and 100 mg/kg) for 15 days protected the gastric mucosa against the damage induced by indomethacin (50 mg/kg). The incidence of ulceration in the control group was 100%. The best results were found in a dosage of 50 mg/kg in sour summer cultivar which inhibited the peptic ulcerin comparison with indomethacin induced gastric ulcer group. Lowest ulcer index (5.4 ± 0.55), an apparent decrease in the infiltration of polymorphonuclear leukocytes and hemorrhage were observed after administration of sour summer extracts (50 mg/kg). In conclusion present study showed that pomegranate peel extract, especially sour summer, has curative potential as an antiulcer, possibly via its high antioxidant activity. These results from pomegranate peel extract can provide an extra income and may contribute about good nutritional values of this product.

Keywords: Anti-Inflammatory; Indomethacine; Peptic Ulcer Disease

1. Introduction

A peptic ulcer, also known as PUD or peptic ulcer disease, is the most common ulcer of an area of the gastrointestinal tract. Peptic ulcers occur worldwide and gastric cancer is the second commonest cause of death from malignant disease [1]. It is defined as mucosal erosions equal to or greater than 0.5 cm. Almost all ulcers are associated with Helicobacter pylori, a spiral-shaped bacterium that lives in the acidic environment of the stomach. They are caused by many factors such as drugs, stress or alcohol, due to an imbalance between offensive acid-pepsin secretion and defensive mucosal factors like mucin secretion and cell shedding [2].

A number of drugs including prostaglandins analogs, histamine receptor antagonists, proton pump inhibitors, and cytoprotective agents are available for the treatment of peptic ulcer. By the way various side effects of these products such as, hepatotoxicity and anaphylaxis, are not totally managed yet, so medicinal plant as an alternative treatment always has been the focus of many studies [3,4].

A long-term use of non-steroidal anti-inflammatory drugs (NSAIDs) causing inflammation of the gastric mucosa, gastrointestinal associated with a range of toxicity, and may finally cause ulceration, bleeding and changes with a very high morbidity and mortality [5-7].

Currently due to several adverse effects of chemical drugs, herbal medicines are generally used in treatment. Several natural products have been reported to poses anti-ulcerogenic activity by virtue of their predominant effects on mucosal defensive factors including apple bananas, papeeta, and brindleberry [8,9].

Pomegranate (*Punica granatum* L.) is a well-known

---

*Corresponding author.

table fruit of tropical and subtropical regions of the world. Some botanists place it in the family Lythraceae, of the peculiar type of fruit, called as balausta, most authorities make it the only genus in the family Punicaceae. It belongs to genera *Punica* and family Punicaceae [10]. It is mainly grown in Mediterranean regions and is one of the major cultivated productions of Iran and also in Afghanistan [11,12].

The goal of many studies has been to identify the therapeutic constituents of pomegranate. Commonly found in many plants, ellagic acid exhibits powerful anticarcinogenic and antioxidant properties, propelling it to the forefront of pomegranate research. The pericarp part of pomegranate showed more antioxidant content including phenolic punicalagins; gallic acid and other fatty acids; catechin, quercetin, rutin; flavones, flavonones; anthocyanidins [13-15]. In addition to its ancient historical uses, pomegranate is used in several systems of medicine for a variety of ailments. It used as an antiparasitic agent, a "blood tonic" in addition to heal aphthae, diarrhea, and ulcers. Pomegranate also serves as a remedy for diabetes [15,16].

Hence the aim of the current study is to investigate whether pomegranate peel extract consumption has curative effect toward gastric ulcers and may be helpful in the prevention of indomethacin-induced ulcerogenesis.

2. Material and Method

2.1. Plant Material

There are many varieties of pomegranate in Iran, from very sour to very sweet in taste and from white, yellow, pink, red, purple and even black in color. Three fresh pomegranate cultivars were harvested randomly in September 2010 from different mature trees (14-year-old). The average temperature, the amount of rainfall, relative humidity, and the soil pH in growing season were 28.65°C, 20 mm, 26% and 7.21, respectively. The trees were spaced 6 and 3 m between and along the rows, respectively. Trees were grown under traditional irrigation and routine cultural practices suitable for commercial fruit production. All cultivars were grown under the same geographical conditions and with the same applied agronomic practices. According to the list of Iranian pomegranate cultivars studied by Noormohammadi et al. [17] which studied on the Iranian pomegranate genotypes in two garden of Saveh germ plasm, the North white peel (Poost-Sefid-Shirin), Sour summer (TabestaniTorsh) and Black peel (PoostSiyahTorsh) are almost cultivated in Saveh (Markazi), and Ardakan (Yazd) respectively. Three cultivars of pomegranate including 1) North white peel, 2) Sour summer and 3) Black peel, were donated from Saveh Agricultural Investigation Center. Pomegranate fruits (three numbers of each cultivar) were collected and

washed three times with distilled water.

The black pomegranates are almost rare and more expensive than others, since it is a natural medicine for various diseases. The North white peel, are the accessible cultivar in Iran and according to the Tarighi et al. [18], the north white peel mean length, width and thickness are 82.62 mm, 83.45 mm and 81.31 mm, respectively. Another prevalent pomegranate cultivar in Iran is Sour summer. The weight of each 100 seeds is 4.12 g and the ratio of its seed to its pulp is 55.

To prepare pomegranate extract, fresh fruits were peeled and 30 g of peels were weighted and extracted separately for 4 h by Soxhlet apparatus with methanol 80%. The extract of each cultivar were mixed and the extracts were concentrated under control reduced pressure at 40°C to obtain the methanolic extracts.

2.2. Chemicals

All chemicals used in the experiments were of analytical grade. All reagents and solvents were purchased from Merck (Darmstadt, Germany) and Sigma (St. Louis, MO).

For laboratory experimentation, indomethacin, and cimetidine were obtained from Daroupakhsh Pharmaceutical Company (Tehran, Iran).

2.3. Pharmacological Experiments

2.3.1. Animals
Male Wistar rats weighing 175 - 220 g (Pasteur Institute, Tehran, Iran) were used in the study. The animals were fed under normal conditions (22°C) in 9 separate groups consisting of 5 rats. Animal experiments were performed in accordance with national guidelines for the use and care of laboratory animals and approved by the local animal care committee of Tehran University of Medical Sciences.

2.3.2. Preparation of Test Samples for Bioassay
Test samples were administrated in traperitoneally (i.p.) after dissolving in saline (NaCl 0.9%). The control group animals received the same experimental handling as those of the test groups except that the extract treatment was replaced by administration of appropriate volumes of the dosing vehicle. The histamine-receptor type-2 (H2) blocker, cimetidine (100 mg/kg, i.p.), was used as a reference compound. Peptic ulcer induced by indomethacin. Test samples were given 1h before the i.p. administration of indomethacin [a non-steroidal anti-inflammatory drug] suspended in saline (50 mg/kg, 2 ml) to a group of five rats. Oral pretreatment with peel extracts (25, 50 and 100 mg/kg) for 15 days protected the gastric mucosa against the damage induced by indomethacin (50 mg/kg). Four hours later, the stomachs were removed and inflated with

10 ml of formalin solution and immersed in the same solution to fix the outer layer of stomach. Each stomach was then opened along the greater curvature, rinsed with tap water to remove gastric content and blood clots and examined under the dissecting microscope to assess the formation of ulcers [19].

The ulcer index (UI) and inhibition percentage was calculated according to previous reported [19].

2.4. Statistical Analyses

All data were analyzed by one-way ANOVA using SPSS version 16 (SPSS Inc) software. Differences among groups were obtained using the LSD option and Duncan test, and significance was declared at $p < 0.05$. The results were expressed as percentage and as mean ± standard deviation (mean ± SD).

3. Result and Discussion

Nowadays gastric ulcer is one of the most important concerns as a result of many factors especially widespread using of NSAIDs. Because of poorly understanding the pathophysiology of this disease [20], studies investigating new active compounds are needed. As well, various pharmaceutical products currently used for treatment of gastric ulcers are not completely efficient and cause many adverse side effects [3]. Consequently, it is necessary to develop more effective agents that are also less toxic, with medicinal plants being an attractive source for the development of new drugs because of their wide array of active ingredients [21].

In this study we used the model of indomethacin induced ulcer. As a matter of fact indomethacin is a NSAID and this class of drugs are widely used in clinical practice due to their efficacy and various therapeutic effects, on the other hand acute gastrointestinal lesions are the most serious and frequent side effects of NSAIDs, making them the most common cause of gastro duodenal ulcers in Western countries [22,23]. In present study indomethacin induced ulcerations in 100% of the rats. Rats in the control group suffered from very severe lesions, as shown in representative dissected stomach sections (**Figure 1**). The body weight of the rats was a homogenous parameter. There were no significant differences between rat groups.

Our data shows the anti-ulcer activity of methanol extract of pomegranate peel in experimentally indomethacin induced gastric ulcers in rats. The results of the initial trials carried out with different extracts (25, 50, 100 mg/kg) are presented in **Table 1**. Microscope views of dissected stomachs representing antiulcer effects of sour summer and in black peel extracts compared to indomethacine.

Statistical analysis showed a significant difference between the ulcer indexes between the cultivar (*p*-value < 0.05). From the **Table 1**, it is evident that 50 mg/kg dosage of sour summer cultivar markedly inhibited the peptic ulceras compared with indomethacin induced gastric ulcer. In comparison with positive control, mentioned dosage of sour summer inhibited indomethacin-induced

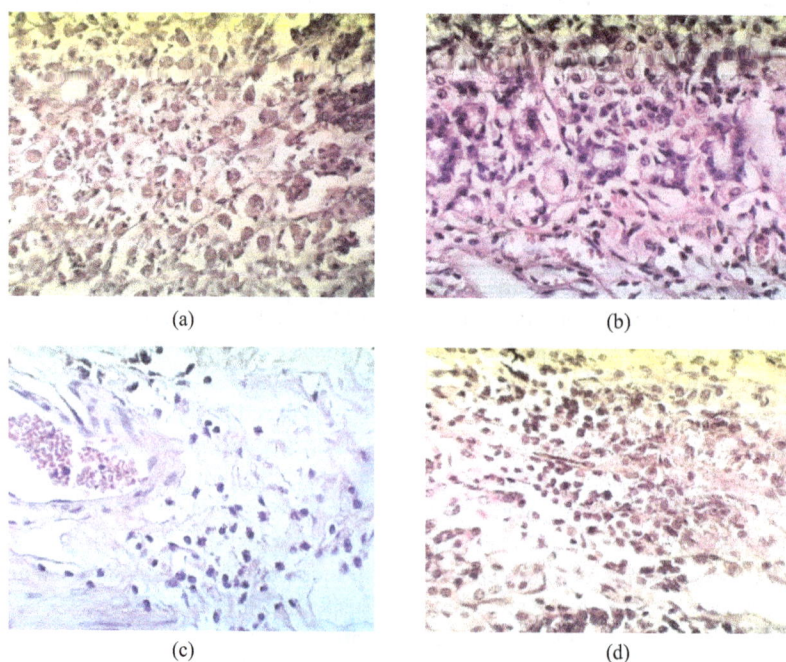

(a)

(b)

(c)

(d)

Figure 1. Microscope views of dissected stomachs representing antiulcer effects of pomegranate peel methanol extracts of (a) sour summer (50 mg/kg); (b) North white peel (25 mg/kg); (c) Indomethacin (50 mg/kg); and (d) Cimetidine (100 mg/kg). All photomicrographs hematoxylin and eosin stained 400×.

Table 1. Anti-ulcer effects of three different pomegranates peel methanolic extractions in rat.

Treatment (mg/Kg)	Ulcer Index (Mean ± SD)	Ulcer Inhibition (%)
Control	63.57 ± 3.2	-
Cimetidine—100	12.07 ± 1.04	81.01
1*—25	61.82 ± 1.62	2.75
1—50	31.14 ± 2.84	51.01
1—100	27.11 ± 1.71	57.35
2**—25	28.93 ± 0.98	54.49
2—50	5.40 ± 0.55	91.50
2—100	10.26 ± 2.97	52.39
3***—25	14.63 ± 1.84	76.98
3—50	8.20 ± 1.03	87.10
3—100	21.36 ± 3.29	19.20

1*: North white peel, 2**: Sour summer and 3***: Black peel.

peptic ulcer more potently than cimetidine. There was an apparent decrease in the infiltration of polymorphonuclear leukocytes and hemorrhage after administration of sour summer extracts (50 mg/kg) (**Figure 1**).

Using the sour summer and black peel cultivar with the dosage of 100 mg/kg decreased the level of ulcer index significantly ($p < 0.05$). Furthermore oral pretreatment with north white peel extract (25 mg/kg), significantly ($p < 0.05$) prevented the adverse changes and maintained the rats at near indomethacin induced status, while the group pretreated by the dosage of 50 and 100 mg/kg of same cultivar extract had a marginal with a little increase in ulcer inhibition. The submucosa was edematous and there was visible infiltration of polymorphonuclear leukocytes (**Figure 1**).

There are no related reports which describe antiulcer effect of pomegranate peel but in other studies fruit parts are assessed. Although the exact mechanism of the anti-ulcer activities of pomegranate peel has not been clearly delineated, it contains some active constituents which ulcer protective properties have been identified based on their antioxidant activity. Pomegranate is a rich source of polyphenols. It contains antioxidants like soluble polyphenols, tannins, and anthocyanins which possessing anti-atherosclerotic properties. The *in vitro* evaluation of the pomegranate extracts and some selected medicinal plants on H pylori activity demonstrate that the extract of pomegranate has remarkable anti-*H. pylori* function [24].

Alam *et al*. [25] claimed that in pylorus-ligated rats the ulcer lesion index, gastric volume, and total acidity significantly reduced by oral administration of an aqueous methanol extract of pomegranate fruit. It prevented ulceration by increasing the pH and mucus secretion in pylorus-ligated rats. Lai *et al*. [26] observed the antiulcer

effects of pomegranate tannins in animal models. Pomegranate tannins play a protective role against gastric ulcer. It's antiulcer effect is related to the increasing secretion of adherent mucus and free mucus from the stomach wall. This may inhibit generation of oxygen-derived free radicals, decrease the consumption of glutathione peroxidase and superoxide dismutase, and maintain the content of nitric oxide at a normal level [26].

In another research, the inhibition of gastric mucosal injury was evaluated by Ajaikumar *et al*. [27] Administration of 70% methanol extract of *Punica granatum* fruit rind revealed the gastro-protective activity of the extract through antioxidant mechanism. The *in vivo* antioxidant levels were increased and found in the normal values in treated groups of animals. All the histopathological examination of the stomach of the ulcerated animals which showed severe erosion of gastric mucosa, sub-mucosal edema and neutrophil infiltration were found to be normal in treated groups.

Based on Shams Ardekal *et al*. [12] and Sadeghi *et al*. [14] study, the antioxidant activity of sour summer as a suitable source for extraction and purification of phenolic and flavonoid compound was higher than north white peel and black peel. The results of present study which is in line with previous data, show that pomegranate peel extract of pomegranate specially sour summer possess good potential as an antiulcer agent too. Additionally, no adverse effects have been reported on consuming pomegranate and its constituents, since time immemorial, animal studies have failed to report any toxicities at doses conventionally used in the conventional system of medicine [28].

4. Conclusion

It is worthy to note that the antioxidant capacity of pomegranate peel extract is 10 times higher than the pulp extract. Also a large quantity of pomegranate peel could be easily collected from the pomegranate processing industries or from the waste products originating. This suggests that the most extracts of pomegranate peel are both anti-inflammatory and anti-ulcerogenic. Taken together, the results can provide an extra income and may contribute to have good nutritional values of this product.

5. Acknowledgments

This work was supported by the grant (No: 88-01-33-8530) from the research council of Tehran University of Medical Sciences, Tehran, Iran.

REFERENCES

[1]   R. L. Zhang, W. D. Luo, T. N. Bi and S. K. Zhou, "Evaluation of Antioxidant and Immunity-Enhancing Activi-

ties of Sargassumpallidum Aqueous Extract in Gastric Cancer Rats," *Molecules*, Vol. 17, No. 7, 2012, pp. 8419-8429.

[2] M. Shakeerabanu, K. Sujatha, C. Praveen-Rajneesh and A. Manimaran, "The Defensive Effect of Quercetin on Indomethacin Induced Gastric Damage in Rats," *Advances in Biological Research*, Vol. 5, No. 1, 2011, pp. 64-70.

[3] F. K. Chan and W. K. Leung, "Peptic-Ulcer Disease," *Lancet*, Vol. 21, No. 360, 2002, pp. 933-941.

[4] L. K. Kohn, C. L. Queiroga, M. C. Martini, L. E. Barata, P. S. Porto, L. Souza, *et al.*, "*In Vitro* Antiviral Activity of Brazilian Plants (*Maytenus ilicifolia* and *Aniba rosaeodora*) against Bovine Herpesvirus Type 5 and Avian Metapneumovirus," *Pharmaceutical Biology*, Vol. 50, No. 10, 2012, pp. 1269-1275.

[5] M. Toborek, A. Malecki, R. Garrido, M. P. Mattson, B. Hennig and B. Young, "Arachidonicacidinduced Oxidative Injury to Cultured Spinal Cord Neurons," *Journal of Neurochemistry*, Vol. 73, No. 2, 1999, pp. 684-692.

[6] S. Somasundaram, G. Sigthorsson, R. J. Simpson, J. Watts, M. Jacob and I. A. Tavares, *et al.*, "Uncoupling of Intestinal Mitochondrial Oxidative Phosphorylation and Inhibition of Cyclooxygenases Are Required for the Development of NSAID Enteropathy in the Rat," *Alimentary Pharmacology & Therapeutics*, Vol. 14, No. 5, 2000, pp. 639-650.

[7] C. J. Hawkey, "Non-Steroidal Anti-Inflammatory Drugs and Peptic Ulcers," British Medical Journal, Vol. 300, No. 6720, 1990, pp. 278-284.

[8] M. Umashanker and S. Shruti, "Traditional Indian Herbal Medicine Used as Antipyretic, Antiulcer, Anti-Diabetic and Anticancer: A Review," *International Journal of Research in Pharmacy and Chemistry*, Vol. 1, No. 4, 2011, pp. 1152-1159.

[9] D. Shirode, T. Patel, S. Pal-Roy, T. M. Jyothi, S. V. Rajendra, K. Prabhu, *et al.*, "Research Article Anti-Ulcer Properties of 70% Ethanol Extract of Leaves of Albizzialebbeck," *Pharmacognosy Magazine*, Vol. 4, No. 15, 2008, pp. 228-231.

[10] D. Chatterjee and G. S. Randhawa, "Standardized Names of Cultivated Plants in India. Fruits," *Indian Journal of Horticulture*, Vol. 9, No. 2, 1952, pp. 24-36.

[11] P. R. Bhandari, "Pomegranate (Punicagranatum L) Ancient Seeds for Modern Cure? Review of Potential Therapeutic Applications," *International Journal of Nutrition, Pharmacology, Neurological Diseases*, Vol. 2, No. 3, 2012, pp. 171-184.

[12] M. R. Shams-Ardekani, M. Hajimahmoodi, M. R. Oveisi, N. Sadeghi, B. Jannat, A. M. Ranjbar, *et al.*, "Comparative Antioxidant Activity and Total Flavonoid Content of Persian Pomegranate (Punicagranatum L) Cultivars," *International Journal of Pharmaceutical Research*, Vol. 10, No. 3, 2011, pp. 519-524.

[13] M. V. Dassprakash, R. Arun, S. K. Abraham and K.

Premkumar, "*In Vitro* and *in Vivo* Evaluation of Antioxidant and Antigenotoxic Potential of *Punica granatum* Leaf Extract," *Pharmaceutical Biology*, Vol. 50, No. 12, 2012, pp. 1523-1530.

[14] N. Sadeghi, B. Jannat, M. R. Oveisi, M. Hajimahmoodi and M. Photovat, "Antioxidant Activity of Iranian Pomegranate (*Punica granatum* L) Seed Extracts," *Journal of Agricultural Science and Technology*, Vol. 11, 2009, pp. 633-638.

[15] M. T. Julie Jurenka, "Therapeutic Applications of Pomegranate (*Punica granatum* L)," *Alternative Medicine Review*, Vol. 13, No. 2, 2008, pp. 128-144.

[16] S. A. Naqvi, M. S. Khan and S. B. Vohora, "Antibacterial, Antifungal, and Antihelminthic Investigations on Indian Medicinal Plants," *Fitoterapia*, Vol. 62, No. 3, 1991, pp. 221-228.

[17] Z. Noormohammadi, A. Fasihee, S. Homaee-Rashidpoor, M. Sheidai, S. Ghasemzadeh-Baraki S, A. Mazooji and S. Z. Tabatabaee-Ardakani, "Genetic Variation among Iranian Pomegranates (*Punica granatum* L.) Using RAPD, ISSR and SSR Markers," *Australian Journal of Crop Science*, Vol. 6, No. 2, 2012, pp. 268-275.

[18] J. Tarighi, S. Dadashi, M. Abbass-Ghazvini and A. Mahmoudi, "Comparison of Physical and Hydrodynamic Properties of Two Iranian Commercial Pomegranates," *Agricultural Engineering International: CIGR Journal*, Vol. 13, No. 3, 2011, pp. 1-5.

[19] M. Khanavi, R. Ahmadi, A. Rajabi, S. Jabbari-Arfaee, G. Hassanzadeh, R. Khademi, A. Hadjiakhoondi and M. Sharifzadeh, "Pharmacological and Histological Effects of *Centaurea bruguierana* ssp. *belangerana* on Indomethacin-Induced Peptic Ulcer in Rats," *Journal of Natural Medicines*, Vol. 66, No. 2, 2012, pp. 343-349.

[20] P. Malfertheiner, F. K. Chan and K. E. McColl, "Peptic Ulcer Disease," *Lancet*, Vol. 374, No. 9699, 2009, pp. 1449-1461.

[21] C. Bonacorsi, L. Marcos, D. Fonseca, M. Stella, G. Raddi, R. R. Kitagawa, M. Sannomiya, *et al.*, "Relative Antioxidant Activity of Brazilian Medicinal Plants for Gastrointestinal Diseases," *Journal of Medicinal Plants Research*, Vol. 5, No. 18, 2011, pp. 4511-4518.

[22] G. B. Glavin and S. Szabo, "Experimental Gastric Mucosal Injury: Laboratory Models Reveal Mechanisms of Pathogenesis and New Therapeutic Strategies," *The FASEB Journal*, Vol. 6, No. 3, 1992, pp. 825-831.

[23] Y. Yuan, I. T. Padol and R. H. Hunt, "Peptic Ulcer Disease Today," *Nature Clinical Practice Gastroenterology & Hepatology*, Vol. 3, No. 2, 2006, pp. 80-89.

[24] M. Hajimahmoodi, M. Shams-Ardakani, P. Saniee, F. Siavoshi, M. Mehrabani, H. Hosseinzadeh, *et al.*, "*In Vitro* Antibacterial Activity of Some Iranian Medicinal Plant Extracts against *Helicobacter pylori*," *Natural Product Research*, Vol. 25, No. 11, 2011, pp. 1059-1066.

[25] M. S. Alam, M. A. Alam, S. Ahmad, A. K. Najmi, M.

Asif and T. Jahangir, "Protective Effects of *Punica granatum* in Experimentally-Induced Gastric Ulcers," *Toxicology Mechanisms and Methods*, Vol. 20, No. 9, 2010, pp. 572-528.

[26] S. Lai, Q. Zhou, Y. Zhang, J. Shang and T. Yu, "Effects of Pomegranate Tannins on Experimental Gastric Damages," *Zhongguo Zhong Yao ZaZhi*, Vol. 34, No. 10, 2009, pp. 1290-1294.

[27] K. B. Ajaikumar , M. Asheef , B. H. Babu and J. Padik-

kala, "The Inhibition of Gastric Mucosal Injury by *Punicagranatum* L (Pomegranate) Methanol Extract," *Journal of Ethnopharmacology*, Vol. 96, No. 1-2, 2005, pp. 171-176.

[28] A. Vidal, A. Fallarero, B. R. Peña, M. E. Medina, B. Gra, F. Rivera, *et al.*, "Studies on the Toxicity of Punicagranatum L (Punicaceae) Whole Fruit Extracts," *Journal of Ethnopharmacology*, Vol. 89, No. 2-3, 2003, pp. 295-300.

23

Effects of Pressurized Argon and Krypton Treatments on the Quality of Fresh White Mushroom (*Agaricus bisporus*)

Camel Lagnika[1,2], Min Zhang[1*], Mohanad Bashari[1], Fatoumata Tounkara[1]

[1]State Key Laboratory of Food Science and Technology, School of Food Science and Technology, Jiangnan University, Wuxi, China;
[2]Ecole Nationale des Sciences et Technique de Conservation et Transformation des Produits Agricoles de Sakété, Université d'Agriculture de Kétou, Kétou, Benin.

ABSTRACT

Effects of argon, krypton and their mixed pressure treatments on the quality of white mushrooms were studied during 9 days of storage at 4°C. Among all treatments in this study, the minimum respiration rate, polyphenoloxidase activity, retained color change, antioxidants and delayed pseudomonas growth were observed with pressure argon (5 MPa) followed by mixing argon and krypton (2.5 MPa each) treatments. Respiration rates after 9 days of storage were 5.35%, 6.20%, 7.50%, 7.60%, 7.91% and 8.95% for HA5, HAK, HA2, HK5, HK2 and control, respectively. DPPH inhibition percentages of free radical for HA5, HAK, HK5, HA2, HK2 and control mushrooms were 28.03%, 25.24%, 24.96%, 21.87%, 20.56% and 19.06%, respectively, after 9 days of storage. The pressurized argon treatment was the most effective compared to pressurized krypton. Thus, application of pressurized argon and krypton treatments could extend the storage life of white mushrooms to 9 days at 4°C.

Keywords: Pressurized Argon; Krypton; Clathrate Hydrates; White Mushroom; Storage

1. Introduction

Production and consumption of mushrooms have been gaining substantial ascendency in many parts of the globe due to their deliciousness, flavor and overall nutritional value. White mushroom (*Agaricus bisporus*) is rich in acidic polysaccharides, dietary fiber, and antioxidants including vitamins (C, B12, and D), folate, ergothioneine and polyphenol [1,2]. Due to these nutrients, consumption of white mushrooms may have potential anti-inflammatory, hypoglycemic and hypocholesterolemic consequences. Unfortunately, fresh mushrooms have very short shelf life (3 to 4 days) compared to most other vegetables at room temperature [3]. This might be due to the fact that mushrooms do not have cuticles to protect them from physical or microbial attack or water loss [4] also because of their high respiratory rate. Postharvest browning of *Agaricus bisporus* is a severe problem that reduces the shelf-life. The most important factors that determine the rate of enzymatic browning are the con-

centrations of active polyphenol oxidase (PPO) and phenolic compounds present [5].

Moulds, bacteria, enzymatic activity and biochemical changes can cause spoilage during storage. Gram-negative microorganisms, such as Pseudomonas bacteria, have been associated with mushroom spoilage.

Thus, since mushrooms are highly perishable, they need special care, especially during harvesting and storage to retain freshness and overall quality.

Parameters such as visual appearance, respiration rate, color, microbial growth and weight loss are usually used to determine the quality of mushrooms [6]. Moreover, the antioxidant status of fruits and vegetables is related to its shelf life and may provide a useful indicator of the quality during storage [7]. Various studies have demonstrated that shelf life of fruits and vegetables is modulated by antioxidants [8,9].

Recently, a great deal of interest has been shown in the potential benefits of using argon in food preservation. Many studies on the application of pressurized inert gases in preserving fresh fruits and vegetables have been

*Corresponding author.

published [10-16]. Use of argon, a major component of the atmosphere in modified atmosphere packaging (MAP) has been reported to reduce microbial growth and improve product overall quality retention [12,17]. Noble gases dissolved in water under appropriately selected temperature and pressure conditions, could result in the formation of highly ordered "iceberg-like" structures (called gas hydrate or clathrate) around solute molecules in aqueous solution due to hydrophobic hydration [18]. At 0°C, argon and nitrogen clathrate hydrates can form and remain stable at more than 8.7 and 14.3 MPa respectively [19]. Zhan and Zhang, 2005 [11] observed clathrate hydrates (structure-type I) using a mixture of argon and xenon at a pressure range of 0.4 - 1.1 MPa in cucumber samples. Ando et al., [20] examined the formation of the hydrate crystals of fresh-cut onions, which were preserved under Xe pressure up to 0.8 MPa at 5°C for few hours. Purwanto et al., [21] found the formation of gas hydrate in distilled water and coffee solutions at 8°C and 0.70 MPa.

Fresh-cut vegetables and fruits pressurized in the presence of inert gases under appropriate conditions of pressure and temperature, cause the inert gases to form hydrate in these fresh-cut material's tissue and lower the activity of intracellular water and inhibit the enzymatic reactions. The combination of these two phenomena contributes to the reduction of metabolism of fruits and vegetables [10,12].

To the best of our knowledge, there are no reports of scientific research works on the effects of combined pressure argon and krypton treatments on the shelf life of white mushrooms. Therefore, the present research was designed to investigate the effects of pressurized argon and krypton, as well as, the mixture of the two on the physico-chemical, microbiological properties and sensory quality of mushrooms during cold storage.

2. Materials and Methods

High-pressure equipment HCYF-3 (HuaAn Scientific Instruments Co. Ltd., Jiangsu-China), commercially available argon and krypton of 99.7% purity (Wuxi Xinnan Gas Co. Jiangsu-China) were used. Freshly harvested, white mushrooms (Agaricus bisporus) were purchased from a local market at Wuxi, China. All other chemicals and solvents used were of analytical grade.

The freshly harvested mushrooms were transported to the laboratory and selected base on uniformity of shape and colour and free from mechanical damage. Mushrooms obtained were randomly divided into six groups, and each group was samples (65 ± 5 g) at least three times using glass jars. The different groups were subjected to the following treatments:

- Control (C): mushrooms washed with distilled water to remove soil then storage at 4°C;

- pressurised argon (HA2): mushroom treated with Ar under pressure 2.5 MPa at 4°C for 1 h;
- pressurised argon (HA5): mushroom treated with Ar under pressure 5 MPa at 4°C for 1 h;
- pressurised krypton (HK2): mushroom treated with Kr under pressure 2.5 MPa at 4 °C for 1 h;
- pressurised krypton (HK5): mushroom treated with Kr under pressure 5 MPa at 4°C for 1 h;
- pressurised mixed argon and krypton (HAK): mushroom treated with mixed Ar and Kr under pressure 2.5 MPa each gas at 4°C for 1 h.

Fresh mushrooms were placed in a high pressure chamber, and then argon or/and krypton was passed into the chamber after the evacuation time. After the pressurized argon or/and krypton treatments, all samples were stored at 4°C with 90% relative humidity for 9 days.

Measurements and analyses of the mushrooms were performed on the following days of storage period; 0, 3rd, 6th, and 9th day. Twelve replicates were included in each treatment group, and subsequently every 3 days, three replicates from each treatment group were analyzed. All measurements were done in triplicates.

2.1. CO$_2$ Production

Mushrooms (65 ± 5 g) were placed in 500 mL glass jars and sealed with high gas barrier film then stored at 4°C for 9 days. Carbon dioxide production was measured on the 3rd, 6th and 9th day of storage period using an O$_2$ and CO$_2$ Analyser (Cyes-II, Jiading federation Instrument, Shanghai, China). Gas samples were taken from the jars with a 20 mL syringe. Carbon dioxide production (ΔCO$_2$) was calculated as follows:

$$\Delta CO_2 \left(\%\right) = CO_{2f} - CO_{2i} \qquad (1)$$

where, CO$_{2i}$ is the gas concentration on the first day and CO$_{2f}$ is the gas concentration on the final day of storage.

2.2. Weight Loss

Weight losses were determined by weighing of all mushrooms contained in one package (initially 65 ± 5 g) before and after the storage period, which was expressed as weight loss percentage with respect to the initial weight.

$$\text{Weight loss} \left(\%\right) = \left(W_0 - W_f\right)/W_0 \times 100 \qquad (2)$$

where, W$_0$ is the weight on the first day and W$_f$ the weight on final storage day.

2.3. Color

Surface color of mushrooms was measured with a Minolta spectrophotometer (CR-400, Konica Minolta Sensing, Tokyo, Japan) using CIE color parameters L^* (light/dark), a^* (red/green) and b^* (yellow/blue) values.

Three readings were taken at three equidistant points on each mushroom cap. Numerical values of L^* and color difference (ΔE) were considered for the evaluation of color modification of fresh mushroom. The value ΔE defines the magnitude of total color difference and is expressed by the equation [15]:

$$\Delta E = \left[\left(L_t^* - L_i^* \right)^2 + \left(a_t^* - a_i^* \right)^2 + \left(b_t^* - b_i^* \right)^2 \right]^{1/2} \quad (3)$$

where ΔE indicates the degree of overall color change in comparison to color values of an ideal mushroom, L_i^*, a_i^* and b_i^* represented the reading of fresh mushroom without any treatments, and L_t^*, a_t^* and b_t^* referred to the instantaneous individual readings during storage time after the mushrooms were treated.

2.4. Polyphenoloxidase (PPO) Activity

Polyphenoloxidase (PPO, E.C. 1.14.18.1) activity in mushroom, during the storage period was determined according to the method proposed by Pizzocaro et al. [22] with slight modifications. Fresh mushroom (10 g) was ground in 10 mL of McIlvaine citric-phosphate buffer, pH 6.5. The homogenate was centrifuged at 3000 × g at 4°C for 30 min. The supernatant obtained was filtered with Whatman no. 4 filter paper and analyzed for PPO activity at 25°C afterward. A 2 mL of catechol solution (0.1%) and 2 mL of McIlvaine buffer pH 6.5 were added to 0.1 mL of PPO extract. PPO activity was assayed in triplicate using a spectrophotometer (UV-visible 2600, Precision Science Instrument, Shanghai, China) at 420 nm and calculated on the basis of the slope from the linear portion of the curve plotted with ΔA_{420}. One unit of PPO was defined as the amount of enzyme present in the extract that resulted in an absorbance increase of 0.001 units per minute. The activity was expressed in units of PPO per minute and gram ($U \cdot min^{-1} \cdot g^{-1}$) of fresh mushroom.

2.5. DPPH Free Radical-Scavenging Assay

The determination of free radical scavenging effect on 1, 1-diphenyl-2-picrylhydrazyl (DPPH) radical was carried out according to the method of Alothman et al., [23] with slight modifications. Mushroom samples (2 g) were homogenised with a mortar and pestle in 10 mL of methanol and centrifuged at 6000 × g for 15 min at 4°C and filtered through a Whatman No 1 paper. Aliquots of 0.05 mL of the supernatant were mixed with 1 mL of DPPH and 1.5 mL of Tris buffer. The homogenate was shaken vigorously and kept in darkness for 30 min. The absorption of the samples was measured with the UV-visible spectrophotometer at 517 nm against methanol as blank. Results were expressed as percentage of inhibition of the DPPH radical. Percentage of inhibition of the DPPH

radical was calculated according to the following equation:

DPPH radical scavenging activity (%)

$$= \left[\left(A_0 - A \right) / A_0 \right] \times 100 \quad (4)$$

where, A_0 is the absorbance of DPPH solution without extracts and A is the absorbance of the mushroom extract.

2.6. Total Phenolic and Flavonoids Contents

Total phenolic contents were measured according to Singleton and Rossi [24]. Mushroom samples (5 g) were crushed and homogenised in 50 mL methanol. The mixture was centrifuged at 3000 × g for 30 min at 4°C, filtered with Whatman no. 4 filter paper and the mushroom extract was collected. Mushroom extract (200 µL) was mixed with 1.80 mL distilled water then 1 mL of Folin and Ciocalteu's phenol reagent was added. After 2 min, 2 mL of 20% sodium carbonate solution (Na_2CO_3) was added. Thereafter, the reaction was allowed to proceed in the dark for 90 min and absorbance was then read at 750 nm using the spectrophotometer. Gallic acid was used to calculate the standard curve and the results were expressed as mg of gallic acid equivalents (GAE) per g of extract fresh weight.

Flavonoids were extracted and determined according to the methods of Barros et al. [25] with slight modifications. Namely, 1.8 mL of mushroom extract was added to 20 µL distilled water and 75 µL of 5% sodium nitrite ($NaNO_2$) then allowed standing for 6 min. Thereafter, 150 µL of 10% aluminium chloride ($AlCl_3$) was added. After standing for another 5 min, 2 mL of 1 $mol \cdot L^{-1}$ sodium hydroxide (NaOH) was added to the mixtures and immediately their absorbance (pink in colour) was determined at 510 nm. Rutin was used to establish the standard curve and the total flavonoids of mushroom were calculated and expressed on a fresh weight as mg Rutin equivalents (RUE) per g.

2.7. Microbiological Analysis

All samples were analyzed for the pseudomonas bacteria counts. Mushrooms samples (10 g) were removed aseptically from each pack and diluted with 90 mL of 0.1% sterile peptone water. The samples were homogenised by a stomacher at high speed for 2 min. Serial dilutions (10^{-1} - 10^{-8}) were made in tubes (1.0 mL with 9.0 mL of 0.1% peptone water). Pseudomonas bacteria were counted on cephaloridin fucidin cetrimide agar (CFC; Difco), with selective supplement SR 103 (Oxoid). The plates were incubated for 48 h at 25°C and the number of colony forming units per gram ($CFU \cdot g^{-1}$) of mushroom was determined.

2.8. Sensory Analysis

Sensory analysis of the mushroom was evaluated according to Abdallah *et al.* and Conesa *et al.* [26,27] with slight modifications on days 3, 6 and 9 by 10 semi-trained recruited among students of the Food Science and Technology, Jiangnan University. Sensory evaluation was performed based on four aspects (color, aroma, texture and overall acceptability). The aspects were evaluated on a scale of 9-1, where 9—excellent, 8—very good, 7—good, 6—fairly good, 5—satisfactory and limit of marketability, 4—fair and limit of usability, 3—bad, 2—very bad and 1—extremely bad and inedible.

2.9. Statistical Analysis

Data were expressed as mean ± standard deviation (SD). The Tukey's test and one-way analysis of variance (ANOVA) were used for multiple comparisons by the SPSS 17.0 (SPSS, Chicago, Illinois, USA). Difference was considered to be statistically significant if $P < 0.05$.

3. Results and Discussion

3.1. CO_2 Production and Weight Loss

Changes of the CO_2 production during storage under different treatments at 4°C are shown in **Figure 1(A)**. As can be seen from the figure a progressive increase in CO_2 production during the entire storage period was observed with all samples. At the end of storage time, CO_2 productions were 5.35%, 6.20%, 7.50%, 7.60%, 7.91% and 8.95% for HA5, HAK, HA2, HK5, HK2 and control samples, respectively. The results also demonstrated that, argon or krypton treatment at 5 MPa showed lower CO_2 production compared to that at 2 MPa. However, argon treatments were significantly more efficient than krypton treatments ($P < 0.05$).

The weight loss of the control sample was the highest (1.04%) among the six treatments during the storage time (**Figure 1(b)**). Samples treated by HA5 had significantly lower weight loss than the other samples throughout the storage ($P < 0.05$). The weight loss of the HAK, HA2, HK5 and HK2 treatments increased progressively during storage to a maximum of 0.95% for HK2 sample after 9 days, without significant differences among the HAK and HA2 treatments. In our results, we observed a correlation between CO_2 production and weight loss. HAK treated ones showed reduced weight loss and respiration rate compared to samples with 2.5 MPa treatments and the control. Therefore, the lowest CO_2 production and water loss observed in argon treatment could be related to the highest solubility nature of argon which caused the highest capability of gas hydrate formation compared to krypton [28]. Argon hydrate is the most fundamental clathrate hydrate in the sense that argon is spherical and

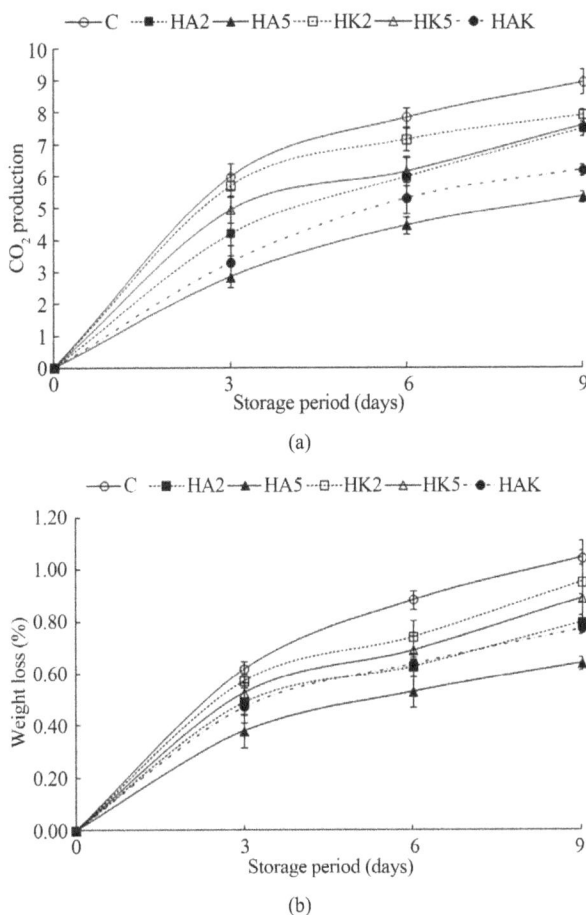

Figure 1. Changes in (a) CO_2 production and (b) Weight loss in white mushrooms during storage at 4°C for 9 days under different treatments (n = 3).

the smallest of the molecules which can be accommodated in the clathrate cages and therefore, its interaction with the lattice is the weakest [29].

3.2. Color

Browning after harvest is a common and economically detrimental phenomenon in the mushroom industry, which may have negative effect not only on the appearance quality, but also on the flavor and nutrient composition.

All mushroom samples showed a decrease in whiteness (L^*), however, the color difference (ΔE) increased during storage (**Table 1**). Compared to the control mushrooms, those treated had a higher L^* ($P < 0.05$) and lower ΔE ($P < 0.05$) values. Mushroom with HA5 and HAK treatments followed by HK5, HA2 and HK2, respectively, had a higher L^* value than that of control during the 9 days of storage.

Color difference (ΔE) during storage differed among treated ones with HA5 and HAK samples recording lower color difference in contrast with HK5 followed by

Table 1. Colour changes of white mushrooms at different types of treatment during storage at 4°C for 9 days.

Days	0	3	6	9
	L			
C	91.02 ± 1.66^a	85.30 ± 0.66^a	82.40 ± 0.33^a	63.28 ± 1.43^a
HA2	90.40 ± 0.69^a	87.47 ± 1.66^b	86.18 ± 0.84^c	71.51 ± 1.03^b
HA5	90.78 ± 1.69^a	90.78 ± 1.43^e	87.29 ± 1.06^d	84.61 ± 1.21^e
HK2	90.92 ± 0.27^a	86.85 ± 0.33^b	83.41 ± 1.16^b	70.58 ± 0.16^b
HK5	90.70 ± 0.97^a	87.51 ± 1.04^c	86.57 ± 1.34^c	77.58 ± 1.36^c
HAK	91.04 ± 0.52^a	88.73 ± 1.31^d	87.00 ± 0.96^d	80.64 ± 2.01^d
	ΔE			
C	3.89 ± 1.32^a	9.52 ± 0.98^a	13.08 ± 0.53^a	30.09 ± 1.72^a
HA2	4.20 ± 1.06^a	8.42 ± 1.03^b	8.62 ± 1.01^c	25.30 ± 2.03^b
HA5	4.35 ± 1.35^a	4.73 ± 1.23^e	7.49 ± 1.03^d	11.03 ± 1.54^e
HK2	4.44 ± 0.37^a	8.46 ± 0.17^b	12.07 ± 0.69^b	25.62 ± 1.05^b
HK5	4.44 ± 0.17^a	7.40 ± 61.06^c	8.30 ± 1.09^c	19.13 ± 0.87^c
HAK	4.33 ± 0.35^a	6.54 ± 1.64^d	7.94 ± 1.48^d	15.80 ± 1.08^d

Values are mean ± standard deviation of triplicates. Data in same column with different letters are significantly different ($P < 0.05$). Control (C): mushrooms washed with distilled water to remove soil then storage at 4°C; (HA2): 2.5 MPa pressure argon treatment at 4°C for 1 h; (HA5): 5 MPa pressure argon treatment at 4°C for 1 h; (HK2): 2.5 MPa pressure krypton treatment at 4°C for 1 h; (HK5): 5 MPa pressure krypton treatment at 4°C for 1 h; (HAK): mixing argon and krypton treatment at 2.5 MPa pressure each at 4°C for 1 h.

HA2 and HK2 and control. Previously reported relationship between different quality levels in mushrooms (*A. bisporus*) and Hunter *L*-value provided a criterion for classification [6,30]. Mushrooms with *L*-values greater than 93 were classified as excellent sample, however, that with L-values ranging between 90 to 93, 86 to 89, 80 to 85, and 69 to 79 were classified as very good, good, fair and poor sample, respectively.

This criterion can be used as an indicator of mushroom shelf life; for example mushrooms with an *L*-value less than 80 would not be acceptable at wholesale level [31]. This grading method is the most frequently used indicator of mushroom shelf-life both in the industry and research [30]. According to that criterion [31], except the HA5 and HAK samples, other samples had to be rejected.

3.3. Polyphenoloxidase (PPO) Activity

PPO plays an important role in the browning process of many fruits and vegetables. Browning reactions are generally assumed to be a direct consequence of PPO actions on polyphenols to form quinones, which ultimately polymerize to produce the browning appearance of fruit and vegetable [32]. PPO activity in white mushrooms increased on the first day of storage and reached maximum value on day 6 for all treatments and decreased during

the latter period afterward (**Table 2**). The lowest activity was observed in HA5, HAK followed by HK5, HA2 and HK2 treatment as compared to control. Mobility of water is restricted by the formation of clathrate hydrates [33]. The lowest PPO activity observed with treated samples compared to control could be attributed to the formation of clathrate hydrate. This implied that, structured water contributed into low water mobility which consequently delayed the enzymatic browning. It's also might be due to the capacity of these gases to dissolve in the aqueous layer of the mushroom through the cells of the flesh. Therefore, they can inactivate some chemically-active sites on the enzymes and/or reduce the level of dissolved oxygen, whose presence is necessary for oxidative enzymes to catalyze metabolic reactions. Behnke [34] demonstrated that high pressure inert gases inhibited tyrosinase systems by decreasing oxygen availability rather than by physically altering the enzyme. When noble gases dissolve in water, enzymatic reactions are inhibited, resulting in restrained vegetable metabolism [12].

3.4. DPPH Free Radical-Scavenging Assay

Changes during storage in the percentage of inhibition of DPPH radical by antioxidants present in white mushroom, are shown in **Table 2**. DPPH scavenging power of the mushroom generally showed a reduction trend over

Table 2. Changes in polyphenoloxidase (PPO) activity and free radical scavenging effect on 1,1-diphenyl-2-picrylhydrazyl (DPPH) of white mushrooms at different types of treatment during storage at 4°C for 9 days.

Days	0	3	6	9
		PPO activity $(U \cdot min^{-1} \cdot g^{-1})$		
C	3132 ± 3.56^a	3851 ± 8.02^a	4496 ± 7.23^a	4275 ± 9.20^a
HA2	3132 ± 3.56^a	3673 ± 3.24^b	4229 ± 4.67^b	3863 ± 2.65^c
HA5	3132 ± 3.56^a	3210 ± 1.45^c	3809 ± 2.56^c	3155 ± 5.63^f
HK2	3132 ± 3.56^a	3826 ± 2.34^a	4449 ± 5.65^a	4143 ± 4.65^b
HK5	3132 ± 3.56^a	3394 ± 6.12^c	4200 ± 4.34^b	3781 ± 2.64^d
HAK	3132 ± 3.56^a	3488 ± 5.32^d	4037 ± 1.54^c	3451 ± 6.34^e
		DPPH (%)		
C	40.61 ± 0.64^a	31.29 ± 0.81^a	27.51 ± 0.76^a	19.06 ± 0.53^a
HA2	40.61 ± 0.64^a	32.02 ± 0.68^b	30.29 ± 0.81^c	21.87 ± 0.44^d
HA5	40.61 ± 0.64^a	37.55 ± 0.37^d	34.30 ± 0.27^f	28.03 ± 0.76^f
HK2	40.61 ± 0.64^a	31.24 ± 0.63^a	28.54 ± 0.57^b	20.56 ± 0.16^b
HK5	40.61 ± 0.64^a	31.94 ± 0.78^b	31.18 ± 0.14^d	24.96 ± 0.84^c
HAK	40.61 ± 0.64^a	35.02 ± 0.68^c	32.22 ± 0.68^e	25.24 ± 0.59^e

Values are mean ± standard deviation of triplicates. Data in same column with different letters are significantly different ($P < 0.05$). Control (C): mushrooms washed with distilled water to remove soil then storage at 4°C; (HA2): 2.5 MPa pressure argon treatment at 4°C for 1 h; (HA5): 5 MPa pressure argon treatment at 4°C for 1 h; (HK2): 2.5 MPa pressure krypton treatment at 4°C for 1 h; (HK5): 5 MPa pressure krypton treatment at 4°C for 1 h; (HAK): mixing argon and krypton treatment at 2.5 MPa pressure each at 4°C for 1 h.

the 9 days storage time in all samples but at different extent. However, all the treated samples delayed in the decrease but at different degrees, with significant differences ($P < 0.05$) in DPPH scavenging power between the treated samples and the control. After 9 days of storage, percentage of inhibition of DPPH radical for HA5, HAK, HK5, HA2, HK2 and untreated mushrooms were 28.03%, 25.24%, 24.96%, 21.87%, 20.56% and 19.06% respectively. It was observed that, high pressure argon could delay the reduction of antioxidant capacity of mushroom during the refrigerator storage, probably due to noble gas hydrate formation and residual gas in micropore of fruit tissue.

3.5. Total Phenolic and Flavonoids Contents

In this study, total phenolics levels declined in all treatments during the 9 days of storage (**Table 3**). However, pressurized argon (5 MPa) was more effective in delaying decrease of phenolics than other samples. Mushrooms treated samples presented a higher level of total phenolics, compared to control.

Similar to total phenolics, all the treatments showed a reduction in flavonoids contents during storage (**Table 3**). However, HA5 and HAK treatments appeared to be significantly ($P < 0.05$) efficient in delaying the reduction in

flavonoids in the white mushrooms as compared to other treatment. Whereas, untreated white mushrooms presented a lower level of flavonoids, all treatments affected significantly ($P < 0.05$) the flavonoid content in white mushrooms during the 9 days.

It demonstrated that, HA5 and HAK treatments were significantly effective in maintaining the total phenolic and flavonoids compounds, which might be due to clathrate hydrate formation that inhibited the enzyme activity of phenolic and flavonoids compounds degradation. The pressurized argon treatment was the most effective in delaying the reduction in total phenolics and flavonoids content compared to pressurized krypton may be due to the high solubility of argon.

3.6. Microbiological Analysis

Figure 2 presents growth of pseudomonas bacteria (expressed as log $CFU \cdot g^{-1}$) of fresh mushroom during 9 days of storage at 4°C. Gradual growth of microorganisms was seen during storage in all samples. However, some treatments retarded the microbial growth more than others. The highest amount of microorganisms was observed in control samples. Pressurized argon samples (5 Mpa) followed by combined of argon-krypton and HK5 samples, were found to be effective in delaying pseudo-

Table 3. Changes in functional components in white mushrooms during storage at 4°C for 9 days under different treatments.

Days	0	3	6	9
	Total phenolics (min·g^{-1})			
C	1.03 ± 0.92^a	0.90 ± 0.63^a	0.62 ± 0.10^a	0.43 ± 0.18^a
HA2	1.03 ± 0.92^a	0.99 ± 0.17^{bc}	0.75 ± 0.25^b	0.63 ± 0.16^d
HA5	1.03 ± 0.92^a	1.02 ± 0.15^c	0.89 ± 0.34^e	0.76 ± 0.32^f
HK2	1.03 ± 0.92^a	0.97 ± 0.23^b	0.77 ± 0.32^c	0.50 ± 0.71^b
HK5	1.03 ± 0.92^a	0.98 ± 0.43^{bc}	0.77 ± 0.45^c	0.58 ± 0.58^c
HAK	1.03 ± 0.92^a	1.01 ± 0.68^c	0.85 ± 0.30^d	0.74 ± 0.63^e
	Total flavonoids (min·g^{-1})			
C	0.71 ± 0.44^a	0.50 ± 0.36^a	0.38 ± 0.46^a	0.25 ± 0.36^a
HA2	0.71 ± 0.44^a	0.57 ± 0.56^c	0.46 ± 0.34^c	0.33 ± 0.32^c
HA5	0.71 ± 0.44^a	0.66 ± 0.14^d	0.54 ± 0.17^e	0.40 ± 0.37^d
HK2	0.71 ± 0.44^a	0.51 ± 0.44^{ab}	0.41 ± 0.32^b	0.29 ± 0.43^b
HK5	0.71 ± 0.44^a	0.53 ± 0.65^b	0.43 ± 0.45^{bc}	0.30 ± 0.48^b
HAK	0.71 ± 0.44^a	0.61 ± 0.35^{cd}	0.50 ± 0.33^d	0.34 ± 0.23^c

Values are mean ± standard deviation of triplicates. Data in same column with different letters are significantly different ($P < 0.05$). Control (C): mushrooms washed with distilled water to remove soil then storage at 4°C; (HA2): 2.5 MPa pressure argon treatment at 4°C for 1 h; (HA5): 5 MPa pressure argon treatment at 4°C for 1 h; (HK2): 2.5 MPa pressure krypton treatment at 4°C for 1 h; (HK5): 5 MPa pressure krypton treatment at 4°C for 1 h; (HAK): mixing argon and krypton treatment at 2.5 MPa pressure each at 4°C for 1 h.

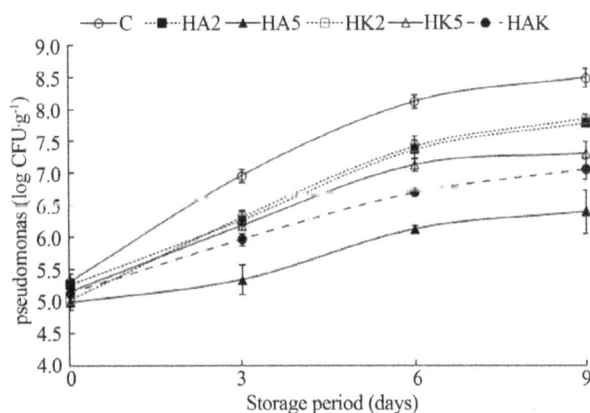

Figure 2. Growth of pseudomonas bacteria of fresh mushroom during 9 days of storage at 4°C under different treatments (n = 3).

monas bacteria growth in mushroom during the 9 days of cold storage. For HA2 and HK2 treatments, no significant ($P < 0.05$) difference was observed during the storage. In food, microbial growth is closely related to the water activity of those products and can be delayed by pressurized inert gases. The inhibitory effect of pressurized gases treatment on microbial growth in white mushroom might be owned to clathrate hydrates formation, which reduced water activity and remained gas in the micropore mushroom to reduce the growth of micro-organism.

3.7. Sensory Analysis

Figure 3 shows the sensory evaluation including color, aroma, texture and overall preference of the six treatments for the three typical storage days. On day 3, all the treatments showed a moderate decrease in the overall quality. As expected, color, aroma, texture and overall acceptability significantly changed ($P < 0.05$) with storage time, supporting the validity of using these parameters as indicators of mushroom deterioration. As the storage time progressed to day 6, there was a continued decrease in sensory quality. On day 9 of storage, considering the development of the evaluated sensory attributes, HA$_5$ mushrooms showed the lowest deterioration rate, followed by HAK, HK5 and HA2. On the other hand, control and HK$_2$ samples reached a score lower than 5 a value that is below the borderline of acceptability and marketability.

4. Conclusion

Compared to the untreated (control) samples, treated samples had significantly ($P < 0.05$) longer shelf-life. The argon treatment delayed quality deterioration, reduced the loss of water, exhibited the smallest respiration rate, retained mushrooms color change, showed smaller

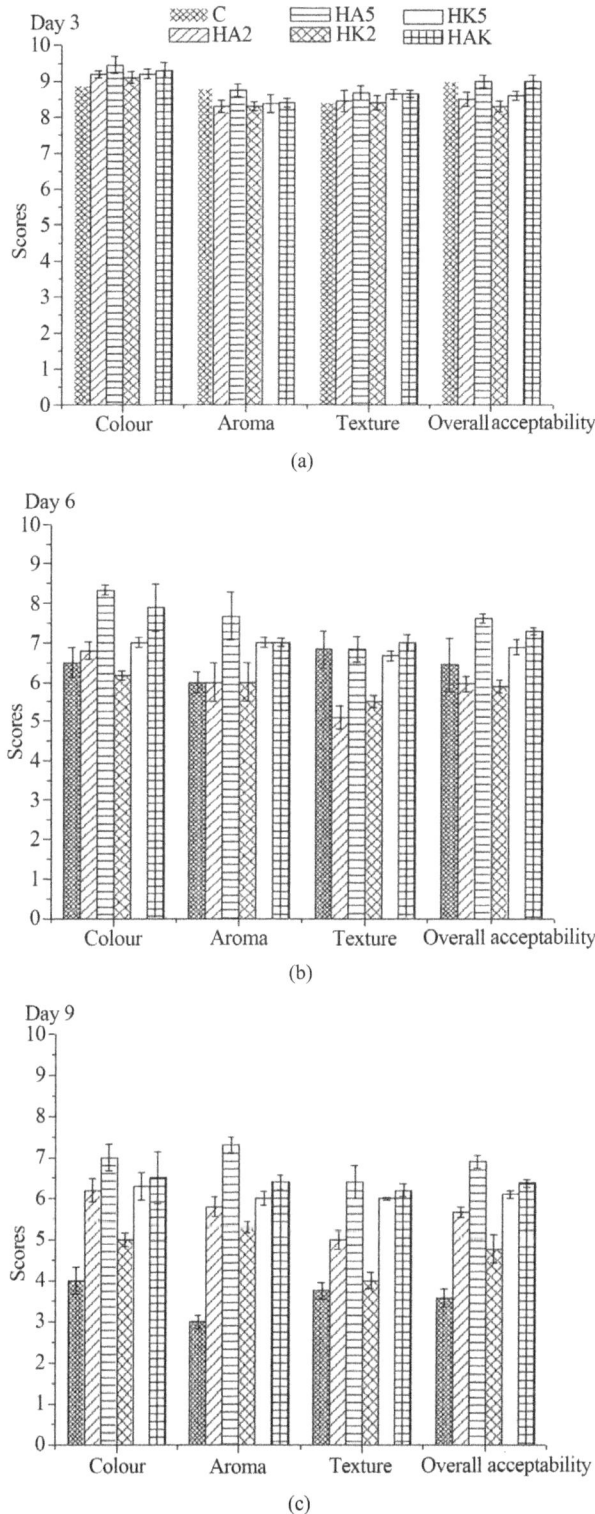

Figure 3. Sensory characteristics of fresh mushroom of six treatments stored for 3, 6 and 9 days at 4°C. Values represent the means of the replicates and error bars represent the standard error of the means (n = 3).

polyphenoloxidase activity, retained antioxidants, delayed pseudomonas growth and maintained sensory qual-

ity compared to krypton treatment. Our research showed that pressurized argon and mixed argon-krypton may be a useful way of maintaining quality and extending the shelf-life of white mushroom.

5. Acknowledgements

The authors are grateful to China National Natural Science Foundation for supporting this research under contract No. 30972058.

REFERENCES

[1] M. Fukushima, M. Nakano, Y. Morii, T. Ohashi, Y. Fujiwara and K. Sonoyama, "Hepatic LDL Receptor mRNA in Rats Is Increased by Dietary Mushroom (*Agaricus bisporus*) Fiber and Sugar Beet Fiber," *Journal of Nutrition*, Vol. 130, No. 9, 2000, pp. 2151-2156.

[2] P. Mattila, K. Könkö, M. Eurola, J. M. Pihalava, J. Astola, L. Vahteristo, *et al.*, "Contents of Vitamins, Mineral Elements, and Some Phenolic Compounds in Cultivated Mushrooms," *Journal of Agricultural and Food Chemistry*, Vol. 49, No. 5, 2001, pp. 2343-2348.

[3] S. Preeti, L. Horst-Christian, A. W. Ali and S. Sven, "Recent Advances in Extending the Shelf Life of Fresh Agaricus Mushrooms: A Review," *Journal of Science Food and Agricultural*, Vol. 90, No. 9, 2010, pp. 1393-1402.

[4] B. Martine, L. P. Gaëlle and G. Ronan, "Post-Harvest Treatment with Citric Acid or Hydrogen Peroxide to Extend the Shelf Life of Fresh Sliced Mushrooms," *LWT-Food Science* and *Technology*, Vol. 33, No. 4, 2000, pp. 285-289.

[5] P. V. Mahajan, F. A. R. Oliveira and I. Macedo, "Effect of Temperature and Humidity on the Transpiration Rate of the Whole Mushrooms," *Journal of Food Engineering*, Vol. 84, No. 2, 2008, pp. 281-288.

[6] M. Taghizadeh, A. Gowen, P. Ward and C. P. O'donnell, "Use of Hyperspectral Imaging for Evaluation of the Shelf-Life of Fresh White Button Mushrooms (*Agaricus bisporus*) Stored in Different Packaging Films," *Innovative Food Science* and *Emerging Technologies*, Vol. 11, No. 3, 2010, pp. 423-431.

[7] E. Aguayo, C. Requejo-Jackman, R. Stanley and A. Woolf, "Effects of Calcium Ascorbate Treatments and Storage Atmosphere on Antioxidant Activity and Quality of Fresh-Cut Apple Slices," *Postharvest Biology and Technology*, Vol. 57, No. 1, 2010, pp. 52-56.

[8] D. M. Hodges, "Postharvest Oxidative Stress in Horticultural Crops," Food Production Press, New York, 2003.

[9] C. T. Wang, Y. P. Cao, M. J. R. Nout, B. G. Sun and L. Liu, "Effect of Modified Atmosphere Packaging (MAP) with Low and Superatmospheric Oxygen on the Quality and Antioxidant Enzyme System of Golden Needle Mush-

rooms (*Flammulina velutipes*) during Postharvest Storage," *European Food Research and Technology*, Vol. 232, No. 5, 2011, pp. 851-860.

[10] S. Oshita, Y. Seo and Y. Kawagoe, "Relaxation Time of Protons in Intracellular Water of Broccoli," *Proceedings of the CIGR International Symposium*, Kyoto, October 2000, pp. 77-82.

[11] Z. G. Zhan and M. Zhang, "Effects of Inert Gases on Enzyme Activity and Inspiration of Cucumber," *Journal of Food Biotechnology*, Vol. 24, No. 3, 2005, pp. 16-18.

[12] M. Zhang, Z. G. Zhan, S. J. Wang and J. M. Tang, "Extending the Shelf Life of Asparagus Spears with a Compressed Mix of Argon and Xenon Gases," *LWT-Food Science and Technology*, Vol. 41, No. 4, 2008, pp. 686-691.

[13] L. Wang, H. Ando, Y. Kawagoe, Y. Makino and S. Oshita, "Basic Study of the Storage of Agricultural Produce by Using Xenon Hydrate," *Cryobiology*, Vol. 59, No. 3, 2009, p. 408.

[14] C. Lagnika, M. Zhang and S. Wang, "Effect of High Argon Pressure and Modified Atmosphere Packaging on the White Mushroom (*Agaricus bisporus*) Physico-Chemical and Microbiological Properties," *Journal of Food Nutrition and Research*, Vol. 50, No. 3, 2011, pp. 167-176.

[15] Z. S. Wu, M. Zhang and S. Wang, "Effects of High-Pressure Argon and Nitrogen Treatments on Respiration, Browning and Antioxidant Potential of Minimally Processed Pineapples during Shelf Life," *Journal of Science Food and Agriculture*, Vol. 92, No. 11, 2012, pp. 2250-2259.

[16] Z. S. Wu, M. Zhang and S. Wang, "Effects of High Pressure Argon Treatments on the Quality of Fresh-Cut Apples at Cold Storage," *Food Control*, Vol. 23, No. 1, 2012, pp. 120-127.

[17] P. Jamie and M. E. Saltveit, "Postharvest Changes in Broccoli and Lettuce during Storage in Argon, Helium, and Nitrogen Atmospheres Containing 2% Oxygen," *Postharvest Biology and Technology*, Vol. 26, No. 1, 2002, pp. 113-116.

[18] D. W. Davidson, "Clathrate Hydrates," In: F. Franks, Ed., *Water: A Comprehensive Treatise*, Plenum Press, New York, 1973, pp. 115-234.

[19] Y. F. Makogon, "Natural Gas Hydrates—A Promising Source of Energy," *Journal of Natural Gas and Science Engineering*, Vol. 2, No. 1, 2010, pp. 49-59.

[20] H. Ando, S. Takeya, Y. Kawagoe, Y. Makino, T. Suzuki and S. Oshita, "*In Situ* Observation of Xenon Hydrate Formation in Onion Tissue by Using NMR and Powder X-Ray Diffraction Measurement," *Cryobiology*, Vol. 59, No. 3, 2009, p. 405.

[21] Y. A. Purwanto, S. Oshita and Y. Seo, "Concentration of Liquid Foods by the Use of Gas Hydrate," *Journal of Food Engineering*, Vol. 47, No. 2, 2001, pp. 133-438.

[22] F. Pizzocaro, D. Torreggiani and G. Gilardi, "Inhibition of Apple Polyphenoloxidase (PPO) by Ascorbic Acid, Citric Acid and Sodium Chloride," *Journal of Food Process and Preservation*, Vol. 17, No. 1, 1993, pp. 21-30.

[23] M. Alothman, R. Bhat and A. A. Karim, "Antioxidant Capacity and Phenolic Content of Selected Tropical Fruits from Malaysia, Extracted with Different Solvents," *Food Chemistry*, Vol. 115, No. 3, 2009, pp. 785-788.

[24] V. L. Singleton and J. A. Rossi, "Colorimetry of Total Phenolics with Phosphomolybdic-Phosphotungstic Acid Reagents," *American Journal of Enology and Viticulture*, Vol. 16, 1945, pp. 144-158.

[25] L. Barros, S. Falcão, P. C. Baptistafreire, M. Vilas-Boas and I. C. F. R. Ferreira, "Antioxidant Activity of *Agaricus* sp. Mushrooms by Chemical, Biochemical and Electrochemical Assays," *Food Chemistry*, Vol. 111, No. 1, 2008, pp. 61-66.

[26] M. A. S. Abdallah, O. A. Ali, A. A. I. Sharaf Eldeen and A. Y. A. Mohammed, "A Comparative Study on Red and White Karkade (*Hibiscus sabdariffa L.*) Calyces, Extracts and Their Products," *Pakistan Journal of Nutrition*, Vol. 10, No. 7, 2011, pp. 680-683.

[27] A. Conesa, F. Arteshernandez, S. Geysen, B. Nicolai and F. Artes, "High Oxygen Combined with High Carbon Dioxide Improves Microbial and Sensory Quality of Fresh-Cut Peppers," *Postharvest Biology and Technology*, Vol. 43, No. 2, 2007, pp. 230-237.

[28] D. L. Lide, "CRC Handbook of Chemistry and Physics," 76th Edition, CRC Press, Boca Raton, 1995.

[29] O. Yamamuro, M. Oguni, T. Matsuo and H. Suga, "Calorimetric Study on Pure and KOH-Doped Argon Clathrate Hydrates," *Journal of inclusion phenomena*, Vol. 6, No. 3, 1988, pp. 307-318.

[30] L. Aguirre, J. M. Frias, C. B. Ryan and H. Grogan, "Assessing the Effect of Product Variability on the Management of the Quality of Mushrooms (*Agaricus bisporus*)," *Postharvest Biology and Technology*, Vol. 49, No. 2, 2008, pp. 247-254.

[31] G. López-Briones, P. Varoquaux, G. Bureau and B. Pascat, "Modified Atmosphere Packaging of Common Mushroom," *International Journal of Food Science* and *Technology*, Vol. 28, No. 1, 1993, pp. 57-68.

[32] Z. Chen, C. H. Zhu, Y. Zhang, D. B. Niu and J. H. Du, "Effects of Aqueous Chlorine Dioxide Treatment on Enzymatic Browning and Shelf-Life of Fresh-Cut Asparagus Lettuce (*Lactuca sativa L.*)," *Postharvest Biology and Technology*, Vol. 58, No. 3, 2010, pp. 232-238.

[33] S. A. Yoshioki, "Note on Transformation between Clathrate Hydrate Structures I and II," *Journal of Molecular Graphics and Modelling*. Vol. 29, No. 2, 2010, pp. 290-

294.

[34] E. A. Behnke, "Enzyme-Catalysed Reactions as Influenced by Inert Gases at High Pressures," *Journal of Food*

Science, Vol. 34, No. 4, 1969, pp. 370-375.

24

The Effect of Seeds and Fruit Pulp of *Adansonia digitata* L. (Baobab) on Ehrlich Ascites Carcinoma

Fahmy G. Elsaid[1,2]

[1]Zoology Department, Faculty of Science, Mansoura University, Mansoura, Egypt; [2]Biology Department, Science College, King Khalid University, Abha, KSA.

ABSTRACT

The anti-tumor effect of *Adansonia digitata* on Ehrlich ascites carcinoma cells (EAC) is still novel talk. This study is focusing on the role of the extracts of seeds and the fruit pulp of *Adansonia* on the antioxidants activity and the molecular changes of pro-apoptic and anti-apoptic genes expression before and after the treatment of EAC cells bearing mice. Adult female BALB/C mice were used in this study; subgrouped randomly into four groups: control group (non-tumorized); EAC tumorized group, mice was i.p. inoculated with 2.5×10^6 of EAC cells; EAC+ extract of seeds group, tumorized mice was inoculated with 2.5×10^6 of EAC cells and i.p. administered with the extract of *Adansonia* seeds (300 mg/kg b. wt.); EAC+ fruit pulp group, tumorized mice was inoculated with 2.5×10^6 of EAC cells and i.p. administered with the extract of *Adansonia* fruit pulp (300 mg/kg b. wt.). The antioxidant enzymes were inhibited in EAC cells and in ascetic fluid of tumorized mice. Also the oxidative stress was increased significantly in EAC cells bearing mice. The liver was affected with the transplantation of EAC cells as reflected by the imbalance in the antioxidants and oxidants in the EAC cells bearing mice. Moreover, the molecular changes in *p53* and B-cell lymphoma (*Bcl-2*) genes expression were recorded in EAC cells bearing mice. The extracts of *adansonia* have a promising role as antioxidant action due to their antioxidant effect as they ameliorate the imbalance in antioxidants and oxidants balance. The plant extract has anti-apoptosis role by restoring the P53 and Bcl-2 genes expression. Also the plant has antitumor action as they restore tumor markers levels such as α-l-fucosidase and arginase to the normal levels.

Keywords: Ehrlich Ascites Carcinoma; Tumor Markers; Antioxidants; Oxidative Stress; Gene Expression; *Adansonia digitata*; Baobab

1. Introduction

Ehrlich ascites carcinoma (EAC) is transplantable tumors and they were taken into concern in the last 2 to 3 decades. Experimental tumors have vital significance for the purposes of modeling, and EAC is one of the top and famed cases. EAC is indicated to as an undifferentiated carcinoma, and has high transplantable capability, noregression, rapid proliferation, shorter lifespan and 100% malignancy [1]. EAC is used as ascites or a solid form due to these purposes, that is, if ascites fluid contains the tumor cell that injects intraperitoneal, the ascites form is obtained, but if it contains subcutaneously, a solid form is obtained [2]. In 4 or 6 days after inoculation, the ascites fluid is formed and a total of 5 or 12 mL ascites fluid is accumulated [3]. Following the inoculation into the peritoneal cavity of mice, EAC cells proliferating by which the number of cells increases exponentially,

and a plateau phase followed by are sting period, in which a number of cells stay almost constant [4]. Bulan [5] reported that the number of EAC cells increased exponentially in the 9th day after intraperitoneal transplantation of 3×10^6 EAC cells. After inoculation of EAC, the host animal died due to the pressure exerted by the tumor volume and/or the damage that resulted from the tumor [6]. For the accumulation of ascites fluid, whether or not the tumor cells secrete a vascular permeability factor that stimulated the accumulation of ascites fluid was investigated that damage changed the vascular permeability [7]. Altun [8] in another study investigated the liver regeneration in mice with EAC and reported that tumor growth stimulated the regenerative growth.

Oxidative stress exerts toxic effects on cells and is potent in inducing apoptosis [9]. The inhibition of superoxide dismutase (SOD) and catalase (CAT) activities as a result of EAC growth was reported in liver [10]. Aktas [6] and Lobo [11] showed that the antioxidant level such as glutathione (GSH) concentration decreased in EAC cells. Similar decrease in the activity of SOD, CAT, GSH contentand increase in lipid peroxidation was observed in the EAC and the liver of EAC-bearing mice [12]. Overproduction of reactive oxidative stress in SOD antisense-transfected cells was reported to increase pro-apoptotic proteins Bad and Bax and to decrease anti-apoptotic proteins Bcl-2 and Bcl-xL by regulating their phosphorylation and ubiquitination [13]. However, Xu [14] found that the expression of some antioxidant enzymes such as SOD presented a more complex pattern, which differed from the expression of Bcl-2. Among the positive and negative regulators of apoptosis, p53, the tumor suppressor gene, has an important role against cancer as it suppresses tumor growth through two mechanisms, cell cycle arrest and apoptosis [15]. Cell growth and its proliferation are two tightly linked processes. Both are characterized by increasing protein synthesis which are achieved by changes in the activity of ribosomal biogenesis and p53 expression [16]. Bax, thepro-apoptotic member of Bcl-2 family, is a p53 target and istransactivated in a number of systems during p53-mediate-dapoptosis [17]. The up-regulation of Bax expression and down-regulation of Bcl-2 have been well demonstrated during apoptosis [18]. Also Cho [19] showed that H_2O_2-induced apoptosis was accompanied by changes in apoptosis-related factors such as the Bcl-2 family of regulatory proteins. The intracellular concentration of the anti-apoptotic protein Bcl-2 acts as a molecular indicator that influences whether a cell lives or dies [20]. All these studies demonstrated that pro- and anti-apoptotic proteins are the important factors deciding the fate of neoplastic cells.

Some researchers reported that some plant extracts were effective against EAC [21]. Although there are a lot of floristic studies, approximately 10% of the 250,000 complex plant species only were investigated at their chemical and pharmacological sites [7]. The exploring of new toxic agents from natural products has been concerned in collaboration with scientists worldwide [22]. Kamatou [23] reported that it was an excellent antioxidant due to the vitamin C content which was seven to ten times higher than the vitamin C content of oranges. Kamatou [23] suggested that consumption of A. digitata fruit might play an important role in resistance to liver damage in areas where baobab was consumed [24]. Phytochemical investigation showed the presence of flavonoids, phytosterols, amino acids, fatty acids, vitamins and minerals. The seeds are a source of significant quantities of lysine, thiamine, calcium and iron [23]. Recently, the European Commission authorised the import of baobab fruit pulp as a novel food [25] and it was approved in 2009 by the Food and Drug Administration as a food ingredient in the United States of America [26]. Due to the high demand for commercial baobab products in EU and United States, this tree with its edible fruits needs to be conserved and treasured [27].

Upon of these, the study aimed to investigate the role of Adansonia digitata on the biochemical changes associated with the EAC transplantation in mice. Moreover, this is may be the first topic that deals with the vital role of methanolic extracts of seeds and dry powder fruitpulp of Adansonia digitata in regulating the role of pro- and anti-apoptotic genes in tumor cells.

2. Material and Methods

2.1. About Plant

Baobab (Adansonia digitata L.; Bombacaceae) is a large tree with an unusual shape growing in the drier parts of Africa (Sudan). The plant has a wide range of uses, not only as a food and beverage, but also medicinally to treat fevers and dysentery [28]. In recent years, there has been an upsurge of interest in the development of baobab fruits as a botanical dietary supplement in the United States. The fruit pulp affords high levels of vitamin C (range 2.8 - 3 g/kg) and has also been documented as having high antioxidant potency [29]. An aqueous extract of A. digitata fruit pulp has shown anti-inflammatory and analgesic effects in rat models, but at quite a high dose range (400 - 800 mg/kg, p.o.) [30]. The various plant parts of baobab have been subjected to relatively few studies phytochemically, and, for example, a number of proanthocyanidins of previously known structure were reported as major constituents from an 80% methanol extract of the fruit pulp [31]. The acute toxicity of baobab fruit pulp extract was tested in vivo on rats and the results showed that the LD50 was 8000 mg/kg following parenteral administration suggesting low toxicity [30].

2.2. Plant Extract

The air-dried and powdered of seeds and fruit pulp (10 g) were extracted with 200 ml methanol by Soxhlet extraction for 18 hrs. The residue was dried overnight and then extracted with 250 ml water by using a shaking water bath at 70°C for 2 hrs. The obtained methanolic and water extracts were filtered and evaporated by using a rotator evaporator. The dried extracts of seeds and fruits pulp were stored at −80°C until use. The method of extraction is used herein with a little modification as Mothana [32].

2.3. Tumor Cell Line

A line of Ehrlich Ascites Carcinoma (EAC) obtained from Egyptian National Cancer Institute, Cairo, Egypt.

The tumor line was maintained in female BALB/C mice by weekly intraperitoneal injection of 2.5×10^6 cells/mouse/0.25 ml of phosphate buffer saline (PBS). EAC cells counts were done in a Neubauer hemocytometer using the trypan blue dye. Cell viability was always found to be 95% or more. Tumor cell suspensions were prepared in PBS. EAC cell line is characterized by its reasonable rapid growth which could not kill the animal due to the accumulation of Ascites.

2.4. Animal Grouping

The experiment has been used three month old female BALB/c mice 40 ± 5 g. The local committee approved the design of the experiments and the protocol conforms to the guidelines of the National Institutes of Health (NIH). All measures were taken to minimize the number of mice used and their suffering. Animals were kept in vivarium, housed in polycarbonate cages, 10 mice per cage, at a constant room temperature of $23°C \pm 1°C$ with a 12-h light/dark cycle (light on from 7:00 AM till 7:00 PM). Standard rat chow diet and water were available *ad libitum*. Animals were divided into four groups (Ten/group); Control group: mice were received no treatment; Ehrlich Ascites Carcinoma (EAC) group: mice were inoculated intraperitoneally with of 2.5×10^6 cells/mouse/0.25 ml of PBS; EAC+ seed extract group: Tumorized mice were daily intraperitoneally administered with seed extract of *Adansonia* at 300 mg/kg b. wt. for 14 days; EAC+ fruits extract group: Tumorized mice were daily intraperitoneally administered with fruits extract of *Adansonia* at 300 mg/kg b. wt. for 14 days. At the end of the experimental period mice were sacrificed by decapitation then blood, ascites fluid, EAC cells, and liver tissue were collected from each group. EAC cells from EAC bearing-mouse were homogenized then centrifuged at 2000 rpm for 10 min at 4°C. The ascetic fluid obtained was used for the estimation of the antioxidant enzymes such as SOD, GSH-Px in different animal groups. Also the EAC cells pellet and liver tissues were immediately homogenized in a phosphate buffer solution pH 7.4 and centrifuged at 2000 rpm for 10 min at 4°C. The liver samples were then rinsed in ice-cold normal saline followed by cold 0.15 M Tris-HCl (pH 7.4), blotted dry, and weighed. A 10% w/v homogenate was prepared in 0.15 M Tris-HCl buffer; of which a portion was utilized for the estimation of lipid peroxidation and H_2O_2 and a second portion was used for the estimation of GSH after precipitating proteins with TCA. The supernatant of liver tissues, EAC cells, ascites fluid and sera were stored at $-80°C$ until be used in the biochemical assay. The EAC-cells that be used in the molecular study was stored at $-80°C$ until be further analyzed.

2.5. Enzymatic Assay

2.5.1. Superoxide Dismutase (SOD) Activity

This assay relied on the ability of the enzyme to inhibit the phenazinemethosulphate-mediated reduction of nitrobluetetrazolium dye [33].

2.5.2. Glutathione Peroxidase (GSH-Px) Activity

The activity was measured by the method described by Ellman [34]. Briefly, the reaction mixture contained 0.2 ml of 0.4 M phosphate buffer (pH 7.0), 0.1 ml of 10 mM sodium azide, 0.2 ml of supernatant (homogenate on 0.4 M phosphate buffer, pH 7.0), and 0.2 ml glutathione, 0.1 of 0.2 mM H_2O_2. The reaction content was incubated at 37°C for 10 min. The reaction was arrested by adding 0.4 ml of 10% TCA and then centrifuged. The molar extinction coefficient for NADPH is 6220 $mM^{-1} \cdot cm^{-1}$ at 340 nm.

2.5.3. GSH (GSH) Content

GSH served as an antioxidant, reacting with free radicals and organic peroxides, in amino acid transport, and as a substrate for the GSH-Px and glutathione-S-transferase in the detoxification of organic peroxide and metabolism of xenobiotics, respectively. A tissue was homogenized in 10 ml cold buffer (50 mM potassium phosphate, pH 7.5, 1mM EDTA) per gram tissue. Centrifuge at 100,000 x g for 15min at 4°C was performed. The supernatant was transferred into a new sterile tube for assay and the rest of the samples stored at $-80°C$ for further work. The method based on the reduction of 5,5'-Dithiobis(2-nitrobenzoic acid) (DTNB) with GSH to produce a yellow compound. The reduced chromogen directly proportional to GSH concentration and its absorbance can be measured at 405 nm [35].

2.5.4. Malondialdehyde (MDA) Level

The thiobarbituric acid reactive substances (TBARS) as malondialdehyde were estimated by the method of Ohkawa [36]. Briefly, to 0.2 ml of homogenate, 0.2 ml of 40% sodium dodecylsulphate, 1.5 ml of 20% acetic acid (prepared in 0.27 M of HCl) and 1.5 ml of 0.5% thiobarbituric acid were mixed together. The mixture was heated for 60 min at 95°C in a water bath to give a pink color. The mixture was then centrifuged at 3500 rpm for 10 min. Finally absorbance of the supernatant layer was read spectrophoto-metrically at 532 nm, the molar extinction coefficient factor equal 1.56×10^5 $M^{-1} \cdot cm^{-1}$.

2.5.5. Hydrogen Peroxide (H_2O_2) Level

Its concentration was measured according to the method of Aebi [37]. The principle of this method based on that, in the presence of peroxidase, H_2O_2 reacts with 3,5-dichloro-2-hydroxy benzene sulfonic acid (DHBS) and 4-aminophenazone (AAP) to form a chromophore. The absorbance was read spectrophotometrically at 510 nm.

2.5.6. Bilirubin Content

The reaction between bilirubin and the diazonium salt of sulphanilic acid produced azobilirubin which shows a maximum absorption at 535 nm in an acid medium [38].

2.5.6. Statistical Analysis of Biochemical Data

The biochemical data recorded were expressed as mean ± SD and statistical and correlation analyses were undertaken using the One-way ANOVA followed by a post-hoc LSD (Least Significant Difference) test. A P value < 0.05 was statistically significant. A Statistical analysis was performed with the Statistical Package for the Social Sciences for Windows (SPSS, version 10.0, Chicago, IL, USA).

2.5.7. Extraction of Total RNA from EAC Homogenates

Total RNA was isolated from EAC pellets using RNeasy Mini Kit according to manufacturer's instructions (QIAGEN, Germany). About 100 µl of homogenate was subjected to RNA extraction and the resultant RNA was dissolved in DEPC-treated water, quantified spectrophotometrically and analyzed on 1.2% agarose gel. RNAs inhibitors were added to the samples during the RNA extraction process.

2.5.8. Real Time PCR and Gene Expression

1) For Pro-Apoptic and Anti-Apoptic Genes (*P*53 and *Bcl*2)

The extracted RNA was subjected to Real Time PCR reaction to examine the expression *p*53 and *Bcl*2 genes in EAC, EAC+ Seeds extract and EAC+ fruits extract of *Adansonia* using specific primers in the presence of glyceraldehyde-3-phosphate dehydrogenase (GPDH) as a housekeeping gene. The Real time reaction consists of 12.5 µl of 2X Quantitech SYBR® Green RT Mix (Fermentaz, USA), 2 µl of the extracted RNA (50 ng/µl), 1 µl of 25 pM/µl forward (F) primer, 1 µl of 25 pM/µl reverse (R) primer (**Table 1**), 9.5 µl of RNAase free water for a total of 25 µl. Samples were spun before loading in the rotor's wells. The real time PCR program was performed as follows: initial denaturation at 95°C for 10 min.; 40 cycles of 64°C for 15 sec for *p*53, annealing at 63°C for 30 sec for *Bcl*2 and extension at 72°C for 30 sec. Data acquisition performed during the extension step. This reaction was performed using Rotor-Gene 6000 system (QIAGEN, USA).

2.5.9. Molecular Data Analysis

Comparative quantitation analysis was performed using Rotor-Gene-6000 Series Software based on the following equation:

$$\text{Ratio target gene expression} = \frac{\text{Fold change in target gene expresion}\left(\dfrac{\text{sample}}{\text{control}}\right)}{\text{Fold change in reference gene expresion}\left(\dfrac{\text{house keeping gene}}{\text{control}}\right)}$$

Real-time PCR data of all samples were analyzed with appropriate bioinformatics and statistical program for the estimation of the relative expression of genes using real-time PCR and the result normalized to its gene (Reference gene). The data were statistically evaluated, interpreted and analyzed using Rotor-Gene-6000 version 1.7.

3. Results

As in **Table 2**, the sera bilirubin and tumour markers such as arginase and alpha-l-fucosidase levels were significant increase in the transplant EAC group ($p \leq 0.001$)

when compared with the control group. The extracts of seeds and pulp fruits of *Adansonia* showed significant decrease in the tumor markers when compared with EAC group. Moreover, the level of bilirubin was ameliorated with the administration of the extracts of seeds and fruits of *Adansonia*.

The EAC group in **Table 3**, showed very high significant decrease ($P \leq 0.001$) in the sera superoxide dismutase and glutathione peroxidase activities and reduced glutathione content when compared with control group. The oxidative stress was determined as an increment in

Table 1. Primers nucleotides sequence used in this study.

Primers name	Primer sequence from 5'-3'	Annealing temp.
*P*53	F-AGGGATACTATTCAGCCCGAGGTG R-ACTGCCACTCCTTGCCCCATTC	64°C
*Bcl*2	F-ATGTGTGTGGAGAGCGTCAACC R-TGAGCAGAGTCTTCAGAGACAGC	63°C
GPDH (Housekeeping gene)	F-ATTGACCACTACCTGGGCAA R-GAGATACACTTCAACACTTTGACCT	60°C - 65°C

GPDH: Glyceraldehyde-3-phosphate dehydrogenase.

Table 2. Serum bilirubin and tumor markers of different treated groups.

Groups Parameters	Control	EAC	EAC+ Seeds Extract	EAC+ Fruit Extract
Bilirubin (U/dl)	0.51 ± 0.08	$1.1 \pm 0.12^{***\,\text{I II}}$	$0.87 \pm 0.14^{***\,\text{I II}}$	$0.79 \pm 0.02^{***\,\text{I II}}$
Arginase (U/dl)	31.8 ± 2.2	$110.6 \pm 5.1^{***}$	$63.6 \pm 2.5^{***\,\text{I II}}$	$72.0 \pm 1.6^{***\,\text{I II}}$
L-Fucosidase (U/dl)	21.8 ± 10.4	$75.1 \pm 6.9^{***}$	$56.9 \pm 11.2^{***\,\text{II}}$	$53.0 \pm 3.9^{***\,\text{I II}}$

All data ± standard deviation (SD), each group equal 5 rats, $p \leq 0.001^{***}$, represented the comparison of EAC, EAC+ Seeds Extract and EAC+ Fruit Extract with the control group; $p \leq 0.001^{\text{I II}}$, represented the comparison of EAC+ Seeds Extract and EAC+ Fruit Extract with the EAC group.

Table 3. The serumantioxidants activities and oxidative stress in different treated groups.

Groups Parameters	Control	EAC	EAC+Seeds Extract	EAC+Fruit Extract
SOD (U/ml)	3.5 ± 0.1	$1.3 \pm 0.2^{***}$	$2.6 \pm 0.4^{***\,\text{I II}}$	$2.3 \pm 0.4^{***\,\text{I II}}$
GSH-Px (U/ml)	162.6 ± 19.7	$55.2 \pm 7.8^{***}$	$74.1 \pm 5.7^{***\,\text{I}}$	$77.6 \pm 5.9^{***\,\text{II}}$
GSH (mg/mg protein)	28.7 ± 3.3	$14.1 \pm 0.87^{***}$	$18.4 \pm 1.78^{***\,\text{II}}$	$18.1 \pm 1.5^{***\,\text{II}}$
MDA (n mol/ml)	82.7 ± 5.3	$124.2 \pm 9.1^{***}$	$101.8 \pm 9.1^{***\,\text{I II}}$	$98.2 \pm 5.9^{***\,\text{I II}}$
H_2O_2 (U/ml)	71.1 ± 10.7	$146.8 \pm 7.4^{***}$	$128.1 \pm 3.8^{***\,\text{I II}}$	$118.9 \pm 3.6^{***\,\text{I II}}$

All data ± standard deviation (SD), each group equal 5 rats, $p \leq 0.001^{***}$, represented the comparison of EAC, EAC+ Seeds Extract and EAC+ Fruit Extract with the control group; $p \leq 0.01^{\text{II}}$, $p < 0.001^{\text{I II}}$, represented the comparison of EAC+ Seeds Extract and EAC+ Fruit Extract with the EAC group.

TBARS and H_2O_2 in the EAC group. The methanolic extracts of seeds and fruits of *Adansonia* ameliorate the deflection of antioxidants status as they increase the antioxidant enzymes as well as decrease the TBARS and H_2O_2 in the EAC+ Seeds Extract and EAC+ Fruit Extract groups when compared with EAC group.

As in **Table 4** the EAC+ Seeds Extract and EAC+ Fruit Extract groups showed high significant increase in the Ascites fluid superoxide dismutase and glutathione peroxidase activities and reduced glutathione content when compared with EAC group. The extracts of seeds and fruits of *Adansonia* showed high significant decrease the TBARS and H_2O_2 in the EAC+ Seeds Extract and EAC+ Fruit Extract groups when compared with EAC group.

As in **Table 5** the EAC+ Seeds Extract and EAC+ Fruit Extract groups showed high significant increase in the EAC cells superoxide dismutase and glutathione peroxidase activities and reduced glutathione content when compared with EAC group. The extracts of seeds and fruits of *Adansonia* ameliorate the imbalance in antioxidants status as they increase the antioxidant enzymes as well as decrease the TBARS and H_2O_2 in the EAC+ Seeds Extract and EAC+ Fruit Extract groups when compared with EAC group.

The EAC group in **Table 6**, showed very high significant decrease ($P \leq 0.001$) in the liver superoxide dismutase and glutathione peroxidase activities and reduced glutathione content when compared with control group. The oxidative stress was estimated as an increment in TBARS and H_2O_2 in the EAC group. The extracts of seeds and fruits of *Adansonia* ameliorate the imbalance

Table 4. The Ascites fluid antioxidants activities and oxidative stress in different treated groups.

Groups Parameters	EAC	EAC+ Seeds Extract	EAC+ Fruit Extract
SOD (U/ml)	90.4 ± 9.7	$180.3 \pm 14.5^{***}$	$193.6 \pm 12.1^{***}$
GSH-Px (U/ml)	101.48 ± 6	$117 \pm 5.5^{**}$	$124.6 \pm 10.4^{***}$
GSH (mg/ml)	19 ± 0.98	$29.1 \pm 1.3^{***}$	$31.1 \pm 3.3^{***}$
MDA (nmol/ml)	145.6 ± 8.3	$111.8 \pm 13.8^{***}$	$126.4 \pm 11.1^{*}$
H_2O_2 (U/ml)	17.8 ± 1.2	$11.8 \pm 0.91^{***}$	$12.1 \pm 0.61^{***}$

All data ± standard deviation (SD), each group equal 5 rats, $p < 0.05^{*}$, $p \leq 0.01^{**}$, $p \leq 0.001^{***}$, represented the comparison of EAC+ Seeds Extract and EAC+ Fruit Extract with the EAC group.

Table 5. The EAC cells antioxidants activities and oxidative stress in different treated groups.

Groups Parameters	EAC	EAC+ Seeds Extract	EAC+ Fruit Extract
SOD (U/g protein)	1.6 ± 0.13	$2.1 \pm 0.12^{**}$	$1.9 \pm 0.46^{*}$
GSH-Px (U/g protein)	92 ± 4.9	$104.5 \pm 8.4^{**}$	$110.9 \pm 7.4^{***}$
GSH (U/g protein)	32.9 ± 17.3	$46.6 \pm 27.7^{***}$	$43.9 \pm 29.4^{***}$
MDA (n mol/g protein)	148.6 ± 11.9	$121.8 \pm 6.8^{***}$	$128.4 \pm 14.6^{**}$
H_2O_2 (U/g protein)	15.7 ± 1.8	$10.4 \pm 0.76^{***}$	$11.1 \pm 1.5^{***}$

All data ± standard deviation (SD), each group equal 5 rats, $p < 0.05^{*}$, $p \leq 0.01^{**}$, $p \leq 0.001^{***}$, represented the comparison of EAC+ Seeds Extract and EAC+ Fruit Extract with the EAC group.

antioxidants status as they increase the antioxidant enzymes as well as decrease the TBARS and H_2O_2 in the EAC+ Seeds Extract and EAC+ Fruit Extract groups when compared with EAC group.

The real time PCR results (**Figure 1**) showed that the expression of *p53* and *Bcl2* genes was decreased in EAC+ Seeds extract group when compared with the EAC one. Also the extract of fruit extract decreased the expression of both genes EAC+ Fruit Extract treated group when compared with the EAC group.

4. Discussion

The transplantation of EAC neoplastic cells into the mice causes deflection in the balance of antioxidant/oxidative stress system both in EAC and liver tissue. The inhibition of SOD and GSH-Px activities as a result of EAC growth was reported in liver [39]. Corresponding decrease in the levels of SOD and GSH was observed in the liver of EAC-bearing mice [12]. The decrease in the activities of antioxidants may be due to the increase of oxidative stress induced by the inoculation of EAC. There was increase in the levels of thiobarbituric acid reactive substance representing by malondialdehyde (MDA) and hydrogen peroxide in the EAC cells, ascetic fluid and liver homogenates in EAC group when compared with the

EAC treated groups. The extracts of seeds and fruits pulp of *Adansonia* moderate the levels of MDA and H_2O_2 in EAC cells, ascetic fluid and liver homogenates. This amelioration may be due to the antioxidant effect of *Adansonia*. Vertuani [40] showed that the fruit pulp exhibited strong anti-oxidant activity corresponding to 6 - 7 mmol/g of Trolox, in comparison to the fruit pulps of orange (0.1 mmol/g), strawberry (0.90 mmol/g), apple (0.16 mmol/g) and kiwi (0.34 mmol/g).The antioxidant activity of fresh ripe fruit of *A. digitata* was 1000 mg AEAC/100g (ascorbic acid equivalent antioxidant content) [41]. So the significant increase in the antioxidant enzymes activity such as SOD, GSH-Px in EAC and liver due to the higher antioxidant effect of *Adansonia*. Nhukarume [42] investigated the ability of solvent extracts of various fruits including baobab to inhibit the peroxidetion of lipids. This may explain the significant decrease in the MDA and H_2O_2 levels after the treatment with the extract of *Adansonia*. The hepatoprotective activity of a water extract of the fruit pulp was evaluated in vivo against chemical-induced toxicity with CCL4 in rats [23]. In the present study, EAC-bearing mice showed to be under higher oxidative stress than control animals indicated by elevated lipid peroxidation and reduced endogenous antioxidants in the liver. Tumors in the human body or in

Table 6. The liver antioxidants activities and oxidative stress in different treated groups.

Groups Parameters	Control	EAC	EAC+Seeds Extract	EAC+Fruit Extract
SOD (U/g tissue)	32.4 ± 2.2	$14.4 \pm 7.04^{***}$	$22.1 \pm 2.01^{***\text{II}}$	$21.8 \pm 1.9^{***\text{II}}$
GSH-Px (U/g tissue)	107.1 ± 6.5	$86.5 \pm 10.5^{***}$	$103.3 \pm 8.9^{\text{II}}$	$97.6 \pm 11^{*}$
GSH(mg/g tissue)	96.31 ± 29.1	$54.58 \pm 14.3^{***}$	$84.13 \pm 25.1^{***\text{I II}}$	$84.34 \pm 38.6^{***\text{I II}}$
MDA (n mol/g tissue)	92.8 ± 13.7	$148.2 \pm 6^{***}$	$126.3 \pm 2.3^{***}$	$132 \pm 5.2^{***}$
H_2O_2(U/g tissue)	9.0 ± 0.91	$16.8 \pm 2.2^{***}$	$9.9 \pm 0.89^{\text{I II}}$	$11.0 \pm 1.9^{\text{I II}}$

All data $\pm$ standard deviation (SD), each group equal 5 rats, $p < 0.05^{*}$, $p \leq 0.001^{***}$, represented the comparison of EAC, EAC+ Seeds Extract and EAC+ Fruit Extract with the control group; $p \leq 0.01^{\text{II}}$, represented the comparison of EAC+ Seeds Extract and EAC+ Fruit Extract with the EAC group.

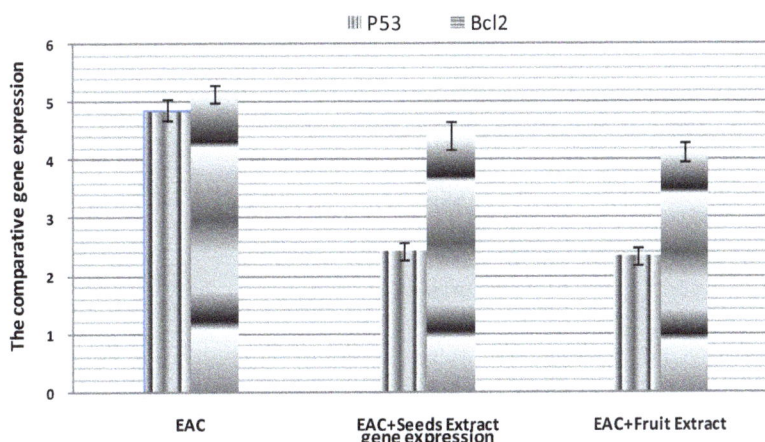

Figure 1. *P53* and *Bcl2* genes expression in Ehrlich Ascites Carcinoma cells of different groups of rats.

experimental animals are known to affect many functions of the vital organs, especially the liver, even when the site of the tumor does not interfere directly with organ function [43]. However, the decrease in bilirubin indicates the improvement of liver function, so the treated mice with *adansonia* improve the liver function. This is may be due totriterpenoids, β-sitosterol, β-amyrinpalmitate, and/or α-amyrin and ursolic acid present in thefruit pulp [24]. The reduction in oxidative stress leads to subsequent decrease in injury and damage of the hepatocytes membranes. It is suggested that if oxidative stress is involved in the origin of EAC-induced liver oxidative injury, then a successful extract with antioxidant potential should protect against that injury. Seeds and fruit pulp of *Adansonia* ameliorated oxidative stress in seeds and fruit pulp + EAC-bearing mice and prevented cellular injury as evidenced by a decrease in levels of TBARS in liver. This is associated with increase in GSH level as well as the activities of SOD, and GSH-Px compared with EAC-implanted mice. The control of endogenous redox state might contribute to the demonstrated increase in the life span of EAC-tumor bearing mice treated with *Adansonia*. Thus, the tumor markers such as α-l-fucosidase and arginase indicate that *Adansonia* exhibited significant anti-tumor and might show potential antioxidant activity in EAC-bearing mice. This might be via changes in the expression of endogenous antioxidants. The administration of seeds and fruit pulp of *Adansonia* significantly increased the levels of GSH, SOD and GSH-Px as well as decreased lipid peroxidation in EAC cells, ascetic fluid and liver homogenate compared with that obtained from control EAC group. This indicates that *Adansonia* extracts altered the redox state in the sera, EAC cells, ascetic fluid and liver tissue.

Apoptosis is a programmed cell death where the cell expends energy towards its own end. It is well known that functional *p53* provides a protective mechanism against tumor growth. It acts as a tumor suppressor through its capacity to induce cell cycle arrest and apoptosis in response to a variety of chemotherapeutic drugs [44]. Cell growth and its proliferation are two tightly linked processes. Both are characterized by increased protein synthesis which is achieved by changes in the activity of ribosomal biogenesis and *p53* expression [16]. It is reported that deregulated signaling in cancer cells drives excessive ribosome biogenesis within the nucleolus, which elicits uncontrolled cell growth and proliferation. It is controlled by a complex interconnection of proteins such as *p53* and is often triggered by oxidative stress and the release of cytochrome *c* from the mitochondria [19]. *Bcl*-2 indicates a family of genes and proteins found in rats that regulates the outer membrane permeability of mitochondria, and most importantly the fate of the cell through programmed cell death, apoptosis.

The increased level of ROS is involved in generating the signal that causes change the permeability of the mitochondrial membrane, and, thus, the release of cytochrome *c* into the cytosol. Once this occurs, the initiation of the cascade of caspases occurs [45]. Activation of caspases will eventually lead to the death of the cell and other surrounding cells. H_2O_2-induced apoptosis is accompanied by changes in apoptosis-related factors such as the *Bcl*-2 family of regulatory proteins [19]. However, ROS such as H_2O_2 act not only as cellular messengers capable of causing oxidative damage to macromolecules, but also as signaling molecules that activate protein kinase cascades [46]. The high expression of *p53* and *Bcl*-2 genes due to their molecular changes accompanied with tumor development in mice. The impairment of nucleolar function might stabilize *p53* by preventing its degradation and, therefore, for the arrest of cell-cycle progression [10]. Therefore, the present finding indicatesthat seeds and fruit pulp of *Adansonia* might be able to attenuate apoptotic mechanisms through the changes of *p53* and *Bcl*-2 gene expression leading to the control of growth and proliferation of EAC cells.

5. Conclusion

The study demonstrated that extracts of seeds and fruit pulp of *Adansonia*could have anti-tumor action through modulation of redox state in sera, EAC cells, ascetic fluid and liver tissue. These extracts were able to restore and modulate the tumor markers levels such as α-l-fucosidase and arginase activity in the sera of EAC bearing mice. Also *Adansonia* attenuates the *p53* and *Bcl*-2 gene expression as pro-apoptic and anti-apoptic genes leading to management of tumor growth. The study initiates human to use *Adansonia* as a good edible stuff that has strong antioxidant action and anti-tumor agent.

6. Acknowledgements

The author appreciates the helpful effort of his colleagues Dr. Ali H. Amin and Dr. Mohammed E. Elbeeh, zoology department, faculty of science, Mansoura University, Egypt, during the practical part of this work.

REFERENCES

[1]   Ö Kaleoglu and N. Isli, "Ehrlich-Lettre Asit Tümörü," *Tıp Fakültesi Mecmuası*, Vol. 40, No. 1, 1977, pp. 978-984.

[2]   H. G. Okay, "Deneysel EAT Olu_turulan Fare Karaciger Plazmasında Nitrik Oksit Metabolizmasının Incelenmesi. Yüksek Lisans Tezi," Istanbul Üniversitesi Saglık Bilimleri Enstitüsü, Biyokimya ABD, Istanbul, 1998.

[3]   H. Gümüshan and I.P., I.V. ve S.C., "Yollarla Uygulanan Adriamycin'in Ehrlich Asit Tümörü (EAT) Tasıyan Fare-

lere Etkileri Üzerine Bir Çalısma," Harran Üniversitesi Fen Bilimleri Enstitüsü, Yüksek Lisans Tezi, Sanlıurfa, 2002.

[4] Z. Song, J. Varani and I. J. Goldstein, "Differences in Cell Surface Carbohydrates, and in Laminin and Fibronectin Synthesis between Adherent and Non-Adherent Ehrlich Ascites Tumor Cells," *International Journal of Cancer*, Vol. 55, No. 6, 1993, pp. 1029-1035.

[5] Ö. Bulan, "Ehrlich Ascites Tümör Hücrelerinde Yaslanmaile Hücre Kinetigi Arasındaki Iliskiler," YüksekLisans Tezi, IstanbulÜniversitesi Fen Bilimleri Enstitüsü, Istanbul, 1990.

[6] E. Aktas, Ehrlich Asit Sıvısının L-Hücrelerinin Çogalma Hızına Etkisi," Yüksek Lisans Tezi, Istanbul Üniversitesi Fen Bilimleri Enstitüsü, Istanbul, 1996.

[7] M. Ozaslan, I. D. Karagoz, I. H. Kilic and M. E. Guldur, "Ehrlich Ascites Carcinoma," *African Journal of Biotechnology*, Vol. 10, No. 13, 2011, pp. 2375-2378.

[8] S. Altun, "Normal, Tümöralve Rejeneratif Büyümeler Arasındaki Kinetik Iliskiler," *Turkish Journal of Biology*, Vol. 20, No. 3, 1996, pp. 153-173.

[9] T. Andoh, P. B. Chock and C. C. Chiueh, "The Roles of Thioredoxin in Protection against Oxidative Stress-Induced Apoptosis in SH-SY5Y Cells," *The Journal of Biological Chemistry*, Vol. 277, 2002, pp. 9655-9660.

[10] M. R. Bleavins, D. A. Brott, J. D. Alvey and F. A. de la Iglesia, "Flow Cytometric Characterization of Lymphocyte Subpopulations in the Cynomolgus Monkey (*Macaca fascicularis*)," *Veterinary Immunology and Immunopathology*, Vol. 37, No. 1, 1993, pp. 1-13.

[11] C. Lobo, M. A. Ruiz-Bellido, J. C. Aledo, J. Marquez, I. N. De Castro and F. J. Alonso, "Inhibition of Glutaminase Expression by Antisense mRNA Decreases Growth and Tumourigenicity of Tumour Cells," *Biochemical Journal*, Vol. 348, 2000, pp. 257-261.

[12] M. A. El-Missiry, A. I. Othman, M. A. Amer and E. Mohamed, "Ottelione a Inhibited Proliferation of Ehrlich Ascites Carcinoma Cells ın Mice," *Chemico-Biological Interactions*, Vol. 200, No. 2-3, 2012, pp. 119-127.

[13] D. Li, E. Ueta, T. Kimura, T. Yamamoto and T. Osaki, "Reactive Oxygen Species (ROS) Control the Expression of Bcl-2 Family Proteins by Regulating Their Phosphorylation and Ubiquitination," *Cancer Science*, Vol. 95, No. 8, 2004, pp. 644-650.

[14] H. Xu, R. J. Steven and X. M. Li, "Dose-Related Effects of Chronic Antidepressants on Neuroprotective Proteins BDNF, Bcl-2 and Cu/Zn-SOD in Rat Hippocampus," *Neuropsychopharmacology*, Vol. 28, 2003, pp. 53-62.

[15] T. Das, G. Sa, P. Sinha and P. K. Ray, "Induction of Cell Proliferation and Apoptosis: Dependence on the Dose of the Inducer," *Biochemical and Biophysical Research Communications*, Vol. 260, No. 1, 1999, pp. 105-110.

[16] G. Donati, L. Montanaro and M. Derenzini, "Ribosome Biogenesis and Control of Cell Proliferation: p53 Is Not Alone," *Cancer Research*, Vol. 72, 2012, pp. 1602-1607.

[17] T. Miyashita and J. C. Reed, "Tumor Suppressor p53 Is a Direct Transcriptional Activator of the Human Bax Gene," *Cell*, Vol. 80, No. 2, 1995, pp. 293-299.

[18] T. Miyashita, S. Krajewski, M. Krajewska, H. G. Wang, H. K. Lin, D. A. Liebermann, B. Hoffman and J. C. Reed, "Tumor Suppressor p53 Is a Regulator of Bcl-2 and Bax Gene Expression *in Vitro* and in *Vivo*," *Oncogene*, Vol. 9, No. 6, 1994, pp. 799-1805.

[19] D. H. Cho, T. N. Tomohiro, J. Fang, P. Cieplak, A. Godzik, Z. Gu and S. A. Lipton, "S-Nitrosylation of Drp1 Mediates β-Amyloid-Related Mitochondrial Fission and Neuronal Injury," *Science*, Vol. 324, No. 5923, 2009, pp. 102-105.

[20] X. R. Cheng, L. Zhang, J. J. Hu, L. Sun and G. H. Du, "Neuroprotective Effects of Tetramethylpyrazine on Hydrogen Peroxide-Induced Apoptosis in PC12 Cells," *Cell Biology International*, Vol. 31, 2007, pp. 438-443.

[21] M. Ozaslan, I. D. Karagöz, M. E. Kalender, I. H. Kilic, I. Sari and A. Karagöz, "*In Vivo* Antitumoral Effect of *Plantago major* L. Extract on Balb/C Mouse with Ehrlich Ascites Tumor," *The American Journal of Chinese Medicine*, Vol. 35, No. 5, 2007, pp. 841-851.

[22] G. M. Cragg and D. J. Newman, "Discovery and Development of Antineoplastic Agents from Natural Sources," *Cancer Investigation*, Vol. 17, No. 2, 1999, pp. 153-163.

[23] G. P. P. Kamatou, I. Vermaak and A. M. Viljoen, "An Updated Review of *Adansonia digitata*: A Commercially Important African Tree," *South African Journal of Botany*, Vol. 77, No. 4, 2011, pp. 908-919.

[24] A. A. Al-Qarawi, M. A. Al-Damegh and S. A. El-Mougy, "Hepatoprotective Influence of *Adansonia digitata* Pulp," *Journal of Herbs Spices and Medicinal Plants*, Vol. 10, No. 3, 2003, pp. 1-6.

[25] C. Buchmann, S. Prehsler, A. Hartl and C. R. Vogl, "The Importance of Baobab (*Adansonia digitata* L.) in Rural West African Subsistence—Suggestion of a Cautionary Approach to International Market Export of Baobab Fruits," *Ecology of Food and Nutrition*, Vol. 49, No. 3, 2010, pp. 145-172.

[26] R. Addy, "Baobab Fruit Approved as Food Ingredient in US," 2009. http://www.nutraingredientsusa.com/content/view/print/2 59574,2009.

[27] A. C. Sanchez, P. E. Osborne and N. Haq, "Identifying the Global Potential for Baobab Tree Cultivation Using Ecological Niche Modeling," *Agroforestry Systems*, Vol. 80, No. 2, 2010, pp. 191-201.

[28] J. Gebauer, K. El-Siddig and G. Ebert, "Baobab (*Adan-*

sonia digitata L.): A Review on a Multipurpose Tree with Promising Future in Sudan," *Gartenbauwissenschaft*, Vol. 67, No. 4, 2002, pp. 155-160.

[29] S. Vertuani, E. Braccioli, V. Buzzoni and V. S. Manfredini, "Antioxidant Capacity of Adansonia Digitata Fruit Pulp and Leaves," *Acta Phytotherapeutica*, Vol. 5, No. 2, 2002, pp. 2-7.

[30] A. A. Shahat, "Procyanidins from *Adansonia digitata*," *Pharmaceutical Biology*, Vol. 44, No. 6, 2006, pp. 445-450.

[31] R. A. Mothana, U. Lindequist, R. Gruenert and P. J. Bednarski, "Studies of the *in Vitro* Anticancer, Antimicrobial and Antioxidant Potentials of Selected Yemeni Medicinal Plants from the Island Soqotra," *BMC Complementary & Alternative Medicine*, Vol. 9, 2009, pp. 7-18.

[32] M. Nishikimi, N. A. Rao, *et al.*, "The Occurrence of Superoxide Anion in the Reaction of Reduced Phenazine Methosulfate and Molecular Oxygen," *Biochemical and Biophysical Research Communications*, Vol. 46, No. 2, 1972, pp. 849-854.

[33] G. L. Ellman, "Tissue Sulfhydryl Groups," *Archives of Biochemistry and Biophysics*, Vol. 82, No. 1, 1959, pp. 70-77.

[34] E. Beutler, O. Duron, B. M. Kelly, "Improved Method for the Determination of Blood Glutathione," The *Journal of Laboratory and Clinical Medicine*, Vol. 61, No. 1, 1963, pp. 882-888.

[35] H. Ohkawa, N. Ohishi, *et al.*, "Assay for Lipid Peroxides in Animal Tissues by Thiobarbituric Acid Reaction," *Analytical Biochemistry*, Vol. 95, No. 2, 1979, pp. 351-358.

[36] H. Aebi, "Catalase *in Vitro*," *Methods in Enzymology*, Vol. 105, 1984, pp. 121-126.

[37] M. Walter and H. Gerade, "An Ultramicromethod for the Determination of Conjugated and Total Bilirubin in Serum or Plasma," *Microchemical Journal*, Vol. 15, No. 2, 1970, pp. 231-243.

[38] A. Lamien-Meda, C. E. Lamien, M. M. Y. Compaoré, R.

N. T. Meda, M. Kiendrebeogo, B. Zeba, J. F. Millogo and O. G. Nacoulma, "Polyphenol Content and Antioxidant Activity of Fourteen Wild Edible Fruits from Burkina Faso," *Molecules*, Vol. 13, No. 3, 2008, pp. 581-594.

[39] M. Gupta, U. K. Mazumder, R. S. Kumar, T. Sivakumar and M. L. Vamsi, "Antitumor Activity and Antioxidant Status of Caesalpinia Bonducella against Ehrlich Ascites Carcinoma in Swiss Albino Mice," *Journal of Pharmacological Science*, Vol. 94, 2004, pp. 177-184.

[40] A. Lamien-Meda, C. E. Lamien, M. M. Compaoré, R. N. Meda, M. Kiendrebeogo, B. Zeba, J. F. Millogo and O. G. Nacoulma, "Polyphenol Content and Antioxidant Activity of Fourteen Wild Edible Fruits from Burkina Faso," *Molecules*, Vol. 13, No. 3, 2008, pp. 581-594.

[41] L. Nhukarume, Z. Chikwambi, M. Muchuweti and B. Chipurura, "Phenolic Content and Antioxidant Capacities of *Parinari curatelifolia*, *Strychnos spinosa* and *Adansonia digitata*," *Journal of Food Biochemistry*, Vol. 34, Suppl. 1, 2008, pp. 207-221.

[42] N. Senthilkumar, S. Badami, S. H. Dongre and S. Bhojraj, "Antioxidant and Hepatoprotective Activity of the Methanol Extract of *Careya arborea* Bark in Ehrlich Ascites Carcinoma-Bearing Mice," *Journal of Natural Medicines*, Vol. 62, No. 3, 2008, pp. 336-339.

[43] C. P. Rubbi and J. Milner, "P53: Guardian of a Genome's Guardian? *Cell Cycle*, Vol. 2, No. 1, 2003, pp. 20-21.

[44] M. P. Mattson, "Apoptosis in Neurodegenerative Disorders," *Nature Reviews. Molecular Cell Biology*, Vol. 1, No. 2, 2000, pp. 120-129.

[45] R. B. Petersen, A. Nunomura, H. G. Lee, G. Casadesus, G. Perry, M. A. Smith and X. Zhu, "Signal Transduction Cascades Associated with Oxidative Stress in Alzheimer's Disease," *Journal of Alzheimer's Disease*, Vol. 11, No. 2, 2007, pp. 143-152.